AF827937

The Notion of Mathematical Proof: Key Rules and Considerations

The Notion of Mathematical Proof: Key Rules and Considerations

Edited by:

Olga Moreira

ARCLER PRESS

www.arclerpress.com

The Notion of Mathematical Proof: Key Rules and Considerations

Olga Moreira

Arcler Press
224 Shoreacres Road
Burlington, ON L7L 2H2
Canada
www.arclerpress.com
Email: orders@arclereducation.com

© 2023 Arcler Press

ISBN: 978-1-77469-498-5 (Hardcover)

This book contains information obtained from highly regarded resources. Reprinted material sources are indicated. Copyright for individual articles remains with the authors as indicated and published under Creative Commons License. A Wide variety of references are listed. Reasonable efforts have been made to publish reliable data and views articulated in the chapters are those of the individual contributors, and not necessarily those of the editors or publishers. Editors or publishers are not responsible for the accuracy of the information in the published chapters or consequences of their use. The publisher assumes no responsibility for any damage or grievance to the persons or property arising out of the use of any materials, instructions, methods or thoughts in the book. The editors and the publisher have attempted to trace the copyright holders of all material reproduced in this publication and apologize to copyright holders if permission has not been obtained. If any copyright holder has not been acknowledged, please write to us so we may rectify.

Notice: Registered trademark of products or corporate names are used only for explanation and identification without intent of infringement.

Arcler Press publishes wide variety of books and eBooks. For more information about Arcler Press and its products, visit our website at www.arclerpress.com

DECLARATION

Some content or chapters in this book are open access copyright free published research work, which is published under Creative Commons License and are indicated with the citation. We are thankful to the publishers and authors of the content and chapters as without them this book wouldn't have been possible.

ABOUT THE EDITOR

Olga Moreira is a Ph.D. and M.Sc. in Astrophysics and B.Sc. in Physics/Applied Mathematics (Astronomy). She is an experienced technical writer and data analyst. As a graduate student, she held two research grants to carry out her work in Astrophysics at two of the most renowned European institutions in the fields of Astrophysics and Space Science (the European Space Agency, and the European Southern Observatory). She is currently an independent scientist, peer-reviewer and editor. Her research interest is solar physics, machine learning and artificial neural networks.

TABLE OF CONTENTS

LIST OF CONTRIBUTORS

Susana Carreira
University of Algarve, 8005-139 Faro, Portugal
UIDEF, Institute of Education, University of Lisbon, 1649-103 Lisboa, Portugal

Nélia Amado
University of Algarve, 8005-139 Faro, Portugal
UIDEF, Institute of Education, University of Lisbon, 1649-103 Lisboa, Portugal

Hélia Jacinto
Institute of Education, University of Lisbon, 1649-103 Lisboa, Portugal

Janka Medová
Department of Mathematics, Faculty of Natural Sciences, Constantine the Philosopher University in Nitra, Tr. A. Hlinku 1, 949 74 Nitra, Slovakia

Kristína Ovary Bulková
Department of School Education, Faculty of Humanities, Tomas Bata University, Štefánikova 5670, 760 01 Zlín, Czech Republic

Soňa Čeretková
Department of Mathematics, Faculty of Natural Sciences, Constantine the Philosopher University in Nitra, Tr. A. Hlinku 1, 949 74 Nitra, Slovakia

Uffe Thomas Jankvist
Danish School of Education, Aarhus University, Campus Emdrup, DK-2400 Copenhagen NV, Denmark

Mogens Niss
Department of Science and Environment, Roskilde University; DK-4000 Roskilde, Denmark

Fauziah Hakim
Program Studi Pendidikan Matematika, Universitas Sulawesi Barat

Murtafiah
Jl. Baurung, Banggae Timur, Kabupaten Majene, Sulawesi Barat

Arta Ekayanti
Faculty of Education and Teacher Training, Universitas Muhammadiyah Ponorogo, Indonesia

Hikma Khilda Nasyiithoh
Faculty of Education and Teacher Training, Institut Agama Islam Negeri Ponorogo, Indonesia

Benjamin Shongwe
Department of Mathematics Education, University of KwaZulu-Natal, South Africa

Vimolan Mudaly
Department of Mathematics Education, University of KwaZulu-Natal, South Africa

Tatag Yuli Eko Siswono
Universitas Negeri Surabaya, Gedung C8 FMIPA Unesa Ketintang, Surabaya, Indonesia

Sugi Hartono
Universitas Negeri Surabaya, Gedung C8 FMIPA Unesa Ketintang, Surabaya, Indonesia

Ahmad Wachidul Kohar
Universitas Negeri Surabaya, Gedung C8 FMIPA Unesa Ketintang, Surabaya, Indonesia

Zakaria Ndemo
Bindura University of Science Education, 741 Chimurenga road, Bindura, Zimbabwe

Rippi Maya
State Islamic University - Sunan Gunung Jati, Bandung

Utari Sumarmo
Indonesia University of Education, Bandung

Syamsuri, Indiana Marethi
FKIPUniversitas Sultan Ageng Tirtayasa, Indonesia

Anwar Mutaqin
FKIPUniversitas Sultan Ageng Tirtayasa, Indonesia

Windia Hadi
University of Muhammadiyah Prof. DR. HAMKA, Indonesia

Ayu Faradillah
University of Muhammadiyah Prof. DR. HAMKA, Indonesia

Aysun Yeşilyurt Çetin
Atatürk University, Erzurum, Turkey

Ramazan Dikici
Mersin University, Turkey

Samsul Maarif
Department of Mathematics Education, Universitas Muhammadiyah Prof. DR. HAMKA

Fitri Alyani
Department of Mathematics Education, Universitas Muhammadiyah Prof. DR. HAMKA

Trisna Roy Pradipta
Department of Mathematics Education, Universitas Muhammadiyah Prof. DR. HAMKA

Muchamad Subali Noto
Universitas Swadaya Gunung Djati, Jl. Perjuangan No.1 Cirebon, Indonesia
Universitas Pendidikan Indonesia, Jl. Dr. Setiabudi No. 229 Bandung 40154, Indonesia

Nanang Priatna
Universitas Pendidikan Indonesia, Jl. Dr. Setiabudi No. 229 Bandung 40154, Indonesia

Jarnawi Afgani Dahlan
Universitas Pendidikan Indonesia, Jl. Dr. Setiabudi No. 229 Bandung 40154, Indonesia

Aloisius Loka Son
Universitas Timor, Kefamenanu, Indonesia
Universitas Pendidikan Indonesia, Bandung, Indonesia

Darhim
Universitas Pendidikan Indonesia, Bandung, Indonesia

Siti Fatimah
Universitas Pendidikan Indonesia, Bandung, Indonesia

Mitchell J. Nathan
University of Wisconsin-Madison, Educational Sciences Building, 1025 West Johnson Street, Madison, WI 53705, USA

Candace Walkington
Southern Methodist University, Dallas, TX, USA

LIST OF ABBREVIATIONS

AQ	Adversity Quotient
ARP	Adversity Response Profile
CFA	Confirmatory Factor Analysis
CFI	Comparative Fit Index
CORE	Connect, Organize, Reflect, and Extend
DC	Didactical Contract
EFA	Exploratory Factor Analysis
FD	Field Dependent
FI	Field Independent
GEFT	Group Embedded Figure Test
GEMC	Grounded and Embodied Mathematical Cognition
LEDs	Light-Emitting Diodes
MPSA	Mathematical Problem-Solving Ability
OECD	Organization for Economic Cooperation and Development
PA	Parallel Analysis
PAF	Principal Axis Factoring
PCA	Principal Components Analysis
PEET	Proof Error Evaluation Tools
PSEP	Perceived Self-Efficacy for Proof
RCGP	Reading Comprehension of Geometric Proof
RFI	Resilience Factor Inventory
RME	Realistic Mathematics Education
RMSEA	Root Mean Square Error of Approximation
STEM	Science, Technology, Engineering, and Mathematics
TIMSS	Trends in International Mathematics and Science Study
TLI	Tucker-Lewis Index

PREFACE

The ability to solve mathematical proof problems is essential in advanced mathematics learning. It requires the development of learning skills such as critical thinking, logical inductive and deductive reasoning, communication and collaboration. Being able to understand, construct and evaluate mathematical proofs, is not only important when it comes to learning mathematics, but it is also fundamental for successfully developing cognitive skills that are considered to play a central role within the teaching and learning in numerous research fields.

In a nutshell, a mathematical proof is a series of logical arguments that must be a valid explanation of the truth of a statement or proposition. Each step in the mathematical proof must be based on previous steps or other facts with guaranteed truth. Thus, schools are transforming mathematics courses by facilitating students to understand formal mathematical language and axiomatic structure.

It is now known that factors such as adversity quotient and resilience are associated with the ability to solve mathematical proof problems; and that deductive reasoning should be cultivated at early ages (at the elementary school level) as it plays a central role in teaching and learning in the field of mathematical proving at high school, undergraduate and postgraduate school level.

The first part of this book (Chapters 1 to 6) focuses on the development of the ability to solve mathematical proof problems. It is focused on the factors that influence such cognitive development, as well as the elements that can play a role in the process of learning and understanding mathematical proof. Elements such as misconceptions, nonproductive mathematics beliefs and myths can become obstacles and result in difficulties in understanding and processing mathematical proofs.

Chapter 1 shows that logical-deductive reasoning is an important element for success in mathematical proof learning. It demonstrates that logical-deductive reasoning should be cultivated and reinforced in schools.

Chapter 2 focuses on specific relations between a student's reasoning skills and their ability to generalize algebraically and then find a solution to a complex open-ended problem. It concludes that algebraic generalization abilities should be the focus of upper-secondary mathematical proof teaching.

Chapter 3 reflects on the influence of students' misconceptions, unproductive beliefs and myths about mathematics on their ability to understand the mathematical proof. It aims to overview possible interventions and targeted efforts in altering students' misconceptions as well as unproductive beliefs, as means of helping them to overcome difficulties and succeed in solving mathematical proof problems.

Chapters 4 and 5 are focused on the students' adversity quotient and resilience influence on their ability to solve mathematical proof problems. Resilience is seen as a person's ability to face a difficult experience and the knowledge of how to deal with or adapt to it. The adversity quotient (AQ) is a person's ability to survive in the face of difficulties and efforts to resolve them. There are 3 categories in AQ: quitter, camper, and climber. Climbers are persons with high AQ who choose to survive and struggle to face the problems, challenges, and obstacles that will continue to hit. Campers persons' with medium AQ who have the willingness to try to deal with problems, challenges, and obstacles but stop when they feel they are no longer able. Quitters are persons with low AQ who lack the ability to accept challenges in life.

Chapter 4 is focused on examining the effect of adversity quotient and resilience on the mathematical proof problem-solving ability, while chapter 5 focused on the analysis of students' errors in producing proof. Each of the categories of students' AQ is identified according to the type of students' error. The type of students' errors used according to Newmann's Error Analysis.

Finally, chapter 6 introduces a new measure which relates to students' perceived self-efficacy for solving a mathematical proof. This refers to a student's mechanism based on the expectation to accomplish proof-related tasks.

The second part of the book (Chapters 7 to 13) focuses on learning and teaching methods for processing mathematical proof. It overviews teaching strategies and measures that help evaluate mathematical proofs. It is aimed at finding tools that can help students and teachers to overcome difficulties in constructing mathematical proofs.

Chapter 7 overviews teachers' difficulties in constructing a mathematical proof. The study uses a descriptive-explorative research design to explore prospective teachers' deductive and inductive reasoning and their difficulties in carrying out a mathematical proving process.

Chapter 8 is focused on improving the teaching and learning of mathematical proof by uncovering the kinds of flaws in postgraduate mathematics education. It also examines how students construct proofs of mathematical statements and their competencies in resolving proof tasks.

Chapter 9 investigates the role of the modified Moore learning approach in improving students' mathematical understanding and proving abilities. Chapter 10 includes a 4 quadrant classification model of students' formal-proof construction. Based on empirical data, the authors develop a quadrant model to describe students' classification of the proof results. It proposes different approaches for dealing with students' proof comprehension in each quadrant: hermeneutics approach, two-column form method, a proof construction learning using a worked-example method, and structural method.

Chapter 11 is focused on determining the effect of the application of discovery learning methods for developing students' mathematical proof ability in trigonometry, while Chapter 12 is focused on identifying the basic components of the mathematical proof process in abstract algebra. Chapter 13 is focused on the implementation of a self-explanation strategy to develop an understanding of proof in geometry. It aims to provide a strategy that can put students in the process of clarifying proof better.

The last 3 chapters are focused on an overview of learning obstacles in transformation geometry (Chapter 14); mathematical problem-solving ability based on teaching model and cognitive skills (Chapter 15); as well as improving mathematical reasoning through action (Chapter 14). This last part of the book aims to reflect on our understanding of mathematical proof and on how the body can support mathematical reasoning.

Chapter 1

VENUES FOR ANALYTICAL REASONING PROBLEMS: HOW CHILDREN PRODUCE DEDUCTIVE REASONING

Susana Carreira [1,2], ORCID,Nélia Amado [1,2] and Hélia Jacinto [3]

[1]University of Algarve, 8005-139 Faro, Portugal

[2]UIDEF, Institute of Education, University of Lisbon, 1649-103 Lisboa, Portugal

[3]Institute of Education, University of Lisbon, 1649-103 Lisboa, Portugal

ABSTRACT

The research on deductive reasoning in mathematics education has been predominantly associated with the study of proof; consequently, there is a lack of studies on logical reasoning per se, especially with young children. Analytical reasoning problems are adequate tasks to engage the solver in deductive reasoning, as they require rule checking and option elimination, for which chains of inferences based on premises and rules are accomplished. Focusing on the solutions of children aged 10–12 to an analytical reasoning

Citation: (APA): Carreira, S., Amado, N., & Jacinto, H. (2020). Venues for analytical reasoning problems: How children produce deductive reasoning. Education Sciences, 10(6), 169. (23 pages).

Copyright: © Creative Commons Attribution 4.0 International License (http://creativecommons.org/licenses/by/4.0/).

problem proposed in two separate settings—a web-based problem-solving competition and mathematics classes—this study aims to find out what forms of deductive reasoning they undertake and how they express that reasoning. This was done through a qualitative content analysis encompassing 384 solutions by children participating in a beyond-school competition and 102 solutions given by students in their mathematics classes. The results showed that four different types of deductive reasoning models were produced in the two venues. Moreover, several representational resources were found in the children's solutions. Overall, it may be concluded that moderately complex analytical reasoning tasks can be taken into regular mathematics classes to support and nurture young children's diverse deductive reasoning models.

Keywords: deductive reasoning; analytical reasoning problem; reasoning models; young students; expression of reasoning; beyond-school mathematics competition; mathematics class

INTRODUCTION

Logical reasoning, namely deductive reasoning, is one of the recognized and celebrated pillars of mathematical reasoning, whether linked to proof and argumentation or communication and problem-solving, which makes it meaningful to school mathematics worldwide (e.g., [1,2]). Despite its relevance in the learning of mathematics from early ages, the difficulties related to the use of deductive reasoning are well known; not only are such difficulties found in students across the various levels of education, but also in the general population when individuals are given logic problem situations from different domains. However, relatively little is known about effective and adequate teaching of logic, and there is much evidence that, even with specific training in mathematics, many students continue to fail in formal proofs [3,4,5,6].

A question that is far from resolved in the available research is whether deductive reasoning has the potential to be cultivated in activities such as word problems or challenging situations, sometimes referred to as puzzles or brain-teasers. Another open question is that of the ability of school-aged children to deal with logical forms of reasoning because it is uncertain when the mastery of different forms of deductive reasoning is achieved [7]. There is, however, a general tendency to look at deductive reasoning as a high-level skill, which will only be available to students of advanced years [8].

Knowing more about the ways that children deal with word problems requiring deductive reasoning and their ability to express that reasoning is important to improve our understanding of children's engagement with that kind of logical thinking from an early age. This interest in children's deductive reasoning stems from an extensive assessment of the solutions to several analytical reasoning problems, which are part of the collection of problems proposed in a web-based problem-solving competition (addressing students aged 10–12) promoted by the University of Algarve, in Portugal [9]. Analytical reasoning problems, also referred to as constraint-satisfaction problems [10,11], have a strong relationship with logical reasoning and hold an important place in various areas of knowledge.

Rationale and Aims

The present study focuses on the deductive reasoning of children aged 10–12 when solving an analytical reasoning problem of a moderately challenging level. Two very different settings were contemplated for data collection. The reason for this decision is related to the contrasting conditions between both, namely the condition of limited against extended time and the condition of possible help from adults against no aid provided. The inclusion of the two settings was designed as a way of ascertaining whether children are capable of producing effective deductive reasoning in both cases, thus proposing two contrasting venues for deductive reasoning within problem-solving activities. In the first case, we gathered students' solutions to an analytical reasoning problem in the course of their participation in a web-based mathematical problem-solving competition, which took place online in a beyond-school environment. In the second case, we collected the solutions produced by other students who solved the same problem in the mathematics classroom during a regular class. In the case of the children who participated in the competition, their success in solving the analytical reasoning problem was evident; they worked on the problem outside of school, mostly at home, with extended time (two weeks) to think about the problem and develop ways of communicating the solution, with possible help and guidance from others (family, teachers, and friends). In the case of children who solved it in the classroom, the work was done individually in a limited time (defined by their teacher, but roughly 30 min), and the answers were written on paper, using only a pencil or pen to express the reasoning. The success rate of the latter was clearly lower, and there were also signs of there being little time to improve the written explanations of the solution process.

The empirical study material is composed of the solutions obtained in the two settings. The research questions were as follows: (i) What are the types of deductive reasoning that can be found in the children's solutions? (ii) How is deductive reasoning expressed by children in their solutions? (iii) Do the same types of deductive reasoning appear in the two settings in which the solutions were created?

THEORETICAL BACKGROUND

From a theoretical point of view, the study acknowledges the research debate between two opposing theories in explaining the psychological mechanisms of deductive reasoning: The mental logic theory and the mental models theory. As they are theories about pure logical reasoning, the theoretical framework also reviews the research on how logical reasoning and mathematical reasoning interrelate, particularly in the case of school-aged children, and finally discusses analytical reasoning problems (also known as relational reasoning problems) as special problems that surround many knowledge domains, from law to mathematics, including many everyday situations that children can be challenged to solve from young ages.

Theoretical Approaches to the Psychology of Deduction

Two major competing theories in the psychology of deduction have been proposed over the last few years: The mental models theory and the mental logic theory. The debate between the proponents of each of them has engendered consecutive empirical investigations attempting to confirm their distinct theoretical assumptions. In the work of Schroyens, Schaeken, and D'Ydewalle [12], a synthesis of the starting points of each of the two opposing theories is presented, as well as the acknowledgement that the results of various investigations could be effectively explained by either theory.

The theorists of the mental logic approach assume that the mental apparatus underlying deduction consists of applying mental inference rules (inference schemata) to the premises. Theorists who support this position postulate that individuals use a mental logic that is similar to propositional calculus. They claim that deductive thinking consists of applying a set of formal rules that allow derivation of a conclusion from given premises, regardless of the meaning of the propositions. In other words, individuals make use of formal inferential rules free of content when reasoning deductively. According to this perspective, when people fail in logical reasoning problems, it is not because they are devoid of mental logic, but because the demands of the situation exceed their logical skills, because they make inferences from non-logical sources, or because they are reasoning on a non-logical matter. Thus, the difficulty of a problem would mainly be a consequence of the number and/or availability of inference rules to be applied. Moreover, the theorists of mental logic consider the importance of the reductio ad absurdum argument in many logic problems, especially in those that include some form of conditional reasoning.

The mental models theory, largely developed by Johnson-Laird and colleagues (e.g., [13,14]), conceptualizes logical reasoning, similarly to other fields of reasoning, as a processing schema in three phases. In the first phase—the construction of the model—various mental models that represent, to the individual, the truth of the premises are constructed. Mental models essentially represent possible worlds, i.e., possibilities of meanings given to the premises in light of relevant knowledge that is triggered during the process of interpretation. In the second phase—the formulation of the conclusion—individuals integrate the mental models of the premises. The inconsistent models are eliminated and the consistent ones are combined in the creation of an integrated model. The resulting model allows people to form a putative conclusion. In the third phase—the validation of the conclusion—people seek alternative models of the premises, which may falsify the putative conclusion. In fact, based on the limitations of the working memory, it is assumed that people do not initially represent all the possibilities consistent with the premises. Developments of the theory related to the validation process suggest that the search for counterexamples is a crucial element, namely in using the reductio ad absurdum form of reasoning, which roughly means avoiding contradictions at all costs. The availability of counterexamples seems to be related to the semantic content of the problem, as the following claim suggests: "The less semantically impoverished the materials are, the stronger a search for counterexamples might be driven by our background knowledge associated with and triggered by the given information" [12] (p. 161). The theory of mental models predicts that the difficulty of a problem increases with the number of models that are compatible with the problem.

Put briefly, a clear distinction between the two opposing theories refers to the semantics of a problem, namely when generating indeterminacy [15], and to its role in the difficulty of a problem: Minimized by the theory of mental logic and emphasized by the theory of mental models.

Logical Reasoning and Mathematics Learning

It is generally accepted that individuals' ability to use logical–deductive reasoning is an important element for success in mathematics learning, and several studies have supported the claim that such an ability can and should be strengthened and developed [6,7,16,17,18,19].

In the field of mathematics education, the study of deductive reasoning has been mainly associated with mathematical proof, leaving the understanding of deductive reasoning, per se, in the background [20,21]. This may have to do with the diverse ways of conceiving deductive thinking by the protagonists involved in mathematics education. The study by Ayalon and Even [20] offers a picture of that diversity. It looked at the core conceptions on deductive reasoning held by a heterogeneous group of people engaged in mathematics education (teachers, mathematicians, researchers in mathematics education, curriculum designers, and teacher educators). Two different approaches to the meaning of deductive reasoning were identified: The systematic approach, which considered deductive reasoning as a systematic problem-solving approach regardless of whether it is used in mathematics or in various other areas, and the logical approach, which emphasizes logic in mathematics as distinct from logic in other domains.

Research on students' logical reasoning is even sparser when it comes to young children. Nevertheless, some studies (namely those revised in Stylianides and Stylianides [7]) have already revealed that children are able to develop logical reasoning in various deductive reasoning tasks and problems. In light of such evidence, the research in mathematics education needs to know more about the ways in which schoolchildren can develop their deductive reasoning, namely under what circumstances and with what kind of tasks.

Authors such as Hoyles and Küchmann [19], Epp [17], Lee [4], Nunes et al. [16], Stylianides and Stylianides [7], and Lommatsch [22], among others, have advocated an explicit teaching of deductive reasoning in school mathematics. In general, they have criticized the teaching of propositional logic as an empty unit, considering the weak transfer of that learning to mathematics and problem-solving.

Hoyles and Küchmann [19] developed a study with a large sample of students in the eighth and ninth grades (ages 12–13) on the use of conditional reasoning in tasks involving properties of number sums and products. Students were presented with two statements: An implication and its converse. They were then asked to: Decide whether the two statements said the same thing, draw a conclusion assuming that one of the statements was true, and assess the validity of each statement in turn. Overall, the research showed that even students with high performance in mathematics cannot fully understand how to determine the truth of "if p then q". This supported a recommendation for school mathematics: The design of activities that focus on giving meaning to the structural properties of logical implication and encourage education to strengthen and refine those meanings over time.

Nunes et al. [16] investigated the causal relationship between the logical–deductive performance of children and their learning of mathematics. In adopting an experimental design with six-year-old children attending the first grade, the researchers implemented an educational program in the regular period of mathematics classes. The results showed a beneficial effect on children's mathematics learning that persisted after several months following the intervention. The teaching on logical principles related to quantities and operations even improved the learning of children whose levels of logical competence were consistent with their difficulties in mathematics. Nunes et al. [16] derived important consequences about the value of time invested in developing logical competence in school mathematics and on the advantage of giving children opportunities to achieve a logical understanding of arithmetic operations.

The study by Stylianides [8] focused on a class of third graders where the use of a representational model—the "building model"—was investigated. The task required the students to find out how many ways there were for a person to get to the second floor, and asked them to prove their answer. Depending on the assumption the student would make, the task would lead to proving the existence of a finite number of ways or infinitely many ways. The results of the study highlighted the fundamental role of assumptions in proving with children from early grades. Evidence was also produced on the abilities of young children in dealing with mathematical situations that are related to assumptions and proofs, which, in many aspects, resembled the way of thinking of mathematicians.

From these studies, it may be inferred that the richness of the tasks (both inquiring and challenging) and the way they are used to trigger students' logical and deductive forms of reasoning are common traits in eliciting and developing the ability to reason deductively. In addition, it seems that students' use of different strategies and reasoning models offers opportunities for them to understand what it means to engage in logical arguments and to explicitly explore such reasoning in mathematics education.

Analytical Reasoning Problems

Among the problems that require deductive reasoning processes are those known as analytical reasoning problems or relational reasoning problems (e.g., determine the layout of seats at a table, subject to a number of conditions and restrictions, such as protocol rules). Such problems involve reasoning deductively from a set of instructions, rules, or principles that describe relationships between people, things, or events. They require the ability to consider a set of facts and conditions, and to determine, based on logical principles, which may or must be true. Sometimes, such problems are also described as puzzles based on constraint satisfaction [10,11,23].

Analytical reasoning problems involve a wide variety of deductive reasoning skills, which include: (i) Understanding the basic structure of a set of relations in order to find a complete solution to a given problem; (ii) using conditional reasoning of the form "if–then" and recognizing logically equivalent formulations of such conditional statements; (iii) deducing what may be true or must be true from certain facts and rules by removing contradictions or generating new information in the form of an additional or substitute rule; (iv) deciding when two statements are logically equivalent in the context by identifying the rule that corresponds to one or more of the original conditions. Not surprisingly, analytical reasoning problems are quite common in many areas of study, such as law, management, planning, military defense, security, network operations, and medicine, not to mention mathematics.

According to the literature, the difficulty of such problems depends on the number of dimensions and on the number of values along each dimension. Problems containing statements in the negative form are also perceived as being more difficult than those in which all the information is given in the affirmative form. Finally, problems involving reasoning with quantifiers were also considered as having a high level of difficulty. The study by Cox and Brna [23] examined the effects of the choice and use of external representations by university students in solving problems of analytical reasoning with different degrees of difficulty. This study revealed several types of external representations performed by the subjects, which were analyzed in detail as to their benefits and limitations. The results showed that: Some individuals may not have ways to represent more abstract relationships, such as those involving quantifiers; the use of multiple representations appears generally associated with good performance (including tables, ordered text or lists, free text, tree diagrams, set diagrams); the tabular representation was the predominant external representation; and the cognitive load associated with the choice and construction of external representations is compensated by the relief of cognitive load in the subsequent use of those representations to infer conclusions.

The selection of one or more external representations emerges as a crucial stage of the reasoning process, as it is often difficult because the requirements of representation vary with the problems. The expressive properties of the external representation chosen should allow representation of the semantics of the problem. Thus, the subject must correctly understand the essential characteristics of the problem, namely the dimensionality and the level of determination, and then select a suitable form of representation from her/his repertoire. A significant amount of information provided in analytical reasoning problems is given implicitly and, therefore, must be inferred before it can be represented. An important function of external representations appears to be the guiding of the search for implicit information. However, the combination of producing representations and creating inferences brings a high cognitive demand to the subject.

The work of English [24] focused on another aspect of solving analytical reasoning problems—the reasoning processes used by young students attending primary school (9–12 years old). Its main purpose was to investigate the reasoning processes of children in a set of problems presented in writing and complemented by manipulatives. The research involved a detailed analysis of the protocols obtained from the verbalization of children's thinking as they solved the problems. The study included a large sample of children from fourth to seventh grade, and this allowed the exploration of all the general trends of their reasoning processes. The theoretical perspective adopted was the mental models theory, according to which a sound logical reasoning is not equal to the application of the formal rules of logic. The results of the study showed a strong correlation between success in the problem-solving and the choice of assumptions that more easily contribute to the construction of an initial model and its development. The less efficient solvers tended to avoid the relational complexity of the premises, and instead used processes of segmentation of the relations.

Newstead et al. [25] argue that analytical reasoning problems are deductive reasoning problems in that they can be solved completely based on the information presented, and their solutions can be obtained and verified by formal logic. However, they are much more complex than the tasks traditionally used by psychologists in the investigation of deductive reasoning and, in a sense, may be more representative of logical reasoning in the real world. Their greater difficulty is often related to the kind of inferences involved: Necessary, possible, or impossible. During the solving process, there are situations in which multiple consequences are possible, thus making the problem more difficult.

Two main strategies were used by the participants in their study, which are described as rule checking and option elimination. These strategies consist of choosing one of the dimensions of the problem and testing all possible assumptions for a given rule, or, systematically, for each of the rules, seeking contradictions in order to discard hypotheses and draw necessary conclusions.

The study involved undergraduate and graduate students in solving complex analytical reasoning problems, which included many rules and conditions to be met. The research was designed to provide a model for the predictive characteristics of the difficulty of analytical reasoning problems. At the same time, it sought to understand which of the two major theories about the nature of logical thinking (mental models theory or mental logic theory) was best suited to assess the difficulty of the problems. From the results, it may be highlighted that the difficulty of the rules is the feature that most contributes to the difficulty of the problem. Moreover, the predominant strategy was the systematic checking of hypotheses against the rules available. This suggested the construction of models of the situation based on the given premises, followed by their validation through a procedural form of checking. In some cases, the subjects also combined rules in order to derive logical conclusions from them. In this particular point, the authors also found a resonance with the mental logic theory.

A relevant discussion for the present investigation is brought by the work of Van der Henst [15], which analyzes the assumptions of each of the theories to explain an element that generates greater difficulty in analytical reasoning problems: The indeterminacy factor. This means the existence of premises that, while not being irrelevant per se, introduce an element of indeterminacy in the conclusions derived from the premises. Thus, proponents of the mental models theory consider that such problems contain a higher level of complexity because the solvers will have more than one possible model to obtain the answer; this could arguably not be explained by the theory of mental logic, which would only take into account the inferential rules derived from the premises, which do not change due to the fact that there is an indeterminacy factor. On the other hand, the theory of mental logic may also explain the higher degree of complexity by claiming that the reasoner does not know in advance what information is necessary for solving the problem (determinate relations) and what information is not (indeterminate relations). Thus, the higher level of complexity could also be explained by the greater number of relationships to be stored mentally. As the author, concludes: "According to MMT [Mental Model Theory], indeterminacy involves the construction of several models. According to a rule approach, it involves more propositional information to store and manipulate in memory. Although empirical data in relational reasoning are neatly described by MMT, they do not exclude a description based on inference rules" [15] (p. 200).

METHODOLOGY

Participants

The study comprises two settings where children aged 10–12 solved an analytical reasoning problem: (i) Beyond-school mathematics and (ii) the regular math class.

In the first case, the students were voluntarily participating in a mathematical problem-solving competition taking place over the Internet. This was an inclusive competition, where non-routine and moderate mathematical challenges were proposed, and students sent their solutions by email. The requirement of explaining the problem-solving process was always emphasized. Participants could compete individually or in pairs. Some of the presented problems were analytical reasoning problems. One of those problems was selected for the present study.

The second setting consisted of regular math classes of students of the same age (attending the fifth or sixth grade). Three mathematics teachers, all from different public middle schools, agreed to propose the chosen analytical reasoning problem to different classes, allowing students the appropriate time for their resolution (which they planned to be around 30–45 min). Across the three schools, two fifth-grade classes and four sixth-grade classes participated. All of the teachers decided to deliver the problem as the final task of a lesson. They also emphasized the requirement of explaining the problem-solving process when they handed the task. The students solved the problem with paper and pencil and delivered their solutions at the end. The problem was solved individually, without any intervention or help from the teacher.

Several important distinctions between the two settings must be noted. In the competition scenario, the participating students had substantial time (two weeks) to solve the problem and submit the solution; the final product was sent in digital format, and the participants were free to choose any digital tools that would suit them to describe and express their ways of reasoning tidily and cleanly; the participants could count on possible help and guidance from parents, teachers, and friends. In the mathematics classroom scenario, students had limited and significantly less time (about half an hour); the solution was produced on paper and pencil, and it was not feasible in the time available to improve the presentation of the reasoning in their final answers; the students worked individually and received no help from the teacher.

Materials and Data Collection

The following is the analytical reasoning problem that was given to both sets of young children (Figure 1):

Four friends went to the movies and decided to buy popcorn. During the film, one of them was frightened by a terrifying scene and spilled his bucket of popcorn. At the break, the movie theatre attendant saw the spilled popcorn all over the floor and wanted to know who had been the guilty one.

- It wasn't me, said Bernardo.
- It was Paulo, said Carlos.
-It was Carlos, said Gabriel.
- Gabriel is not telling the truth, said Paulo.

It is known that only one of them is lying. Who spilled the popcorn?
Do not forget to explain your problem solving process!

Figure 1. Analytical reasoning problem proposed.

The goal is to know which of the friends spilled the popcorn, drawing on various premises and rules:

- There are four friends: Bernardo, Carlos, Gabriel, and Paulo.
- Each of the four friends makes a statement (three refer to who spilled the popcorn and the latter refers to the truth of one of the statements).
- One—and only one—of the friends spilled the popcorn.
- One—and only one—of the friends lied.

Based on the given information, it is possible to isolate the following atomic propositions involved in the problem (Table 1), divided according to the two dimensions of the problem situation (lying and spilling popcorn).

Among the propositions P1, P2, P3, and P4, only one is true, and among the propositions Q1, Q2, Q3, and Q4, also only one is true.

Table 1. Sets of propositions on the dimensions "lying" and "spilling popcorn"

Dim. Lying		Dim. Spilling	
P1:	Bernardo lies (he said it was not him)	Q1:	Bernardo spills popcorn
P2:	Carlos lies (he said it was Paulo)	Q2:	Carlos spills popcorn
P3:	Gabriel lies (he said it was Carlos)	Q3:	Gabriel spills popcorn
P4:	Paulo lies (he said that Gabriel lied)	Q4:	Paulo spills popcorn

According to the rule that only one of the friends lied, one can make an exhaustive analysis, assuming that each of them, in turn, was the one who lied.

Thus, there are four cases, as follows:

(1) Let P1 be true.

If P1, then we have: Q1 and not-P2, not-P3, and not-P4 (Bernardo was the dropper and none of the others lied). However, the two propositions not-P3 and not-P4 are contradictory (if Gabriel is telling the truth, then Paulo is lying, and vice versa). Therefore, P1 is false.

(2) Let P2 be true.

If P2, then we have: Not-Q4, not-P1, not-P3, and not-P4 (Paulo was not the dropper, and none of the others lied). However, again, the two propositions not-P3 and not-P4 are contradictory. Therefore, P2 is false.

(3) Let P3 be true.

If P3, then we have: Not-Q2, not-P1, not-P2, and not-P4 (Carlos was not the dropper, and none of the others lied). Therefore, we conclude: Not-Q1, not-Q2, and Q4 (Bernardo was not the dropper and Paulo was the dropper). As there is only one that is guilty, we may deduce not-Q3. This means that Paulo spilled the popcorn and Gabriel was the liar.

(4) Let P4 be true.

If P4, then we have: Not-P3, not-P1, and not-P2 (Bernardo is true, Carlos is true, and Gabriel is true). Then, we have Q2 and Q4 (Paulo was the dropper and Carlos was the dropper). This is not true because Q2 and Q4 cannot both be true (only one has spilled the popcorn). Therefore, P4 is false.

Similarly, according to the rule that only one of the friends spilled the popcorn, one can make an exhaustive analysis, assuming that each of them, in turn, was the one who spilled it.

The solutions from the participants in the web-based competition were collected in the first place. All of those solutions were received in a digital form via email, and were collected over a period of fifteen days since the problem was posted. The email messages with the solutions all indicated: Name, school, and class. Many of the solutions were presented in the email message window, but some came enclosed in attached files, such as Word, PowerPoint, Excel, or scanned images of paper-and-pencil work. The total of those solutions was 384. The solutions given by the students who solved the problem in the class were collected at a later time. All of them were written on paper and delivered to the teacher after completion. Depending on the schools, the students used a blank A4 sheet or half a sheet to present their solutions. The students only indicated the school and the class on their worksheets. The total of solutions collected was 102 (the amounts per school were 25, 35, and 42). In the study, ethical standards were ensured, namely by keeping the anonymity of the students and teachers involved.

Data Analysis

The data were analyzed based on qualitative content analysis. The method is suitable for the analysis of recorded material, namely written documents or digital files that contain written information, including text and forms of visual communication. The method is equally suited to the research questions that seek to know what the world of children's deductive reasoning "looks like" [26] (p. 29) by identifying and describing, in as efficient and clear ways as possible, the deductive reasoning produced and expressed by young children in solving the problem, both in the context of a beyond-school competition as well as in the classroom context.

The expectation, with regard to the more likely forms of reasoning, was that students would test systematic hypotheses and look for contradictions. In that sense, they would probably elect one of the two dimensions of the problem for their hypothesis testing (as suggested in the presentation of the problem in Section 3.2). Thus, the coding frame used for the preliminary analysis [27] consisted of classifying the solutions under two major concept-driven categories: Focusing on the rule that one—and only one—of the friends lied, or focusing on the rule that one—and only one—of the friends spilled the popcorn. Solutions that did not fit into either category would be cataloged according to their specific approach. After that separation, a data-driven categorization process followed, in which the various solutions from each category were reviewed, and patterns were searched to provide a finer subdivision of the forms of deductive reasoning according to different types of conceptual models or approaches (Figure 2).

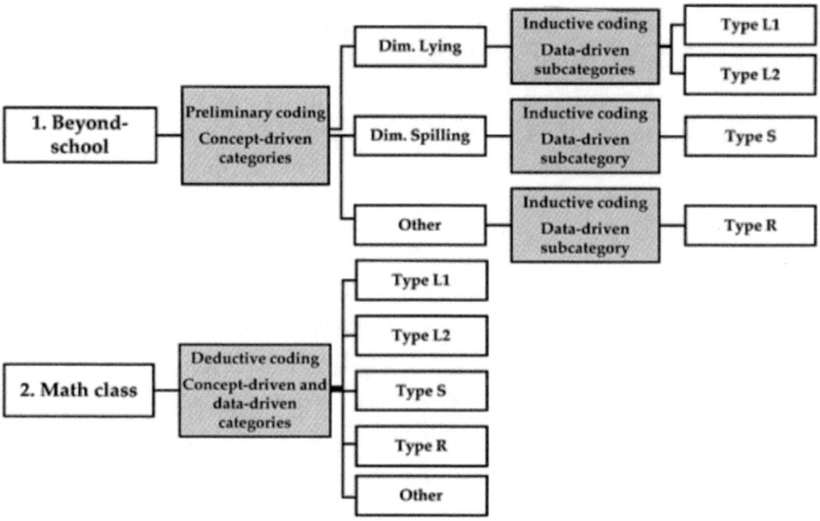

Figure 2. Steps of the qualitative content analysis of each set of data.

Following this procedure, it was possible to identify types and subtypes of solutions to the problem, each of which corresponded to a specific form of deductive reasoning. In a subsequent stage, we chose prototypical examples of the different types of solutions, which were the object of a more detailed analysis and led to the identification of the supporting structures of the deductive reasoning developed. Those prototypes were also scrutinized for the students' kinds of expression of deductive reasoning, including representational features.

The coding process was firstly applied to the body of solutions from the beyond-school scenario because their number was higher and, therefore, they could be more diverse. Next, using the obtained categories and subcategories as the established coding frame, the analysis was applied to the body of solutions from the mathematics class (Figure 2). Finally, a comparison was made between the results from the two bodies of data.

RESULTS AND FINDINGS

In this section, the results are presented and empirical evidence of the findings is provided from a selection of prototypes of solutions pertaining to each type and subtype. The originals in Portuguese were transcribed and translated. The translation aimed to be as faithful as possible to the original text, with the utmost concern of preserving the meaning and the style of the written text used by the students. The language used by most students was quite simple and was usually filled with the very words that the problem itself contains: The names of the characters in the story, the verbs to lie and tell the truth, the adjectives of guilty and liar. Another type of word that frequently appeared included expressions like: Supposing that; we know that; therefore.

First, the results of the analysis of the solutions from the beyond-school setting are presented as a result of which three types of logical reasoning were identified, one of them subdivided into two sub-types. Each form of reasoning is concisely described and illustrated with two examples of students' productions. Next, based on the previous categories, the results from the coding of the classroom solutions are presented, and an example of an answer for each type of reasoning is given.

Results from the Beyond-School Scenario

The first reading of the data allowed identification of the correct, incorrect, and partially correct answers. The latter consisted of solutions that presented an insufficient or unclear explanation of the reasoning that led to the answer to the problem. As can be seen in Table 2, which presents the numbers obtained, the success rate of the students who participated in the problem-solving competition was quite high.

Table 2. Success summary from the beyond-school body of solutions

Completely Correct	Partially Correct or Incorrect	Total
334	50	384
(87%)	(13%)	(100%)

The two main approaches identified in the solutions were each associated with one of the dimensions of the problem: (a) Who lied and (b) who spilled the popcorn (Table 3). Looking for necessary inferences about the one who lied was the most used approach. The two cases involved the testing of hypotheses and the elimination of options due to contradictions with the premises: Only one friend lied; only one friend spilled the popcorn; or one or more of the statements uttered by the characters. In addition to the two main approaches, some solutions adopted an alternative strategy. In a much smaller number, such solutions concentrated on the reduction of the conditions through the previous establishment of relations between the given constraints. In particular, some were based on realizing that two of the boys accused different people, which made it possible to deduce that one of them was necessarily the liar, and some were based on noticing that two of the boys could not both be lying (or telling the truth), as one states that the other lies, which made it possible to infer that one of them was necessarily the liar.

Table 3. Main categories from the beyond-school body of solutions

Reasoning on the Dim. Lying	Reasoning on the Dim. Spilling	Reasoning on Relations	Total
197	99	38	334
(59%)	(30%)	(11%)	(100%)

In many cases, the checking of hypotheses was indicated in an organized manner. The ways in which students solved the problem allowed their classification into three different types: Lying (L), spilling (S), and relations (R). In the case of the deductive reasoning focusing on the dimension "lying", two subtypes of reasoning were then identified in the data (labeled as L1 and L2) and categorized according to the underlying form of logical reasoning.

Dimension "Lying": The L1 Type of Deductive Reasoning

Based on their common features, the first type of solution under the general approach of checking all the options for the dimension "X lies" is illustrated by the Examples 1 and 2, and is characterized as the L1 type of deductive reasoning.

L1 Deductive Reasoning (Example 1)

Hypothesis 1.

Bernardo is lying—then it was Bernardo who spilled the popcorn. This cannot be the case because Gabriel and Carlos would also be lying.

Hypothesis 2.

Paulo is lying—then Gabriel is telling the truth and it was Carlos who spilled the popcorn. This cannot be the case because Carlos would also be lying, and there cannot be two lying.

Hypothesis 3.
Carlos is lying—then Paulo speaks the truth and, therefore, Gabriel has to be lying, which cannot be the case because there would be two of them lying.
Hypothesis 4.
Gabriel is lying—so it was not Carlos [who spilled the popcorn] and, therefore, Paulo speaks the truth. Since Bernardo is also speaking the truth and Carlos too, it was Paulo who spilled the popcorn.

L1 Deductive Reasoning (Example 2)

- If it was Bernardo lying, then he was the one who spilled the popcorn; therefore, Carlos had to be lying, and Gabriel too. This hypothesis is not true because there are three of them lying.
- If it was Gabriel lying, then it was not Carlos [who spilled the popcorn], and the others are telling the truth; therefore, the one who spilled the popcorn was Paulo, and that is the true hypothesis because only one is lying.
- If it was Carlos lying, Paulo did not spill the popcorn on the floor, and then it was Carlos [who spilled the popcorn], because Gabriel is telling the truth. However, Paulo says that Gabriel is not telling the truth, so one just cannot figure out, but there are two of them lying.
- If it was Paulo lying, Carlos and Gabriel were telling the truth, and this cannot be the case because they say different names, and only one spilled the popcorn.

This pattern of deductive reasoning corresponds to trying the falsification of hypotheses on the dimension "X lies" against the premise of having exactly one person lying or by the emergence of conflicting inferences, and can be summarized as follows:

L1 Deductive Reasoning

- Suppose X lies, then…; therefore, there is more than one lying (Eliminate).
- Suppose X lies, then…; therefore, conflicting inferences (Eliminate).
- Suppose X lies, then…; therefore, there is only one liar (Accept).

Dimension "Lying": The L2 Type of Deductive Reasoning

A different version of the reasoning that focused on checking all the values of the dimension "X lies" applies the assumption that only one character is lying, and immediately establishes that all the others tell the truth. Thus, the checking is not concerned with ruling out the cases of more than one liar; instead, the checking is done sometimes against the assumption that only one person spilled the popcorn, and, other times, by the contradiction among conflicting inferences (a statement and its negation). The following two are examples of solutions that illustrate the L2 type of deductive reasoning.

L2 Deductive Reasoning (Example 1)

Scenario 1

- [Bernardo]—lies
- [Carlos]—tells the truth
- [Gabriel]—tells the truth
- [Paulo]—tells the truth

It is impossible because Carlos and Gabriel accuse different people.

Scenario 2

- [Bernardo]—Tells the truth
- [Carlos]—Lies
- [Gabriel]—Tells the truth
- [Paulo]—Tells the truth

It is impossible because Gabriel and Paulo cannot both be telling the truth, since Paulo says that Gabriel is lying.

Scenario 3

- [Bernardo]—Tells the truth
- [Carlos]—Tells the truth
- [Gabriel]—Lies
- [Paulo]—Tells the truth

It is possible because we conclude that it was not Bernardo and it was not Carlos, and Paulo confirms that Gabriel lies. So, it was Paulo.

Scenario 4

- [Bernardo]—Tells the truth
- [Carlos]—Tells the truth
- [Gabriel]—Tells the truth
- [Paulo]—Lies

It is impossible because Carlos and Gabriel accuse different people.

Answer: It was Paulo who spilled the popcorn.

L2 Deductive Reasoning (Example 2)

Hypothesis H1.

(Bernardo lies)

State-ment	*Bernardo*	*False*	*Therefore, he is guilty.*	*Accept*
	Carlos	*True*	*Therefore, it was Paulo.*	*Contradiction*
	Gabriel	*True*	*Therefore, it was Carlos.*	*Contradiction*
	Paulo	*True*	*Therefore, Gabriel is not telling the truth and it was not Carlos.*	*Contradiction*

Conclusion: Hypothesis not accepted.

Hypothesis H2.

(Carlos lies)

Statement	*Bernardo*	*True*	*Therefore, it was not him*	*Accept*
	Carlos	*False*	*Therefore, it was not Paulo*	*Accept*
	Gabriel	*True*	*Therefore, it was Carlos*	*Accept*
	Paulo	*True*	*Therefore, Gabriel is not telling the truth and it was not Carlos*	*Contradiction*

Conclusion: Hypothesis not accepted.

Hypothesis H3.

(Gabriel lies)

Statement	*Bernardo*	*True*	*Therefore, it was not him*	*Accept*
	Carlos	*True*	*Therefore, it was Paulo*	*Accept*
	Gabriel	*False*	*Therefore, it was not Carlos*	*Accept*
	Paulo	*True*	*Therefore, Gabriel is not telling the truth and it was not Carlos*	*Accept*

Conclusion: Hypothesis accepted, that is, Gabriel lied and Paulo spilled the popcorn.

Hypothesis H4.

(Paulo lies)

Statement	*Bernardo*	*True*	*Therefore, it was not him*	*Accept*
	Carlos	*True*	*Therefore, it was Paulo*	*Accept*
	Gabriel	*True*	*Therefore, it was Carlos*	*Contradiction*
	Paulo	*False*	*Therefore, Gabriel is telling the truth and it was Carlos*	*Contradiction*

Conclusion: Hypothesis not accepted.

This form of reasoning also corresponds to the intention of falsifying hypotheses under the dimension "X lies", and can be translated as follows, where the checking is based on the existence of a single guilty person or else on contradictory inferences about who has dropped the popcorn on the floor:

L2 Deductive Reasoning

- Suppose X is the only liar, then…; therefore, there is more than one guilty (Eliminate).
- Suppose X is the only liar, then…; therefore, conflicting inferences (Eliminate).
- Suppose X is the only liar, then…; therefore, there is only one guilty (Accept).

Dimension "Spilling": The S Type of Deductive Reasoning

Another route for solving the problem, corresponding to a smaller percentage of answers, took the dimension "who spilled the popcorn" for the construction of hypotheses. This kind of solution stands out in that it presents a much more condensed explanation and more simple and direct deductions, indicating an apparent simplification of the problem-solving process. Some of these solutions include a double-entry table, where one of the dimensions is attributed to the person that spilled the popcorn and the other dimension to the statement of each of the characters, aiming for cross-checking and determining the option consistent with one single person lying. The following two examples illustrate the reasoning of Type S.

S Deductive Reasoning (Example 1)

Assuming that Bernardo spilled the popcorn, then all lied (L), except Gabriel, who told the truth (T) (see table, first column). If it was Carlos, then Carlos and Paulo were lying. If it was Gabriel who spilled the popcorn, then

Carlos and Gabriel were lying. If it was Paulo, then only Gabriel lied. So, it was Paulo who spilled the popcorn, and Gabriel lied.

	B.	C.	G.	P.
- It was not me, said Bernardo.	L	T	T	T
- It was Paulo, said Carlos.	L	L	L	T
- It was Carlos, said Gabriel.	L	T	L	L
- Gabriel is not telling the truth, said Paulo.	T	L	T	T
N. of lies	3	2	2	1

S Deductive Reasoning (Example 2)

Let us suppose that Bernardo spilled the popcorn. Then, Bernardo lied. Then, Paulo lied. However, as there was only one who lied, IT WAS NOT BERNARDO.

Let us assume that it was Carlos. Then, Carlos lied. Then, Paulo lied. However, as there was only one who lied, IT WAS NOT CARLOS.

Let us suppose that it was Gabriel. Then, Carlos lied. Then, Gabriel lied. However, as there was only one who lied, IT WAS NOT GABRIEL.

As there is only one left, it was Paulo who spilled the popcorn.

Let us check: Bernardo did not lie. Carlos did not lie. Gabriel lied. Paulo did not lie. Conclusion: IT WAS PAULO THAT SPILLED THE POPCORN.

In the case of the S Type of solution, as shown in the examples, the reasoning is built on the falsification of assumptions about the dimension "X spills the popcorn", as these are confronted with the premise that there is only one person who is lying. The hypothesis that leads to only one lying is accepted. The schema of reasoning has a simple structure, and it does not yield contradictory inferences, as summarized below:

S Deductive Reasoning

- Suppose X spills, then…; therefore, there is more than one lying (Eliminate).
- Suppose X spills, then…; therefore, there is only one liar (Accept).

Using Relations between the Premises: The R Type of Deductive Reasoning

The solutions of Type R, even less frequent than the former, draw necessary conclusions from relationships between some conditions of the problem, as illustrated below in Examples 1 and 2. In this type of solution, students use combinations of rules to produce new inferences, thus generating new information from the relations devised among the statements provided.

R Deductive Reasoning (Example 1)

- There is only one liar and only one who spilled the popcorn.
- From the two middle sentences, I concluded the following:

-Carlos says: It was Paulo who spilled the popcorn.

-Gabriel says: It was Carlos who spilled the popcorn.

As only one spilled the popcorn, then Carlos and Gabriel cannot both speak the truth because they say different things. So, one of them lies, and thus the liar is either Carlos or Gabriel. So, Paulo speaks truth.

- Paulo speaks the truth and says that Gabriel is lying; therefore, if Gabriel lies, then Carlos speaks the truth, and he says it was Paulo who spilled the popcorn.

R Deductive Reasoning (Example 2)

The statements of Gabriel and Paulo cannot both be true, which means that one of them is lying. So we raised two hypotheses:

Hypothesis A.

If Gabriel tells the truth, then it was Carlos, and Paulo lies (because he says Gabriel lied); then, Carlos lies because he says that it was Paulo (and it was not, since it was Carlos according to this hypothesis), and Bernardo tells the truth (because he says it was not him).

The Hypothesis A is not right because, according to the data of the problem, only one can lie, but in this case, Paulo and Carlos are both lying.

Hypothesis B.

If Gabriel lies, then it was not Carlos (because he says it was Carlos), and Paulo tells the truth (because he says Gabriel lies); as only one of the four friends is lying, then Carlos tells the truth (so it was Paulo) and Bernardo tells the truth (because it was not him, but Paulo).

The conclusion is that the Hypothesis B is correct, so Gabriel was the one who lied and it was Paulo who spilled the popcorn.

This type of reasoning means checking the validity of one of two mutually exclusive assumptions. In a simplified form, this consists of testing two conflicting hypotheses (in referring to the codes used in Section 3.2, that means testing the assumptions P2 and P3 or the assumptions P3 and P4) by admitting the truth of one and the consequent falsity of the other. Considering P3 and P4, one may conclude that one of them has to be false, since Carlos and Gabriel are indicating different names for the guilty one. Likewise, if we consider P3 and P4, one may deduce that one of them has to be false, since Paulo asserts that Gabriel is lying.

In this case, we may describe the R type of deductive reasoning as follows:

R Deductive Reasoning

- Either P2 or P3; suppose not-P2 and P3…; therefore, conflicting inferences (Eliminate).
- Either P2 or P3; suppose not-P3 and P2…; therefore, one guilty and one liar (Accept).

Or else

- Either P3 or P4; suppose not-P3 and P4…; therefore, one guilty and one liar (Accept).
- Either P3 or P4; suppose not-P4 and P3…; therefore, conflicting inferences (Eliminate).

Results from the School Scenario

In analyzing the data from the mathematics class scenario, the first step was to classify the answers as correct, incorrect, or partially correct. As can be seen from the summary in Table 4, it is clear that the success rate of this group of students in the problem was much lower.

Table 4. Success summary from the school body of solutions

Completely Correct	**Partially Correct or Incorrect**	**Total**
31	71	102
(30%)	(70%)	(100%)

The next step was to examine the 31 correct solutions that were obtained. One aspect that stood out was the fact that several resolutions were apparently produced by the students directly on the answer sheet. The written answers showed numerous erasures and, in some cases, there were signs of a lack of space to make multiple attempts, as well as areas of the paper that were written upon and then crossed out by the student.

This time, the data analysis applied the categories already identified in the analysis of the previous body of solutions, that is, each of the resolutions was assessed for its adjustment to any of the identified types of deductive reasoning, including the two subtypes. After careful reading of the answers, it was observed that every solution could be assigned to one of the previous categories, with no other strategy or alternative resolution scheme having emerged, as shown in **Table 5**.

Table 5. Categories from the school body of solutions

Reasoning on the Dim. Lying		**Reasoning on the Dim. Spilling**	**Reasoning on Relations**	**Total**
11		12	8	31
(35%)		(39%)	(26%)	(100%)
L1 Type	L2 Type	S Type	R Type	Total
3	8	12	8	31
(9%)	(26%)	(39%)	(26%)	(100%)

Interestingly, the results show a more even distribution among the various types of reasoning, the most frequent (albeit with a slight advantage) being the one where students focused on the dimension "who spilled the popcorn", that is, the Type S deductive reasoning. The vast majority of solutions present the reasoning in textual form, with some specific cases using tables and even some quick drawings. Apparent hesitations were observed in the elaboration of the written text (as already stated), with several responses showing crossed-out and then redone text. Prototypical examples were identified and compared with those that had been selected in the analysis of the solutions produced outside of school. The fundamental characteristics of each pattern of deductive reasoning were found and confirmed. Below is an example for each pattern of deductive reasoning selected from the students' solutions to the problem in the mathematics class.

L1 Deductive Reasoning (Example)

If Paulo is lying, then it was Carlos who spilled the popcorn. However, Carlos is telling the truth and says it was Paulo. So it is wrong.

If Gabriel is lying, then it was not Carlos who spilled the popcorn. So, it was Paulo because Carlos is telling the truth. This has no errors (Paulo).

If Carlos is lying, then it was not Paulo. So, it was Carlos because Gabriel is telling the truth. However, Paulo is also telling the truth. So, it is wrong.

If Bernardo is lying, then it was him. However, it was also Paulo and it was also Carlos, because Carlos and Gabriel are telling the truth. So, it is wrong.

L2 Deductive Reasoning (Example)

Bernardo	Carlos	Gabriel	Paulo
He said it was not him	He said it was Paulo	He said it was Carlos	He said that Gabriel lied
True	True	True	Lic
			NO. Carlos is true. Gabriel cannot be true.
True	True	Lie	True
		YES. Bernardo is true. Carlos is true. Paulo is true.	
True	Lie	True	True
	NO. Gabriel is true. Paulo cannot be true.		
Lie	True	True	True
NO. Bernardo is not true. Carlos cannot be true.			

Answer: Gabriel lied and Paulo spilled the popcorn.

S Deductive Reasoning (Example)

It was not Bernardo. It was not Paulo. It was not Carlos. It was Gabriel. This possibility is false because there are two liars here.

It was not Bernardo. It was not Paulo. It was Carlos. It was not Gabriel. This possibility is false because there are two liars here.

It was not Bernardo. It was Paulo. It was not Carlos. It was not Gabriel. This possibility is true because there is only one liar here.

It was Bernardo. It was not Paulo. It was not Carlos. It was not Gabriel. This possibility is false because there are three liars here.

Correct possibility: It was Paulo, because Bernardo was telling the truth, Carlos was telling the truth, Gabriel was lying, and Paulo was telling the truth.

R Deductive Reasoning (Example)

Only one of those four friends lied, and only one of those four friends dropped the bucket of popcorn.

It could only have been Paulo or Carlos because they were both accused and because only one of the accusations is a lie.

It could not be Carlos, because if it were Carlos, two of the friends were lying (Carlos and Paulo).

So, it turns out that it was Paulo; therefore, only one of them was lying (Gabriel).

The four examples displayed above effectively substantiate the four reasoning models that were previously described and outlined, and confirm the logical structure that distinguishes them. This leads us to state with reasonable confidence that 10–12 year old students are able to reason deductively and that several models of deductive reasoning are, in fact, plausible among young children.

Comparative and Interpretative Analysis of the Two Scenarios

When confronting the results obtained in the two scenarios, it is possible to perceive commonalities and dissimilarities.

In terms of common results, it stands out that the four patterns of deductive reasoning were found in both groups of subjects. This indicates that children aged 10–12, in different contexts of activity, reason logically and are able to produce adequate and solid deductive reasoning to solve a moderately complex analytical reasoning challenge. Another aspect to be noted is that the R type of deductive reasoning was observed in both cases, especially focusing on dichotomous conditions that entail the use of exclusive disjunction. A third point to be highlighted has to do with the representational resources that were displayed in both groups. The majority of solutions produced by the children reveal the use of free written text with an argumentative spirit, in which the steps of the reasoning appear in a generally sequential and organized way. The use of text lists was also observed, which usually included abbreviations of sentences and use of expressive linguistic elements (e.g., therefore, however, assume, either, or). Other types of notations were observed; for example, the initials of the names of the friends referred to in the problem, as well as the letters T and F to mean true and false, or the letter L to mean lie or liar. In a smaller number, although in both contexts, children used tables as a way to record and organize the production of inferences and the elimination of options. Some examples of simple drawings were also seen, typically used as a way to express the situation in terms of its logical components (e.g., schematic faces or humans to represent the four characters), which were used to highlight the utterances made or to denote inferences (namely, crossing them out when options were eliminated).

As for the contrasts that emerged in the comparison, one of the most obvious has to do with the success rates in the two groups. The notable difference between the higher success of the participants in the beyond-school competition and the lower success of the students in the classroom leads to the consideration of several important conditions for working on analytical reasoning problems. One of them is the existence of possible aid, ranging from adult guidance to the use of resources in solving and expressing the solution. Students in the beyond-school scenario, in many cases, used digital tools to present their solution process, and were able to take advantage of the affordances of those tools. On the contrary, students in the classroom only had paper and pencils at their disposal. Another important condition is linked to the time available to work on the problem. This condition seems relevant not only to interpret the difference in the number of correct answers, but also to justify the most frequent type of reasoning (focused on the dimension "lying") among students who solved the problem in the beyond-school scenario. As the results demonstrated, both the L1 type and the L2 type of deductive reasoning require a more laborious construction of inferences, not because of the number, but because of its extent and its nature. These are the answers that tended to use more space and that appeared to involve longer solution processes. The apparent prevalence of these types of reasoning among students in the beyond-school scenario may be in line with some of their probable characteristics—they usually like to solve challenges, persist in looking for solutions, invest time and work to arrive at an answer, accept more easily the complexity of a problem, and value the quality of the explanation of the solution. Furthermore, unlike students in the classroom, they had the possibility to do and redo several attempts to solve the problem and to choose what seemed to them the most explicit and complete way of showing the reasoning, which may not be the most shortened one. In the case of the students in mathematics classes, not only the more reduced time, but also the reduced resources may inhibit the development of long or more difficult inferences. They could be more disposed to get easy and swift answers. The expression of reasoning seems to be less essential to them because they had fewer opportunities to do and redo attempts. This was perceptible from some messiness in the students' answers and the lack of space that they possibly struggled with.

DISCUSSION

The present study aimed at getting a fine-grained picture of how young children (aged 10–12), in different scenarios, solve an analytical reasoning problem based on deductive reasoning, and to know about the ways in which that reasoning was expressed. The empirical data consisted of 384 solutions from children participating in a beyond-school problem-solving competition taking place through the Internet, and 102 solutions produced by students from the fifth and sixth grades in their mathematics classrooms without any help from others.

Models of Deductive Reasoning in Problem-Solving

Two main approaches to the problem were identified in the solutions collected. One approach focuses on the dimension "who lied", and the other on the dimension "who spilled the popcorn". It is possible to consider four atomic propositions related to the act of lying and four others related to the act of dropping the popcorn, as well as a set of relations between these, as described earlier. Thus, the approach of looking for who lied directs attention to the goal of determining which of the four friends lied, whereas the approach of looking for who dropped the popcorn sets the objective of finding the dropper, in both cases keeping in mind that only one lied and only one spilled the popcorn. Clearly, in either approach, both dimensions were involved, since the relational structure of the problem entails that.

In those two approaches, the predominant strategy for solving the problem is consistent with the one discussed by Newstead et al. [25], that is, participants showed a systematic use of rule checking and option elimination. So, they worked systematically in testing hypotheses, looking for contradictions, and discarding those generating conflicting inferences. The general structure of the reasoning developed consists of assuming an option and testing it against all the given premises, using different control criteria in searching for contradictions, that is, the hypothesis is excluded when one or more contradictions arise.

Four models of deductive reasoning (L1 Type, L2 Type, S Type, and R Type) were found—the first two belonging to the approach driven by the search for "who lied", the third to the approach centered on finding "who spilled the popcorn", and the fourth based on reformulating the structure of the problem by previously working on relations between the premises. The different models were found in the two bodies of solutions examined, although with different frequencies in the two groups. This result is interesting for several reasons.

On the one hand, it demonstrates that in the two scenarios, namely with and without help, young students generated models of solutions that fit the four types, suggesting that the mental models perspective is useful to conceptualize and understand the possibilities of deductive reasoning within the reach of children. In a nutshell, children build mental models that represent the truth of the premises in the problem situation. Furthermore, it reveals that the different settings are not irrelevant in the children's choice of models. Indeed, our results show that a majority of participants (59%) in the beyond-school scenario used the testing of hypotheses for the dimension "who lied", apparently taking the most arduous approach for representing and checking all the possibilities against the premises. Even more clearly in the case of the L1 Type solutions, the students did not try to reduce the relational complexity and, as such, the reasoning required meticulous analysis of the hypotheses. This result is consistent with the conclusions of English [24], according to which solvers tend to avoid the complexity of relational assumptions by directing their effort to the exhaustive logical analysis of the series of premises. If the students are working on the problem with an extended time to deliver the solution, this seems a rather significant factor. On the contrary, many of the students who solved the problem in the classroom may have struggled with the reduced time to engage in this form of resolution. Thus, this model, which is underrepresented in the second body of solutions (9%), may have been the option of several students who were unable to take it forward.

Moreover, the solutions from both groups where students focused on the dimension "who spilled the popcorn" showed a more direct, expeditious, and clear process of achieving necessary conclusions, but such clear-cut approaches are also known to be less common [24,25]. Hence, it seems that the models L1, L2, and S progress in the increasing control capacity that they offer for hypothesis testing. In particular, in the L1 Type, at the start of the inferential reasoning, students do not keep track of those who told the truth or those who did not spill popcorn; in the L2 Type, they keep track of those who spoke the truth from the beginning; and, in the S Type, they keep track of those who did not spill popcorn and get the number of lies (as in the examples shown).

Finally, the R Type of reasoning is rather dissimilar, and looks like compelling evidence of the theory of mental models [12,24,25,28] in that it suggests the development of a more complete and refined model of the given situation, showing the understanding that some of the premises in the problem may be replaced by other equivalents, thus creating a new "version" of it. In other words, the students who used the R Type of deductive reasoning worked on the premises before moving to the systematic and exhaustive checking of hypotheses. This means that they dealt with the complexity of the relationships expressed in the problem and, in particular, with the indeterminacy involved in the premises [15], which is also related to the semantics of the problem. In fact, it was interesting to find that the frequency of this model was actually higher in the students from the classroom setting. This means they were able to realize that some of the propositions were mutually exclusive. One possible interpretation for this result is related to the contextual nature of the problem and to its semantic richness, one characteristic that is seen as relevant in triggering counterexamples [12,15,29].

Analytical Reasoning Problems as Deductive Reasoning Problems

Given that all the students were attending the fifth and sixth grades, their learning of formal logic in school is inexistent, and logical reasoning is very limited in their mathematics lessons, as it is delayed to more advanced school levels in the Portuguese curriculum. However, over several years of implementation of the mathematical problem-solving competition, it was observed that analytical reasoning problems, compared with several other non-routine problems proposed, were among the competitors' favorites, apparently because they incite their curiosity and appeal to their cleverness [9]. This suggests that such problems deserve new attention from school mathematics concerning the development of deductive reasoning. The importance of knowing how children reason deductively on such contextual logical problems is thus justified. Moreover, since young children do not yet master the language of formal logic, it is also important to understand how they express their deductive reasoning with their language and representational resources.

The view that analytical reasoning problems are substantially deductive reasoning problems [17,25] is clearly conveyed by the analysis of the participants' resolutions to the problem of the spilled popcorn. The children's solutions revealed the strong presence of logical inferences expressed in the process used to solve the problem. With regard to the set of deductive reasoning skills involved in analytical reasoning problems, it seems clear that in the problem of the spilled popcorn, the students reached an understanding of the basic structure of the relations involved in the premises and rules of the problem, made extensive use of inferential reasoning to produce true and necessary inferences based on option elimination, drew additional or substitute rules from some of the initially given conditions, and identified equivalent propositions in the context of the problem situation. As in other studies with older students [11,23,25,30], young students reasoned deductively, using similar processes and language, although adjusted to their level of knowledge and development. Likewise, the results of the study corroborate findings of experiments with younger children, namely with regard to principles that most influence the success of solvers in analytical reasoning problems: (1) Selecting a premise that most readily yields an explicit problem-situation model and (2) integrating premises where appropriate in the existing model [24].

In light of the results, analytical reasoning problems represent a potential kind of challenging task for the development of deductive reasoning. Not only do they seem to encourage the use of logical structures of high complexity, they have been shown to be within reach of the students' ability to reason deductively. The results also lead to the conclusion that this is a valuable resource for the integration of logical and deductive reasoning in the school setting, especially if one embraces the view that logic can and should be driven by the curriculum in order to make it meaningful, and as a way of establishing bridges with deductive reasoning in mathematics topics and in proofs [4,16,17,18,19,22].

As underlined by Newstead et al. [25], analytical reasoning problems differ from those tasks more typically used in the psychology of deductive reasoning. Firstly, they offer contextualized situations; secondly, they put deductive reasoning into action while solving a challenge rather than focusing on determining the truth of an abstract rule. This is in line with claims by several researchers who have studied productive ways of integrating deductive reasoning and logical principles in school mathematics [4,8,16,17,19]. Thus, from the standpoint of supporting students' deductive reasoning and as potential tools for the development of logical competence in school mathematics, analytical reasoning problems are a still-underexplored resource.

Language and Resources in Expressing Deductive Reasoning

The children's linguistic expression of their solutions to the problem seems to simultaneously reflect logical thinking in everyday contexts and deductive reasoning in formal logic problems.

A number of textual formulations were identified, such as: "If this… therefore that", "suppose this… then that", "if this, then that…; hence…". In addition, relevant was the use of the word "however" in expressing the presence of some contradiction that led to the rejection of some hypothesis. That was perceptible within the logical scheme of identifying conflicting inferences, expressed in words like: "However, it cannot be", "however, there is a contradiction", and so forth. Moreover, in the case of the R Type of reasoning, students noticed the presence of mutually exclusive propositions, which they often conveyed through the words: "Either–or", "either this or that", "either this is true or that is true", or "only one of the two can be true". This shows that students efficiently dealt with the "exclusive or" in some of the solutions obtained.

Contrary to what is common in analytical reasoning problems, the use of tables was not the most frequent, although it was a representational resource used by some children in both contexts. The use of tables appears to have proven more effective and also easier in cases where the approach was focused on the dimension of "who spilled the popcorn". In fact, in the approach focused on "who lied", the participants who used the tabular representation had to build a succession of different tables, while in the other approach, a single table proved to be sufficient for the development of the entire reasoning.

To some extent, we may conclude that even young children have the linguistic resources to express their logical reasoning on a contextual problem in meaningful and strongly referential ways, which resonates with the perspective that mental models are not expressed by the rules and much less by the syntax of formal logic, as argued by Markovits and Barrouillet: "We consider it unlikely that reasoning problems are easily represented in even a semiabstract way, particularly among children" [28] (p. 32). This also indicates that analytical reasoning problems are a fruitful resource for the progressive development of logical reasoning, as they represent contextual problems that are, at the same time, intuitive, semi-abstract, and "pre-symbolic"; they are therefore interesting as catalysts in exploring deductive reasoning in mathematics education. The results furthermore corroborate the evidence that some studies have obtained with even younger children [31], namely that they are able to construct mathematical and deductive reasoning in solving non-routine problems and to use more or less conventional forms of representation in expressing their reasoning. One of the crucial functions of the use of expressive resources by young children is to unpack the structure of the problem [31].

CONCLUSIONS

In this study, we examined children's solutions to a moderate mathematical challenge involving analytical reasoning, produced in two different contexts: The classroom and a web-based mathematics competition.

Our study found that children aged 10–12 years are capable of reasoning deductively in solving an analytical reasoning problem and that the deductive reasoning models revealed by their solutions are of four types, each with its own logical structure and corresponding to the dimensions of the problem ("who lied" and "who spilled the popcorn") or a relationship between them. The four types of deductive reasoning appeared in the two settings in which the solutions were created, though with different frequencies.

Children can express deductive reasoning through language and representational resources while testing hypotheses systematically, looking for contradictions, and eliminating options. The textual formulations in their solutions show features of deductive reasoning typical of formal logic, as well as traits of everyday logical thinking. Although less frequent, other forms of representations were used, such as tables, ordered lists, or diagrams. The use of a single table was a powerful resource in implementing the S Type of deductive reasoning, while several tables had to be used for the L Type of approach.

The lowest success rate in the problem of the spilled popcorn was obtained with the students who solved the problem individually, without help, in limited time, and with basic resources. Despite the reduced success, those who successfully completed the task showed deductive reasoning models identical to those built by other students in a beyond-school environment. This leads to the assertion that deductive reasoning should be cultivated and reinforced in the school context, even with young children. Among several factors to consider, students should be given enough time to create their mental models of the situation; the use of resources may be important, namely representational resources, such as tables, ordered text, diagrams, notations, symbols, and registers that help to deal with relational complexity; good contextualized problems and purely mathematical problems with a relational character are desirable (the attractive factor of solving puzzles or enigmas is important in students' engagement); the use of analytical reasoning problems with a slight degree of indeterminacy seems to be useful, as it brings up different types of deductive reasoning that may be beneficially shared in the class; and, finally, solving analytical reasoning problems involves a very significant expression of reasoning and, therefore, it constitutes an opportunity to foster the linguistic and representational abilities of students in expressing logical thinking, namely chains of deductive arguments.

AUTHOR CONTRIBUTIONS

Conceptualization, S.C.; methodology, S.C.; formal analysis, S.C., N.A., and H.J.; writing—Original draft preparation, S.C.; writing—Review and editing, S.C., N.A., and H.J. All authors have read and agreed to the published version of the manuscript.

REFERENCES

1. NCTM. *Principles and Standards for School Mathematics*; National Council of Teachers of Mathematics: Reston, VA, USA, 2000.
2. MOE. *Mathematics syllabus –Primary One to Six*; Ministry of Education—Curriculum Planning and Developmental Division: Singapore, 2012.
3. Dreyfus, T. Why Johnny can't prove. *Educ. Stud. Math.* 1999, *38*, 85–109.
4. Lee, K. Students' Logical Reasoning and Mathematical Proving of Implications. Ph.D. Thesis, Michigan State University, East Lansing, MI, USA, 2011. Available online: https://d.lib.msu.edu/etd/1959 (accessed on 7 January 2019).
5. Inglis, M.; Simpson, A. Mathematicians and the Selection Task. In *Proceedings of the 28th Conference of the International Group for the Psychology of Mathematics Education, Bergen, Norway, 14–18 July 2004*; Høines, M.J., Fuglestad, A.B., Eds.; Bergen University College & PME: Bergen, Norway, 2004; Volume 3, pp. 89–96.
6. Durand-Guerrier, V. Which notion of implication is the right one? From logical considerations to a didactic perspective. *Educ. Stud. Math.* 2003, *53*, 5–34.
7. Stylianides, G.J.; Stylianides, A.J. Proof in school mathematics: Insights from psychological research into students' ability for deductive reasoning. *Math. Think. Learn.* 2008, *10*, 103–133.
8. Stylianides, A.J. Introducing young children to the role of assumptions in proving. *Math. Think. Learn.* 2007, *9*, 361–385.
9. Carreira, S.; Jones, K.; Amado, N.; Jacinto, H.; Nobre, S. *Youngsters Solving Mathematical Problems with Technology*; Springer: New York, NY, USA, 2016.
10. Kaufman, S.B.; Ross, H.J.; Silk, E.M. *A Cognitive Model of Analytical Reasoning Using GRE Problems*; Carnegie Mellon University: Pittsburgh, PA, USA, 2001. Available online: https://scottbarrykaufman.com/wp-content/uploads/2013/01/Kaufman-Etal-2001-GRE-Problems.pdf (accessed on 17 February 2019).
11. Cox, R. Analytical Reasoning with Multiple External Representations. Ph.D. Thesis, University of Edinburgh, Edinburgh, UK, 1996. Available online: https://www.era.lib.ed.ac.uk/bitstream/handle/1842/10550/Cox1996.pdf (accessed on 12 March 2019).

12. Schroyens, W.; Schaeken, W.; D'Ydewalle, G. The processing of negations in conditional reasoning: A meta-analytic case study in mental model and/or mental logic theory. *Think. Reas.* 2001, *7*, 121–172.
13. Johnson-Laird, P.N. Deductive reasoning. *Annu. Rev. Psychol.* 1999, *50*, 109–135.
14. Johnson-Laird, P.N.; Byrne, R. A model point of view. *Think. Reas.* 1995, *1*, 339–350.
15. Van der Henst, J.-B. Mental model theory versus the inference rule approach in relational reasoning. *Think. Reas.* 2002, *8*, 193–203.
16. Nunes, T.; Bryant, P.; Evans, D.; Bell, D.; Gardner, S.; Gardner, A.; Carraher, J. The contribution of logical reasoning to the learning of mathematics in primary school. *Br. J. Dev. Psychol.* 2007, *25*, 147–166.
17. Epp, S. The role of logic in teaching proof. *Am. Math. Mon.* 2003, *110*, 886–899.
18. Bakó, M. Why we need to teach logic and how we can teach it. *Int. J. Math. Teach. Learn.* 2002. Available online: http://www.cimt.plymouth.ac.uk/journal/bakom.pdf (accessed on 12 March 2019).
19. Hoyles, C.; Küchemann, D. Students' understandings of logical implication. *Educ. Stud. Math.* 2002, *51*, 193–223.
20. Ayalon, M.; Even, R. Deductive reasoning: In the eye of the beholder. *Educ. Stud. Math.* 2008, *69*, 235–247.
21. Morsanyi, K.; McCormack, T.; O'Mahony, E. The link between deductive reasoning and mathematics. *Think. Reas.* 2018, *24*, 234–257.
22. Lommatsch, C.W. Learning Logic: A mixed methods study to examine the effects of context ordering on reasoning about conditionals. Ph.D. Thesis, Utah State University, Logan, UT, USA, 2018. Available online: https://digitalcommons.usu.edu/etd/7011 (accessed on 10 February 2020).
23. Cox, R.; Brna, P. Supporting the use of external representations in problem solving: The need for flexible learning environments. *Int. J. Artif. Intell. Educ.* 1995, *6*, 239–302.
24. English, L. Children's reasoning in solving relational problems of deduction. *Think. Reas.* 1998, *4*, 249–281.
25. Newstead, S.; Bradon, P.; Handley, S.; Dennis, I.; Evans, J. Predicting the difficulty of complex logical reasoning problems. *Think. Reas.* 2006, *12*, 62–90.

26. Krippendorff, K. *Content Analysis: An Introduction to its Methodology*, 2nd ed.; Sage Publications: London, UK, 2004.
27. Schreier, M. *Qualitative Content Analysis in Practice*; Sage Publications: London, UK, 2012.
28. Markovits, H.; Barrouillet, P. The development of conditional reasoning: A mental model account. *Dev. Rev.* 2002, *22*, 5–36.
29. Van der Henst, J.-B. Mental model theory and pragmatics. *Behav. Brain Sci.* 2000, *23*, 283–284.
30. Yang, Y.; Johnson-Laird, P.N. Mental models and logical reasoning problems in the GRE. *J. Exp. Psychol. Appl.* 2001, *7*, 308–316.
31. Canavarro, A.P.; Pinto, M.E. O raciocínio matemático aos seis anos: Características e funções das representações dos alunos (Mathematical reasoning at age six: Characteristics and functions of students' representations). *Quadrante* 2012, *21*, 51–79.

Chapter 2

RELATIONS BETWEEN GENERALIZATION, REASONING AND COMBINATORIAL THINKING IN SOLVING MATHEMATICAL OPEN-ENDED PROBLEMS WITHIN MATHEMATICAL CONTEST

Janka Medová [1], Kristína Ovary Bulková [2], and Soňa Čeretková [1]

[1]Department of Mathematics, Faculty of Natural Sciences, Constantine the Philosopher University in Nitra, Tr. A. Hlinku 1, 949 74 Nitra, Slovakia

[2]Department of School Education, Faculty of Humanities, Tomas Bata University, Štefánikova 5670, 760 01 Zlín, Czech Republic

ABSTRACT

Algebraic thinking, combinatorial thinking and reasoning skills are considered as playing central roles within teaching and learning in the field

Citation: (APA): Medová, J., Bulková, K. O., & Čeretková, S. (2020). Relations between Generalization, Reasoning and Combinatorial Thinking in Solving Mathematical Open-Ended Problems within Mathematical Contest. Mathematics, 8(12), 2257. (20 pages).

Copyright: © Creative Commons Attribution 4.0 International License (http://creativecommons.org/licenses/by/4.0/).

of mathematics, particularly in solving complex open-ended mathematical problems Specific relations between these three abilities, manifested in the solving of an open-ended ill-structured problem aimed at mathematical modeling, were investigated. We analyzed solutions received from 33 groups totaling 131 students, who solved a complex assignment within the mathematical contest Mathematics B-day 2018. Such relations were more obvious when solving a complex problem, compared to more structured closed subtasks. Algebraic generalization is an important prerequisite to prove mathematically and to solve combinatorial problem at higher levels, i.e., using expressions and formulas, therefore a special focus should be put on this ability in upper-secondary mathematics education.

Keywords: mathematical problem solving; combinatorial thinking; reasoning; mathematical proof; algebraic generalization

INTRODUCTION

Mathematics education is constantly evolving based on the requirements of contemporary society. Consequently, the curriculum development is strongly influenced by 21st-century learning skills in particular critical thinking, problem solving, communication and collaboration [1]. It is obvious that any problem without some known algorithm or procedural tool can be considered as a challenge [2]. Students intending to inquire as defined in [3] feature the following characteristics: asking questions, attempting and experimenting, identifying the problem, choosing the right strategy in problem solving. The nature of open-ended problems represents the means how to implement the inquiry in mathematics education [4] and help students to learn how to think independently for using the problem solving strategies [5]. Over the course of many years, solving open-ended problems in mathematics has evolved into an activating method that supports the 21st-century learning skills mentioned above.

The ability to solve complex open-ended problems is anchored in some various different factors including the following: (i) metacognitive knowledge [6,7], (ii) positive attitude towards mathematics [8,9,10,11,12], (iii) mastering mathematics processes as are reasoning, generalization, communicating the results or mathematical modeling [13,14,15,16,17,18] and (iv) high level of mathematical proficiency in both, procedures [19,20] and conceptual understanding [21,22,23], including arithmetic, algebraic and combinatorial thinking. These aforementioned factors are not isolated, but they rather support each other mutually in the influence of the ability to solve complex open mathematical problems [24,25].

The distinction between arithmetic and algebraic thinking is sometimes oversimplified as a work with numbers, or a work with variables. In contrast, Radford [26] distinguishes between the arithmetic generalization, algebraic generalization and naive inductive reasoning based on the distinctive ways of thinking. It is more than clear that algebraic generalization builds on the observance of commonalities, by abduction from particulars. Then, a hypothesis, if formulated, and an expression, are or can be produced. On the other hand, just the arithmetic generalization often uses some recurrent relations and, following that, pupils, using this way of thinking are not able to provide a formula for a general case. However, the least sophisticated are the naive inductions based on probable reasoning. The type of generalization can be related to the type of problem occurring, however, the students equipped with well-connected problem solving schemas are not surprisingly more successful in transferring their knowledge [27]. On the other hand, the algebraic generalization raises specific problems for almost every student, even in the course of studying higher grades [28]. Our recent research findings indicate an ability to produce some general algebraic expressions as a crucial factor in solving open-ended non-routine problems in mathematics, even by high-achievers [29].

For the students the use of justification, proving strategies and techniques of various forms of proofs in mathematics, providing them with a crucial importance of mathematical reasoning, is conclusive. Therefore, such abilities clearly involve obtaining strategic knowledge in the specific areas related to the given problem at hand, as well as knowledge and norms specific for proving and reasoning [30], especially in mathematical open-ended problem-solving situations. Pedemonte [31] stresses the structural gap between the argumentation and proof. It is clear that just the argumentation inferences are based on the content, while in case of any proof they follow a deductive scheme. Within the work in the classroom environment, usually, students mostly experience some memorized algorithmic reasoning [32] or with some formal proofs, a sequence of formulas in the formal language [33] instead of the actual creative mathematical reasoning within a complex mathematical inquiry that enables understanding and construction of a new knowledge.

The ability to make some generalizations and to provide appropriate reasoning are both considered as high-level mathematical skills. Pedemonte further [34] studied certain relations between the level of the generalization and reasoning skills. However, generalization as such and concurrently as a process appears to be required for construction of mathematical induction, although, on the other hand, some relation was pointed out as a structural distance between them. Therefore, just the open-ended problems are considered suitable ways to develop required reasoning skills by students. Thus, the abductive reasoning serves for passing from the inductive to deductive reasoning [26]. It is not clear how the classifications or characteristics of generalization and justification arising from the studies that have focused on generalizing a number patterns apply to other mathematics topics or tasks. Nor do these frameworks make clear other reasoning processes that enable generalization and justification to occur [35].

Combinatorics has a special place in the field of mathematics and mathematics education. It is considered as a source of frequent and hard difficulties of students at various levels [15]. The fact is that just the combinatorial thinking presents the process of creating various possible combinations of ideas and cognitive operations [36]. The use of combinatorial thinking in combinatorial situations means developing a special ability to create abstract models and to find the structure of a set of outcomes [37]. When given enough time and hands-on problems, even usually low-achieving students can do well in solving combinatorial problems, and vice-versa, usually high-achieving students can lose track when dealing with novel problems in combinatorics [38]. Jones et al. [39] considered the following four levels of combinatorial thinking development: subjective, transitional, and informal quantitative, and numerical. Lockwood [40] stresses the role of the listing of elements forming a set of outcomes while solving any combinatorial problems. The suggested trajectory of solving the problems from some concrete elements to some general expressions seemed productive in helping students to gain familiarity with a context before generalizing, using variables. Problems in combinatorics provide a reasonable environment for generalization of the solution as "counting problems are often accessible yet challenging and these accessible problems provide a natural structure from which students may generalize" [41] (p. 311). During the process of generalization based on the combinatorial problem-solving any gifted students tried to understand the combinatorial situation more deeply, to identify some assumptions more precisely and a formulated plan was global in nature [42].

Little is known about the connections between the mathematical reasoning and combinatorial thinking. One of a few studies [41] investigating the affordances of combinatorial proof which can be considered as a valuable instrument for development of combinatorial thinking and for their proving skills is worth mentioning. The students participating in the study seem to be more convinced by the algebraic arguments than by the combinatorial reasoning.

The standard memorization of common algorithms and the routine calculations repetition have been slowly abandoned in mathematics education today. Instead, students are motivated to thinking about the given problem more deeply, analyzing and discussing the possibilities of different solutions [43] and looking for other mathematical aspects related to the mathematical content of the given problem. Therefore, any mathematics contest can be considered as a challenge [44] improving the students inquiring and using various problem-solving strategies. The development of the mathematical reasoning and problem solving strategies is supported in any mathematical contest especially aimed at students' mutual collaboration [2].

Group-work in mathematics education represents the learning method based on the definition of collaborative problem solving by PISA (the OECD's Programme for International Student Assessment) [45] as following: "an individual's capacity to use cognitive processes to confront and resolve real, cross-disciplinary situations where the solution path is not immediately obvious and where the content areas or curricular areas that might be applicable are not within a single subject area of mathematics, science or reading" (p. 7). Mathematical open-ended problems requiring the higher mathematical thinking skills, can provide students an opportunity to share their knowledge [46], to discuss the strategies [47] and create the conclusion from the team as the unit [48].

As stated by Ericson and Lockwood [49] in order to prove a combinatorial identity various cognitive models are needed in order to understand the algebraic notations. Furthermore, it seems that the ability to provide plausible reasoning within solving combinatorial problems is influenced by a variety of representations at students' disposal, including different algebraic notation [50]. Therefore, we consider as important to shed some light on the relations between the three mathematical competencies within a solution of complex open-ended mathematical problem by high-achieving students. Furthermore, the relations are usually studied for individual students. However, several researchers [51,52] suggest that especially cognitive abilities develop through social interactions, thus the group solutions should be investigated when producing highly cognitively demanding activity such as solving complex mathematical problems. Collaboration in groups is well suited to accessing skills manifested in problem-solving because it encourages students to explain their thinking and engage in joint elaboration of the final assignment [53].

We are not aware of any study investigating relations between the various mathematical abilities while studying group solutions of highly-demanding complex mathematical problems.

Our research is focused on the analysis of the team mathematical contest known as Mathematics B-day in the two particular perspectives: quantitative and qualitative. The assignment of the contest was analyzed as a coherent set of the mathematical problems aimed at inquiry in mathematics. We analyzed the correctness of the solution of subtasks with the different level of combinatorial thinking, algebraic generalization and mathematical reasoning. Then, the results of the analysis were compared with the authentic solutions of groups consisting of 3–4 students.

The research question was formulated as follows: what are the (implicative) relations between the levels of algebraic generalization, mathematical reasoning and combinatorial thinking manifested in the course of solving non-routine mathematical problems by groups of high-achieving upper-secondary students?

MATERIALS AND METHODS

The Mathematics B-day contest [54] is a unique opportunity for upper secondary students to collaborate on the real mathematical open problem solving and demonstrate their mathematical knowledge and competences by conjecturing and proving in mathematics. In the Slovak Republic the Department of Mathematics of Constantine the Philosopher University in Nitra is an organizer of such an event, although the origin of this contest was in the Netherlands. The essence of the Mathematics B-day contest is based on the educational program of mathematics for the prospective university level of technical studies and studies in science and mathematics.

Students, divided into three or four member teams, are solving the assignment created with an intent motivating inquiry in mathematics. The assignment represents a set of subsequent problems related to the one chosen situation composed in the continuous text.

The goal of Mathematics B-day 2018 assignment [55] was inspired by the mathematical open ended problem called "The Moser's Worm Problem" formulated as follows: "What is the surface of the smallest flat shape that can cover all lines of length 15 cm?" [56] (p.153). The situation was represented for students like a story of a certain snake with its length of 15 cm, that needs a blanket for every position while this snake is sleeping. The assigned problems were aimed at explaining specific relations, while generalizing and proving, and finally, at the mathematical modeling as such. Students could use recommended manipulatives (copper wire, sheet of graph paper, compass, scissors, etc.) for their experimenting in the first phase of the solution process and later the dynamic software GeoGebra was recommended. It is more than obvious that a success solution requires some elemental pieces of knowledge, stemming from geometry (e.g., planar geometry and trigonometry) as well as combinatorics (e.g., enumeration of combinations to produce a pattern).

In 2018, 131 students gathered into 33 teams solved the assignment of the Mathematics B-day contest. The relatively high achievement in mathematics is assumed because only two best reports from each participating school can be submitted for the final national assessment. The actual and real abilities to solve non-routine mathematical problems are one of the basic components of the general problem-solving ability [57]. The non-routine problems will demand a high cognitive load [58], therefore, the high-achievers' solutions need to be analyzed to a greater extent. For this reason, the solvers, who participated in the Mathematical B-day event, can represent an appropriate sample for observing the level of generalization and proving in mathematics.

Assignment of Mathematical B-Day 2018: Snake Nest

The full text of the assignment represents 14 pages of continuous mathematical text, containing several kinds of non-routine problem, aimed at the following topics and themes: experimenting, calculations, and then, explanation, generalizing, as well as proving some relevant and real relations. This study was focused on some necessary subtasks, requiring some higher-order mathematical thinking skills, vested, especially, in the fields of proving and generalizing. The assignment is composed from the 6 tasks further divided into more subtasks characterized like open-ended problems.

The introductory subtasks (1a–1e) were aimed at the shape of a circle covering the all positions of the snake represented by a 15 cm long curve.

Students had to find the smallest possible radius and properties in specific cases. In the subtasks 2a–2e, the positioning strategy of the snake is explained to students based on a special example: the snake bended in a rectangle.

The following tasks 3, 4, and 6, are aimed at discovering the smallest planar shape represented by several shapes like a half and quarter of the circle, triangle or diamond. Subsequently, students are asked to arrange the smallest possible area of a blanket by cutting off any useless parts of the aforementioned shapes. Apparently and definitely, some combinatorial thinking skills are required in the task 5. The problem about the smallest area of the blanket is viewed by means of a simplified model represented by the tetra-snake. The body of any tetra-snake consists of squares and the centers thereof can be connected by lines. Any given broken line can determine any position of the snake according to their shape, as shown in Figure 1.

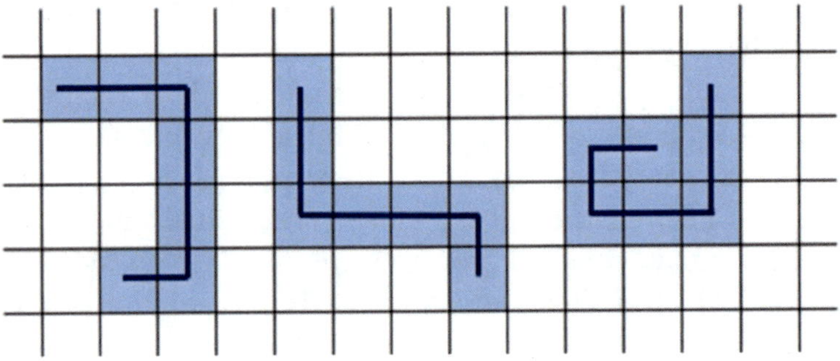

Figure 1. Visual aid for subtask aimed at the combinatorial thinking (adapted from [55]).

The findings from the introductory assignments should be useful for creating a mathematical model of Final assignments assigned as follows: "Design the smallest possible blanket for the 15 cm snake" [55] (p. 14). The proposed approach is generalized in the following six steps:

- Limit what type of snakes and what type of blankets do you are consider.
- Experimenting.
- Give a positioning strategy for the all shapes for those blankets.
- Find the corresponding minimum dimensions.
- Explain that the all forms remain under the blanket with the strategy from the third step.
- Cut bits away from the blanket to make it even smaller.

As mentioned above, the assignment consists of several types of interconnected task characterized as the open-ended problems. The subtasks were sorted based on the level of required abilities into the three groups: (a) experimenting and applying, (b) reasoning and hypothesizing, (c) generalizing and proving. The first group of subtasks requires some basic abilities for solving routine problems aimed at estimating the concrete result or problems with a closed goal but open approach. The other two groups represent problems requiring a higher level of abilities. Figure 2 below shows how the subtasks of the all assignments are interconnected.

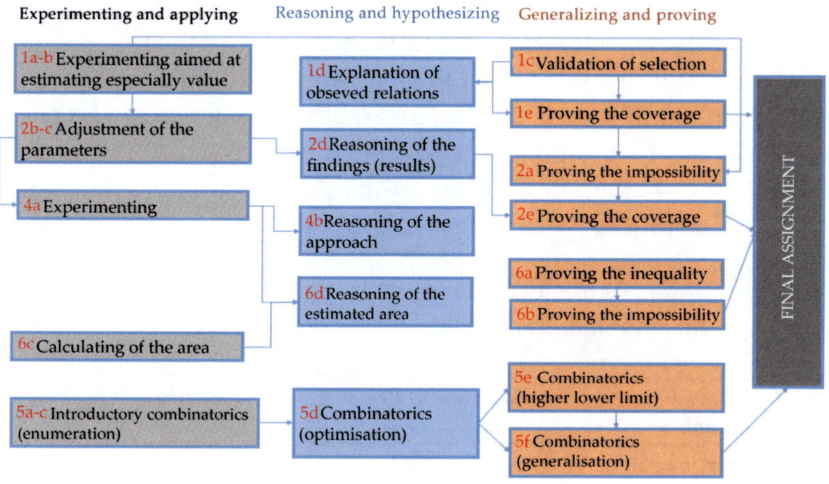

Figure 2. Schema of interconnected subtasks.

The schema (Figure 2) shows that any assignment is composed in the sense of the recommended steps in the final assignment. To achieve success, the use of the students' wide abilities on all levels are required. Task 5 is based on the essence of the main problem to find the smallest blanket for the snake but in view of using combinatorics. Therefore, the subtasks 5a–5f appear to be independent of the whole text of the assignment and the assumption that the previous tasks can influence and affect the success of a solution of the task 5 is omitted. The Moser's Worm Problem can be included in the field of combinatorial geometry. Thus, specifically, we aimed at the subtasks where the investigated skills were supposed to be manifested, particularly the subtasks 1c, 1e, 2a, 2e, 5e, 5f, 6a, 6b (see the full assignment in [55]), while in the case of the subtasks 5a–5f for combinatorial thinking as well as final assignment, being focused mainly on the skills of mathematical modeling.

Statistical Analysis

The descriptive correlational research was used in this study to assess the relationships between students' level of algebraic generalization, mathematical reasoning and combinatorial thinking. The assignment described in Section 2.1 was solved by 33 groups of students from 11 upper-secondary schools in Slovakia within the contest Mathematical B-day 2018. Each group consisted of 3–4 students. The intended output of the contest, final report of each group was assessed and following variables were investigated.

The level of reasoning skills was assigned to every student's solution of the all investigated subtasks (2a, 5f, 6a, 6b, 6d), and the final assignment (FA) as well. The level of algebraic generalization was assigned to every student's solution of the subtask 5f, and the final assignment (FA) as well. The subtasks 5a, 5c, 5f and final assignment were assessed for the level of combinatorial thinking. Table 1 shows the all data covering assessing the mathematical abilities.

Table 1. Description of the levels of algebraic pattern generalization, reasoning and proving, and combinatorial thinking

Level	Algebraic Generalization	Reasoning and Proving	Combinatorial Thinking
0	Observing particular examples	Without any argumentation or reasoning	Listing of the elements in random order, without looking for systematic strategy.
1	Noticing a commonality	Reasoning by one or several particular examples	Using a trial-error strategy, discovery of some generative strategies for small sets of outcomes.
2	Formulating a hypothesis	Correct mathematical evidence, but not formalized in a form of proof	Adopting generative strategies for bigger sets or three- and more-stage case.
3	Producing the expression of p_n	Formal mathematical proof	Applying generative strategies and use of formulas.

Following this, the subsequent statistical analysis of the obtained data was performed, using the software environment R [59], applying the packages RVAideMemoire [60], lsr [61], and rchic [62]. The success-rates in the subtasks of the given assignment were compared by the Cochran Q test that is the generalization of the McNemar test for the two independent samples. The partial problems were considered as independent samples.

Subsequently, the post-hoc analysis, comparing each pair of problems, was performed by the McNemar test. The levels of reasoning skills and algebraic generalization in the analyzed subtasks were compared, and then, post-hoc analyzed by the Friedman test.

The mutual relationships between the assessed features of the group solutions of the complex problem were assessed. Different statistical characteristics were implemented based on the scale used for the particular variable. The relationship between a Boolean variable describing the correctness of the solution of particular subtasks was assessed by φ coefficient. The Spearman's rank correlation coefficient ρ was used to assess a relationship among the level reasoning skills, combinatorial thinking and algebraic pattern generalization manifested in particular subtasks. The corrected Cramer's V was calculated to measure an association between the correctness of a solution of the particular subtask and levels of generalization, reasoning, and combinatorial thinking.

The statistical implicative analysis (SIA) [63] was applied to explore some mutual relations between the defined attributes of assessing and to evaluate relations between the subtasks in the basic assignment and the students' performance in the final assignment.

RESULTS

Out of a total 33 submitted team solutions 27 attempted to solve the final assignment. The relative frequency of investigated items differed significantly ($Q=5.4$, $p=0.020$), a pairwise comparison by the McNemar test is summarized in Table 2. The null hypothesis was formulated for each pair of subtasks as follows: "There is no significant difference between the ratio of incorrectly solved subtasks for the subtask A and subtask B". The levels of reasoning skills varied between the tasks ($chi^2=48.607$, $p<0.001$), contrasting to the level of algebraic generalisation ($chi^2=1.316$, $p=0.251$). The three subtasks were aimed at proving inequality/impossibility (2a, 6a, 6b). The subtask 2a was solved correctly and significantly more often than the other two subtasks ($p_{2a6a}=0.035$; $p_{2a6b}<0.001$; $p_{2a6d}<0.001$). Surprisingly, one of the parts of the assignment with the highest success rate was the final assignment requiring higher order thinking. However, it is an open-ended problem and students were allowed to choose their own objects of investigation according to their involvement and self-efficacy. Furthermore, the analyzed materials were not their hand-out versions, but the final report where they present only chosen results.

Table 2. Success rate and levels of reasoning skills and algebraic generalization manifested in team solutions

Task	Description	Success Rate	Level of Reasoning Skills (M [1] ± SD [2])	Level of Algebraic Generalization (M ± SD)	Level of Combinatorial Thinking (M ± SD)
2a	Proving the impossibility	66% [a]	2.45 ± 1.32 [a]	NA	NA
5a	Introductory combinatorics (enumeration)	55% [agh]	NA	NA	1.42 ± 1.07 [ab]
5b		45% [aehi]	NA	NA	1.00 ± 1.26 [a]
5c		36% [bchijk]	NA	NA	NA
5d	Combinatorics (optimization)	18% [kl]	NA	NA	NA
5e	Higher lower limit	12% [gl]	NA	NA	NA
5f	Generalization	18% [bcdfg]	1.33 ± 1.53 [bce]	1.15 ± 1.37 [a]	1.61 ± 1.25 [b]
6a	Proving the inequality	39% [befj]	1.15 ± 1.39 [b]	NA	NA
6b	Proving the impossibility	21% [cfgi]	1.18 ± 1.48 [bc]	NA	NA
6d	Proving the coverage	9% [dfg]	0.30 ± 0.63 [d]	NA	NA
FA	Mathematical modeling	48% [aej]	0.75 ± 0.90 [bc]	0.66 ± 0.92 [a]	NA

[1] M mean, [2] SD standard deviation. The values assigned by the same letter (a,b,c,… etc.) do not differ significantly based on the McNemar test (correctness of solution) and the median test (levels of reasoning skills and algebraic generalization) ($p \leq 0.05$), e.g. Success rate in 5a does not differ significantly from 5b, as both are assigned by string containing letters a and h, but 5a differs significantly from 5c, as they do not have any letter in common.

The results of SIA (see Figure 3) indicate an evident interconnection between correctness of the investigated subtasks. The correct solution of the subtask 5f aimed at assessing an ability to generalize algebraically implied correct solution of another subtask aimed at assessing the generalization skills, namely the subtask 2e. Although it was implied by the correct solution of the generalization of the subtasks 6b, it was also implied by the high level experimentation and application manifested in the team solution of subtask 6c. Furthermore, the correct solution of 6a implied the correct solution of two generalizations of the subtasks, 2e and 2a. The success in 2a was implied also by the correct solution of FA. Direct implications from the subtasks 6a and 6b to the success in 6d were not confirmed there. It may indicate that just the generalization is a hidden driving ability supporting mathematical reasoning.

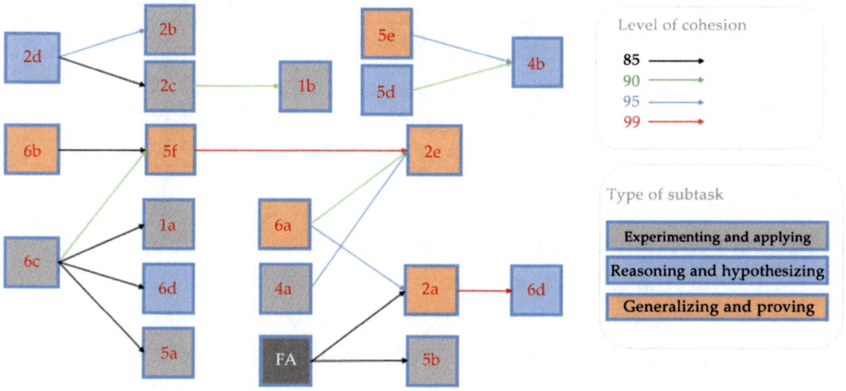

Figure 3. Implication tree of correctness of the results,

The correctness of the team solutions correlated between the introductory combinatorial subtasks (φ_{5a5b}=0.344, p_{5a5b}=0.05; φ_{5b5c}=0.449;p_{5b5c}=0.009) and between combinatorics subtasks aimed at optimization and generalizations (φ_{5d5e}=0.547;p_{5d5e}=0.001). It is worth commenting that the correct solution of the final assignment did not correlate with any of investigated subtasks (Table 3).

Table 3. Correlation matrix of correctness of the team solution in selected subtask (phi coefficient)

		Correctness										
	Task	**2a**	**5a**	**5b**	**5c**	**5d**	**5e**	**5f**	**6a**	**6b**	**6d**	**FA**
Correctness	2a	NA	0.290	0.129	−0.013	0.000	0.066	0.000	0.175	0.367	−0.224	0.043
	5a	0.474	NA	0.344	0.311	0.273	0.339	0.430	−0.011	0.176	−0.077	−0.155
	5b	0.474	0.050	NA	0.449	0.359	0.034	0.043	0.260	−0.027	0.346	0.332
	5c	0.458	0.079	0.009	NA	0.297	0.298	0.297	0.164	0.084	0.199	0.023
	5d	2.000	0.125	0.040	0.093	NA	0.547	0.389	0.263	−0.052	0.124	0.014
	5e	0.717	0.054	0.851	0.092	0.001	NA	0.547	0.081	0.034	−0.117	0.011
	5f	1.000	0.012	0.812	0.093	0.025	0.001	NA	0.414	0.326	0.199	−0.160
	6a	0.329	0.950	0.143	0.362	0.139	0.656	0.016	NA	0.340	0.392	0.211
	6b	0.036	0.327	0.881	0.642	0.772	0.849	0.064	0.053	NA	0.164	0.090
	6d	0.211	0.670	0.448	0.266	0.491	0.515	0.268	0.024	0.362	NA	0.326
	FA [1]	0.813	0.389	0.059	0.899	0.937	0.950	0.372	0.239	0.619	0.064	NA

[1] FA—final assignment.

Based on the results summarized in Table 4 we can claim that the levels of reasoning skill manifested in team solutions of the subtasks 6a, 6b and 6d correlated mutually (ρ_{6ab}=0.814, p<0.001; ρ_{6ad}=0.475, p=0.005; ρ_{6bd}=0.534, p=0.001). On the other hand, the level of algebraic generalizations was not correlated between the subtask 5f and final assignment. The level of combinatorial thinking correlated with the generalization level, manifested in the final assignment (ρ= 0.453, p=0.008).

Table 4. Correlation matrix of the level of reasoning, algebraic generalizations and combinatorial thinking manifested in the team solutions of selected subtasks (Spearman's rho)

		Level of										
		Reasoning					**Generalization**		**Combinatorial Thinking**			
	Item	**2a**	**6a**	**6b**	**6d**	**FA**	**5f**	**FA**	**5a**	**5c**	**5f**	**FA**
Reasoning	2a	NA	0.266	0.210	0.239	0.043	0.081	0.128	0.122	0.259	0.313	0.209
	6a	0.135	NA	0.814	0.475	0.204	0.282	0.089	−0.002	0.229	0.211	−0.020
	6b	0.240	0.000	NA	0.534	0.243	0.292	0.182	0.124	0.181	0.237	0.219
	6d	0.180	0.005	0.001	NA	0.077	0.160	−0.089	−0.011	0.270	0.190	−0.085
	FA	0.813	0.255	0.173	0.668	NA	0.283	0.724	0.140	0.299	0.213	0.247
Gen [1]	5f	0.653	0.112	0.099	0.372	0.111	NA	0.090	0.185	0.536	0.393	0.242
	FA	0.479	0.623	0.311	0.624	0.000	0.617	NA	−0.072	−0.027	0.018	0.453

Combinato-rial thinking	5a	0.499	0.989	0.491	0.952	0.439	0.302	0.689	NA	0.450	0.799	0.095
	5c	0.146	0.201	0.313	0.129	0.091	0.001	0.884	0.009	NA	0.693	−0.141
	5f	0.076	0.239	0.185	0.288	0.233	0.024	0.922	0.000	0.000	NA	0.056
	FA	0.242	0.914	0.220	0.636	0.166	0.175	0.008	0.600	0.434	0.759	NA

[1] Gen—generalization.

The level of combinatorial thinking manifested in the team solutions of the assessed partial subtasks 5a, 5c and 5f was correlated. Furthermore, the combinatorial thinking manifested in the solutions of the subtasks 5c and 5f correlated with algebraic generalization assessed in team solution of the subtask 5f ($\rho_{5c5f}= 0.536$, $p_{5c5f}=0.001$;$\rho_{5f5f}=0.393$, $p_{5f5f}=0.024$). The correlation between the combinatorial thinking manifested in the team solution of the subtask 5c and algebraic generalization in 5f is not caused by a simultaneous use of the two mathematical competencies, but rather in the design of the task as such, when the correct solution of the subtasks 5c was necessary for solving the subtask 5f. On the contrary, in the subtask 5f both competencies were necessary for a successful solution. It seems that contrasting to the mathematical reasoning and algebraic generalization, the ability to use combinatorial thinking appears to be very consistent among the partial subtasks of an investigated complex problem. Due to the special kind of problems as well as high-achieving solvers as participants in the study it is not possible to state whether this consistency is a general characteristic of combinatorial thinking, or it is a feature of the problem or of the solvers.

We tried to relate the correctness of the solution of the particular subtask with the different level of abilities. Correctness of the solution was associated with the reasoning skills manifested in solutions of several investigated subtasks (Table 5). The reasoning skills in solutions of the subtasks 6a, 6b and 6d were associated with correctness of 6a ($V_{6a6a}=0.543$, $p=0.003$; $V_{6a6b}=0.577$, $p=0.014$; $V_{6a6d}=0.501$, $p=0.015$). Furthermore, the subtask 5e was the only subtask in which correctness of the solution was related to the level of reasoning skills in FA ($V=0.206$, $p=0.039$).

Table 5. Correlation matrix of the solution correctness of and the levels of reasoning, algebraic generalization and combinatorial thinking manifested in the team solution of selected subtasks (Cramer's V)

		Level of										
		Reasoning					**Generalization**		**Combinatorial thinking**			
	Item	**2a**	**6a**	**6b**	**6d**	**FA**	**5f**	**FA**	**5a**	**5c**	**5f**	**FA**
2a	V	0.542	0.265	0.351	0.034	0.241	0.079	0.235	0.100	0.205	0.291	0.125
	p	0.046	0.509	0.186	0.930	1.000	0.729	0.537	0.861	0.539	0.358	0.473
5a	V	0.178	0.012	0.072	0.141	0.025	0.463	0.067	0.644	0.291	0.587	0.161
	p	0.653	0.547	0.711	0.805	0.590	0.024	0.596	0.001	0.242	0.001	0.354
5b	V	0.335	0.077	0.177	0.335	0.181	0.303	0.067	0.376	0.388	0.481	0.194
	p	0.222	0.030	0.302	0.037	0.319	0.108	0.199	0.018	0.137	0.012	0.266
5c	V	0.268	0.192	0.121	0.238	0.006	0.335	0.138	0.464	0.602	0.589	0.134
	p	0.115	0.477	0.416	0.394	0.466	0.246	0.103	0.039	0.005	0.008	0.443
5d	V	0.257	0.062	0.166	0.023	0.129	0.412	0.086	0.180	0.250	0.273	0.375
	p	0.367	0.597	0.798	0.504	0.191	0.110	0.126	0.372	0.381	0.398	0.357
5e	V	0.155	0.041	0.046	0.180	0.206	0.439	0.238	0.199	0.296	0.265	0.476
	p	0.357	0.821	0.799	0.542	0.039	0.071	0.045	0.028	0.147	0.435	0.006
5f	V	0.137	0.177	0.102	0.228	0.040	0.645	0.086	0.106	0.188	0.148	0.375
	p	0.439	0.441	0.305	0.373	0.083	0.001	0.126	0.803	0.527	0.242	0.031
6a	V	0.099	0.543	0.577	0.501	0.429	0.414	0.363	0.028	0.198	0.204	0.219
	p	0.774	0.003	0.014	0.015	0.096	0.098	0.166	0.568	0.338	0.366	0.208
6b	V	0.274	0.644	0.745	0.222	0.225	0.326	0.109	0.140	0.059	0.104	0.341
	p	0.307	0.009	0.001	0.089	0.361	0.260	0.268	0.447	0.490	0.189	0.049
6d	V	0.132	0.119	0.105	0.520	0.086	0.199	0.000	0.071	0.336	0.267	0.056
	p	0.936	0.015	0.722	0.002	0.884	0.530	0.838	0.131	0.160	0.490	0.131
FA [1]	V	0.013	0.019	0.128	0.208	0.606	0.160	0.355	0.294	0.290	0.305	0.182
	p	0.083	0.165	0.189	0.383	0.003	0.309	0.235	0.302	0.382	0.167	0.295

[1] FA—final assignment.

Out of 6 teams who were able to produce the correct algebraic expression in the solution of the subtask 5f used an inductive approach, one team did not provide any reasoning for it, and one team used an abductive reasoning approach. The level of generalization was related to correctness of the solution of this subtask ($V=0.645$, $p=0.001$). The level of generalization in 5f was related to correctness of the subtask 5a ($V=0.463$,$p=0.024$). Correctness of the final assignment was related to the reasoning-skills level ($V=0.606$, $p=0.003$), however, it was not related to the level of generalization ($V=0.355$, $p=0.235$), nor with the manifested level of combinatorial thinking in this complex problem (V=0.182,p=0.295). It might again be caused by the different, more open, nature of the final assignment compared to partial subtasks.

Example of the Authentic Team Solutions of the Subtask 2a

The subtask 2a was classified to the third group as aimed at the generalizing and proving, but only the actual levels of reasoning and proving skills were observed in students' solutions of this subtask. The goal of the subtasks is verifying the impossibility of the assigned argument.

Solution 2a.A

"Based on the previous experiment, we found an interval of the angles, for which there is not possible to cover the Lena. There is an interval of angles between the largest angel from the position 1 (41.81°) and the smallest angle from the position 2 (96.38°), which is not possible to fit under the cover. For example, the angle of size 60° creates the isosceles triangle; therefore the lengths of the heights on the triangles sides are equal and greater than 5 cm. Both rectangles, 5 × 15 as well as 5 × 14, are not covering Lena in every case." Students used an experiment as the main method to find a solution (Figure 4). The crucial parameter represents estimated lengths of the highs of the triangles created by a bent line based on the strategies set out in the assignment of the subtask (see the full assignment [55]). Subsequently, students found an interval of the angles that do not match the dimensions of the given rectangle. This approach was aimed at both assigned rectangles. The conclusion of students' experimentation was expressed in the simple statements.

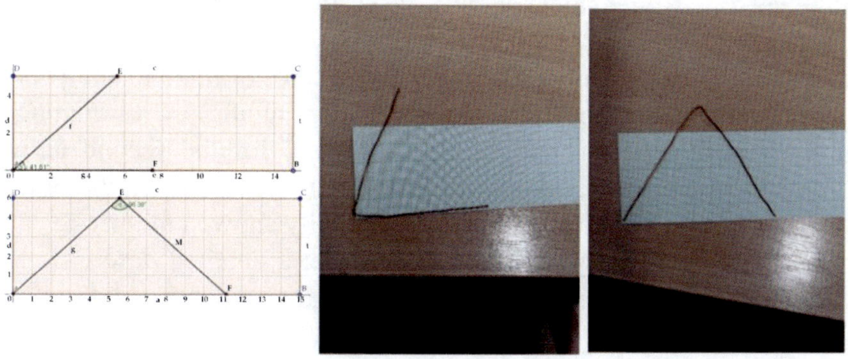

Figure 4. Students' experimenting by the GeoGebra and with the copper wire in solution 2a.A.

Solution 2a.B

"Let us deal with the specific group of snakes, especially with some snakes composed of the two line segments 7.5 cm long with a mutual boundaries point, which we are going to call as the vertex of the snake. The two remaining boundary points of line segments are going to be called the tails of the snake. Let these two line segments form an angle α. We are going to call these snakes laptops, because they remind us of a laptop from their profile.

"First, let us have a look at the rectangle with the dimensions 14 cm and 4 cm. This blanket is not enough for our snake. Therefore, it is obvious that the snake obtains a form of a line segment with its length 15 cm for $\alpha = 180°$. The diagonal represents the longest line segment in the rectangle. And the length of this diagonal can be calculated by the Pythagorean Theorem. In our case the length of diagonal is just $\sqrt{14^2 + 4^2} \doteq 14.56$ cm. The widest part of the blanket is less than our snake for the given angle, and thus for this reason the snake cannot fit under the blanket.

"Let us try the blanket with the dimensions 15 cm and 5 cm. The snake is not going to fit under this blanket. Let us take the snake at 60° angles. This snake has the shape of an equilateral triangle and the whole triangle must be placed inside the rectangle. The shortest distance between the vertex and the point of the opposite side, the height of the triangle, is approximately 6.63 cm long. If we rotate this triangle in the coordinate system, for every turn will be there the y-dimension greater than 5, and so it cannot fit in the rectangle, which has for some turns this dimension equal to 5."

In solution 2a.B, students described their approach of a solution in the two parts for the two assigned rectangles. In the first case of the rectangle with the specific dimensions, students used the Pythagorean theorem. For the conclusion, there is enough to show if the diagonal of the rectangle can cover the whole length of the non-bended snake. Similar to the previous example of the solution, students were approaching by estimating the lengths of triangles created by a bended snake for the second assigned rectangle. However, the approach is not based on the strategies set out in the assignment of the subtask. Students originally showed in the one chosen example the impossibility of the argument, but a detailed description or a picture from experimenting is missing there in the students' solution.

Thus the ways students were thinking in both approaches of the solution were correct. The solution reached the first level of reasoning and proving skills, because their description contains the reasoning by the one particular example.

Example of the Authentic Team Solutions of the Subtask 6d

The following subtask was classified in the second group based on a schema of interconnected subtasks, vested in reasoning and hypothesizing steps and jobs, but this all requires in addition some reasoning and proving skills. The subtask 6d is similar to the previous subtask 2a, but the goal is not to refute an argument. By contrast, students are asked to show that the snake will fit in an estimated area under some specific conditions.

Solution 6d.A

"Let us have a blanket created by joining the two equilateral triangles with the height 7.5 cm long and then we will put the middle of the snake on the common edge of the triangles [Figure 5A].

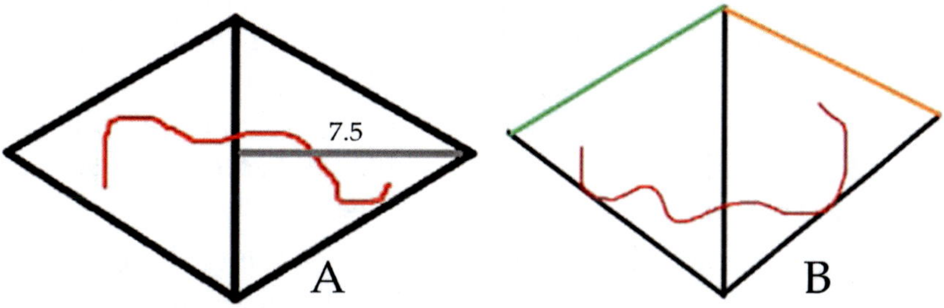

Figure 5. Students' solution A of the subtask 6d.

"We can estimate the side of the triangle x by Pythagorean theorem: $x^2 - \left(\frac{x}{2}\right)^2 = 7.5^2 = 56.25$, so we get by all adjustments, so that $x = \sqrt{75}$. Well, then the area of the one triangle is $\frac{\sqrt{75} \cdot 7.5}{2}$, and so the area of the whole blanket can be estimated as $\sqrt{75} \cdot 7.5$. But is this blanket enough?

"Put the snake into the position, that its middle is placed on the common sides of the two triangles and the snake is touching the two sides of triangles like given in the figure B [Figure 5B] (one half is touching the one side, the second half is touching the second one).

"Then we can see that the one half cannot pass through the two sides, because the heights of the triangle, the distance of these two sides, are equal to the length of the half of the snake."

In the first step of solution 6d.A, students calculated the length of the side of the equilateral triangle for estimating an area of the diamond. Even if the final result of the area is correct, students made a mistake in the notation of the equation. However, the estimation of the diamonds' area was not necessary in the case of a solution of the subtask 6d. Focusing on the reasoning of the argument was more important here. But the students' conclusion does not consist of any mathematical evidence, even any mathematical reasoning which would reach out to the goal of this subtask. The reasoning and proving skills as such manifested in the solution A of the subtask 6d stay on the level 0.

Solution 6d.B

"Both parts of the shape (of the diamond) are the two equilateral triangles, so we just have to look at one of them, because in addition, an equally large piece of snake exceeds upon both sides. Let the equilateral triangle have a height 7.5 cm long. We will choose a point M on the one of the sides of the triangle. We will draw a line from this point, touching on the other side and our goal is to have a snake not exceeding through the third side. For this reason, let us take the shortest possible way. We will find out this way that we will reflect the third side in the axial symmetry based on the second side and extend the perpendicular from the first side.

"In Figure 6 we can see that the green line—a projection of the third side based on the second one —is parallel to the first side—and the point A lies on it. The perpendicular is always the shortest distance, and when a perpendicular line is parallel to another line, that means that the line with the shortest distance will be perpendicular also to the first line. For this reason, the distance is equal with the height of the triangle, therefore 7.5 cm."

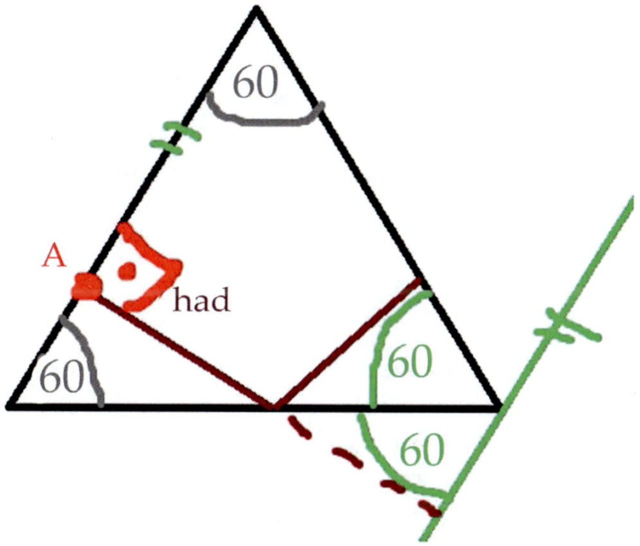

Figure 6. Students' solution B of the subtask 6d.

Students used the proper methods and tools to find the correct mathematical evidence, but the students' description is not clear. At first, the final (underlined) conclusion is written in a confusing way. The description of the sides of triangles is not exactly named. In addition, the added picture does not contain an exact label. For example, there is a point M defined at the beginning of the solution. But in the following explanation as well as in the picture, the point A is marked.

Example of the Authentic Team Solutions of the Subtask 5f

The whole task 5 is aimed at the combinatorial thinking skills. Every single subtask requires some different skills on the different levels. The subtask 5f was chosen as the following example mainly because it required all the observed skills in this research, hence, algebraic generalization, reasoning and proving as well as combinatorial thinking.

Solution A

"If we bend a snake only one time in any point, then the area of the blanket can be calculated as follows $2(n-1)$ for all kinds of snakes, except for the snake with the length of the body 1 because the formula is not valid for it."

The students' solution is giving a short expression, formula, or hypothesis, for the estimating of the wanted area. However, a closer mathematical description or formulation of this hypothesis is missing there. For this reason, students did not put any efforts into the possible combinations of some bended snakes. To sum it up, students manifested the algebraic generalization skill on the level 2, their reasoning and proving skills on the level 0 and their combinatorial thinking is not observable in the solution 5f.A.

Solution B

"Let's have a bended tetra-snake with the length of the body *n*; and *n* cannot be equal to 1 or 2, as the possible smallest bendable tetra-snake has the length of the body 3. In a bent tetra-snake, one square represents a bending point, and other n−1 squares are divided between the two arms. Utmost, the first arm is composed from n−2 square and the second arm from the one square. It represents the most disproportionate distribution. In the next step, we will move the squares from the longer arm to the shorter one. In the most proportionate distribution, if *n* is an even number, there is the same amount of the squares in the arms of the tetra-snake and if *n* is an odd number, there are $\frac{n-1}{2} + 0.5$ squares in the one arm and $\frac{n-1}{2} - 0.5$ squares in the other arm. When we add another square (for example there are arms of a tetra-snake with the length 5 and 3, then 4 and 4, and then 3 and 5), we already have this option and, therefore, we can tilt the tetra-snake very easily.

This means that the one arm of the blanket must have a length of n−2 squares (in the utmost case) and the other arm has a length of $\frac{n-1}{2}$ if n is an even number and $\frac{n-1}{2} - 0.5$ if n is an odd number. Then, the area of the blanket is is $S = n - 2 + 1 + \frac{n-1}{2}$ for n which is even and $S = n - 2 + 1 + \frac{n-1}{2} - 0.5$ for odd number n. The general form of the formula is:

$$S = n - 2 + 1 + \frac{n-1}{2} + ((-(n \bmod 2) + 1) \cdot (-0.5) = \frac{3}{2} \cdot ((-(n \bmod 2) + 1) \cdot (-0.5)."$$

Students offered the general expression of the formula for estimating the area of covering the bent snake, so the algebraic generalization skills were well manifested at the level 3. However, their findings are not formalized in a form of the proof or the theorem. Therefore, students achieved level 2 in the reasoning and proving skills. As far as to the combinatorial thinking skills, level 2 is demonstrated in the solution 5f.B.

Solution C

"Now we are going to focus on the one-bent tetra-snakes, but without a condition that their bending point must be in the middle of the snake. We draw (Figure 7A) for all possible situations for $n \in \langle 1;7 \rangle$.

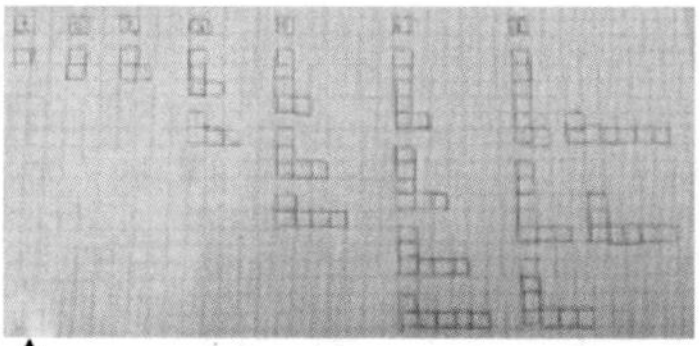

A

N	Obsah obdĺžnikovej prikrývky	Obsah najmenšej prikrývky
3	2*2	2*2 - 1
4	3*2	3*2 - 2
5	4*3	4*3 - 3*2
6	5*3	5*3 - 4*2
7	6*4	6*4 - 5*3
Párne n	(n-1)'(n/2)	$(n-1) \times \frac{n}{2} - (n-2) \times (\frac{n}{2} - 1)$
Nepárne n	(n-1)'[(n/2)+0.5]	$(n-1) \times \left(\frac{n}{2} + \frac{1}{2}\right) - (n-1) \times (\frac{n}{2} - \frac{1}{2})$

B

Figure 7. Students' solution of the subtask 5f (A) the generative strategy for expressing the formula and (B) the table of expressed general formula. Translations: Obsah obdĺžnikovej prikrývky = Area of the rectangular blanket; Obsah najmenšej prikrývky = Area of the smallest blanket; Párne *n* = Even *n*; Nepárne *n* = Odd *n*.

"We will try to generalize the observed properties for getting a formula expressing the smallest possible blanket for the tetra-snake with the length of *n* (Figure 7B).

"Note: we are not going to consider any tetra-snakes with their length of 1 and 2 squares, because these tetra-snakes represent special cases and they cannot be bent."

Definitely, there can be observed some higher skills in the solution 5f.C. Students manifested level 3 in the algebraic generalization skills as well as in combinatorial thinking skills. Based on the chosen approach and tools, students found out a general formula (Figure 7B). However, also in this case it is not formalized in a form of the proof. Students reached level 2 in the reasoning and proving skills.

DISCUSSION

In this paper we tried to investigate specific relations between the ability to generalize algebraically, to use combinatorial thinking and to prove mathematically that were manifested during solving complex open-ended mathematical problems within the team contest. The complex assignment called on a diverse set of knowledge of individual students contributing to the team solutions, as well as on their skills, abilities and expertise [64].

Certain correlations between the algebraic generalization and reasoning skills were confirmed in cases when students solved an open mathematical problem, requiring both, meaning the use of some reasoning skills as well as some generalization skills concurrently. However, the same relation between these two abilities was not confirmed among the all investigated subtasks. This confirms the findings that the mathematical reasoning is not necessarily related to students' achievement in solving problems that do not pose high cognitive demands [65]. Furthermore, Nunes et al. [66] considered mathematical reasoning and algebraic skill as the two separate contributions to the achievement in mathematical problem solving. Relations between several aspects of mathematical problem-solving were investigated and the ability between control of variables and reasoning was confirmed also by [67]. However, Chimoni and Pitta-Pantazi [68] identified the ability to provide reasoning by analogy, serial or deductive reasoning as a significant predictor for lower-secondary students' algebraic thinking. Students have to look beyond the surface characteristics of the problem to be able to construct a mathematical model which is sufficiently general and captures the essence of the situation [69]. The distance between the reasoning skills based on the generalized pattern produced could be an obstacle for students and a possible or probable cause of their failure in the construction of mathematical induction [31]. The correlations between the level of generalization and reasoning skills can be based in the nature of mathematical proof when "construction of deductive proof is performed only when inductive argumentation is based on a process pattern generalization" [70] (p. 147).

The subtasks in combinatorics analyzed were highly intercorrelated and correlated to the level of generalization. However, such correlation is not surprising, whereas the highest level of combinatorial thinking as defined by Jones et al. [39] leads to a solution of the combinatorial problem by the means of an expression or formula [14]. On the other hand, based on findings reported by Lockwood et al. [71] the re-invention of combinatorial formulas puts some higher demands on the level of combinatorial thinking than on algebraic generalization.

The manifested reasoning skills were significantly higher as far as to the subtasks 2a, 6a and 6b compared to the subtask 6d solution. A similar relation can be observed in the case of the correctness of the students' solution. Our results indicate that it is easier to prove that the assigned situation is not possible or equality is not valid than to prove validity of a statement, even though this kind of task is usually unknown or even novel to many students [72]. Whereas the attitudes in mathematics can result from automatizing a repeated emotional reaction, students with a negative attitude towards geometric proofs may attach that same attitude to the all proofs in other mathematical fields [73], hence the use of the aforementioned kinds of proof can reduce negative attitude towards the mathematical proof as such. The seeming independence between ability to provide mathematical reasoning and combinatorial thinking can have a resource in the fact that algebra is typically the first part of mathematics where students are asked to provide abstract mathematical reasoning [74].

The problems investigated were aimed at the reasoning skills and mathematical proof (2a, 6a, 6b and 6d) and at algebraic generalization as such (5f). The subtasks were structured and strongly focused on the particular students' ability. While solving a complex ill-structured open-ended problem aimed at mathematical modeling, i.e., the final assignment, our students have more freedom to investigate mathematically and incorporate the distinct mathematical abilities. This may fulfill the requirement formulated by Novotná [75], showing that the students should have a real chance to experience some pleasure while discovering, but also concurrently having a chance to learn, to acquire some new pieces of knowledge or skills, they would be able to apply in other situations subsequently. Thus, to conclude, any mathematical modeling job allows and requires both, the production of an expression, and proving the validity of the produced result. The interaction between the investigated abilities supports a more sophisticated solution of the problem than each of the skills occurring in isolation from the others [76].

Limitations of the Study

When generalizing the results of the presented study, one should be aware that the sampling was not random, nor stratified. The data came from the results of the contestants of Mathematics B-day what adds self-selection bias to the results. By contrast, any relations between the different sources of mathematical proficiency are most visible when solving a challenging problem by high-achieving students [57]. The groups are usually even more successful when handling complex problems [52], so the relations are more feasible for investigation in this kind of setting. Therefore, the group result was analyzed instead of individual contributions. As the "collaborative partnerships draw on different contributors across different tasks" [64] (p. 254), we focused on the results of shared activity regardless the previous experience with group-work and distribution of workload among the members of the team. Although the analysis of the work of groups of students may appear to be not sufficient to investigate all the cognitive aspects in relationship between the reasoning skills, combinatorial thinking and algebraic generalization, and some underlying variables may perhaps be hidden, some specific relations were observable nevertheless.

As reported in the meta-analysis by Lou et al. [77] the shared group goal (in this case the participation in a mathematical contest) is needed to make the group collaboration effective. Even though the results of this study may be related only to the special group of students, they extrapolate the relations among the ordinary students as solvers of mathematical problems.

CONCLUSIONS

Herein we have explored specific relations between the reasoning skills and the ability to generalize algebraically in the case of a team solution of a complex open-ended problem by high-achieving upper-secondary students. We found out that some relations can be observed mainly in cases of open-ended problem solving that generally require a higher-order level of thinking.

Based on the results of our study, we can suggest using or utilizing more problems aimed at proving an inequality or impossibility in the every-day mathematics classroom environment. Students, it seems to us, are or can be more successful in these kinds of proofs, and it might improve both, their reasoning skills as well as their self-efficacy in the field of mathematical proving. The highest level of combinatorial thinking is related to the work with combinatorial formulas and expressions. When students are asked to solve a combinatorial task using an expression or formula, then they should be able to generalize algebraically. Similarly, to our previous study [**29**] the role of algebraic generalization in solving problems from the different parts of mathematics was confirmed. Thus we can conclude that the algebraic generalization abilities should be in the focus of the upper-secondary mathematics education. It seems worthwhile to study how the generalization and reasoning skills are related to other observable mathematical competencies, e.g., factual combinatorial thinking ability, or factual probability reasoning ability as well.

AUTHOR CONTRIBUTIONS

Conceptualization, S.Č.; methodology, J.M.; software, K.O.B.; validation, J.M., K.O.B. and S.Č.; quantitative analysis, J.M.; qualitative analysis, K.O.B.; investigation, J.M. and K.O.B.; resources, S.Č.; data curation, J.M.

and K.O.B.; writing—original draft preparation, J.M. and K.O.B.; writing—review and editing, J.M. and K.O.B.; visualization, J.M. and K.O.B.; supervision, S.Č.; project administration, S.Č.; funding acquisition, S.Č. All authors have read and agreed to the published version of the manuscript.

REFERENCES

1. National Research Council. *Assessing 21st Century Skills*; National Academies Press: Washington, DC, USA, 2011; p. 154.
2. Powel, A.B. Challenging Tasks and Mathematics Learning. In *Challenging Mathematics in and beyond the Classroom*; Barbeau, E.J., Taylor, P.J., Eds.; ICMI Study Series 12; Springer: Berlin/Heidelberg, Germany, 2009; pp. 133–171.
3. Jaworski, B. Theory and practice in mathematics teaching development: Critical inquiry as a mode of learning in teaching. *J. Math. Teach. Educ.* 2006, *9*, 187–211.
4. Nohda, N. Teaching by Open-Approach Method in Japanese Mathematics Classroom. In Proceedings of the Conference of the International Group for the Psychology of Mathematics Education, Hiroshima, Japan, 23–27 July 2000; Volume 1, pp. 39–53.
5. Frobisher, A.; Frobisher, L. *Didaktika Matematiky II*; Raabe: Bratislava, Slovakia, 2015; p. 169s.
6. Artzt, A.F.; Armour-Thomas, E. Development of a Cognitive-Metacognitive Framework for Protocol Analysis of Mathematical Problem Solving in Small Groups. *Cogn. Instr.* 1992, *9*, 137–175.
7. Veenman, M.V.J. The role of intellectual and metacognitive skills in math problem solving. In *Metacognition in Mathematics Education*; Desoete, A., Veenman, M., Eds.; Nova Science: Haupauge, NY, USA, 2006; pp. 35–50.
8. Chytry, V.; Rican, J.; Korinkova, R. The influence of a teacher′s innovation on a pupil′s relationship to mathematics. In Proceedings of the ICERI19, 12th International Conference of Education, Research and Innovation, Seville, Spain, 11–13 November 2019; Chova, L.G., Martinez, A.L., Torres, I.C., Eds.; IATED-Int Assoc Technology Education & Development: Valencia, Spain, 2019; pp. 128–133.
9. Mason, L. High school students′ beliefs about maths, mathematical problem solving, and their achievement in maths: A cross-sectional study. *Educ. Psychol.* 2003, *23*, 73–85.
10. Pajares, F.; Miller, M.D. Role of self-efficacy and self-concept beliefs in mathematical problem solving: A path analysis. *J. Educ. Psychol.* 1994, *86*, 193.
11. Stage, F.K.; Kloosterman, P. Measuring beliefs about mathematical problem solving. *Sch. Sci. Math.* 1992, *92*, 109–115.

12. Švecová, V. Mathematics Anxiety and Other Psycho Didactic Aspects in University Students. *J. Educ. Soc. Policy* 2018, *5*, 246–250.
13. Kosyvas, G. Levels of arithmetic reasoning in solving an open-ended problem. *Int. J. Math. Educ. Sci. Technol.* 2015, *47*, 356–372.
14. Lockwood, E. Thc Strategy of Solving Smaller, Similar Problems in the Context of Combinatorial Enumeration. *Int. J. Res. Undergrad. Math. Educ.* 2015, *1*, 339–362.
15. Lockwood, E.; Wasserman, N.H.; McGuffey, W. Classifying Combinations: Investigating Undergraduate Students′ Responses to Different Categories of Combination Problems. *Int. J. Res. Undergrad. Math. Educ.* 2018, *4*, 305–322.
16. Mousoulides, N.G.; Christou, C.; Sriraman, B. A modeling perspective on the teaching and learning of mathematical problem solving. *Math. Think. Learn.* 2008, *10*, 293–304.
17. Samková, L. Investigating the Variety and Usualness of Correct Solution Procedures of Mathematical Word Problems. *J. Effic. Responsib. Educ. Sci.* 2020, *13*, 10–26.
18. Wong, T.T.-Y. Is conditional reasoning related to mathematical problem solving? *Dev. Sci.* 2017, *21*, e12644.
19. Abramovich, S.; Pieper, A. Fostering recursive thinking in combinatorics through the use of manipulatives and computing technology. *Math. Educ.* 1995, *6*, 4–12.
20. Swafford, J.O.; Langrall, C.W. Grade 6 Students′ Preinstructional Use of Equations to Describe and Represent Problem Situations. *J. Res. Math. Educ.* 2000, *31*, 89–112.
21. Kazemi, E. Discourse that promotes conceptual understanding. *Teach. Child. Math.* 1998, *4*, 410.
22. Mula, M.; Hodnik, T. The PGBE Model for Building Students′ Mathematical Knowledge about Percentages. *Eur. J. Educ. Res.* 2020, *9*, 257–276.
23. Schüler-Meyer, A.; Prediger, S.; Kuzu, T.; Wessel, L.; Redder, A. Is formal language proficiency in the home language required to profit from a bilingual teaching intervention in mathematics? A mixed methods study on fostering multilingual students′ conceptual understanding. *Int. J. Sci. Math. Educ.* 2017, *17*, 317–339.
24. Cheng, S.C.; She, H.C.; Huang, L.Y. The impact of problem-solving instruction on middle school students′ physical science learning:

Interplays of knowledge, reasoning, and problem solving. *Eurasia J. Math. Sci. Technol. Educ.* 2017, *14*, 731–743.

25. Burch, L.; Pinhero, W.A.; Tillema, E. Opportunities for generalising within pre-service secondary teachers′ symbolisation of combinatorial tasks. In *Proceedings of the Forty-First Annual Meeting of the North American Chapter of the International Group for the Psychology of Mathematics Education*; Otten, S., Candela, A.G., de Araujo, Z., Haines, C., Munter, C., Eds.; University of Missouri: St Louis, MO, USA, 2019.
26. Radford, L. Iconicity and contraction: A semiotic investigation of forms of algebraic generalizations of patterns in different contexts. *ZDM* 2007, *40*, 83–96.
27. Hodnik-Cadez, T.; Manfreda-Kolar, V. Comparison of types of generalizations and problem-solving schemas used to solve a mathematical problem'. *Educ. Stud. Math.* 2015, *89*, 283–306.
28. Ayalon, M.; Watson, A.; Lerman, S. Progression towards Functions: Students' Performance on Three Tasks about Variables from Grades 7 to 12. *Int. J. Sci. Math. Educ.* 2015, *14*, 1153–1173.
29. Bulková, K.; Medová, J.; Čeretková, S. Identification of Crucial Steps and Skills in High-Achievers′ Solving Complex Mathematical Problem within Mathematical Contest. *J. Effic. Responsib. Educ. Sci.* 2020, *13*, 67–78.
30. Robinson, J.A. Proof = guarantee + explanation. In *Intellectics and Computational Logic*; Hölldobler, S., Ed.; Kluwer: Dordrecht, The Netherlands, 2000; pp. 277–294.
31. Pedemonte, B. Argumentation and algebraic proof. *ZDM* 2008, *40*, 385–400.
32. Lithner, J. A research framework for creative and imitative reasoning. *Educ. Stud. Math.* 2007, *67*, 255–276.
33. Andersen, L.E. Acceptable gaps in mathematical proofs. *Synthese* 2018, *197*, 1–15.
34. Pedemonte, B. How can the relationship between argumentation and proof be analysed. *Educ. Stud. Math.* 2007, *66*, 23–41.
35. Vale, C.; Widjaja, W.; Herbert, S.; Bragg, L.A.; Loong, E.Y.-K. Mapping Variation in Children's Mathematical Reasoning: The Case of 'What Else Belongs'? *Int. J. Sci. Math. Educ.* 2017, *15*, 873–894.

36. Wigfield, A.; Eccles, J.S.; Pintrich, P.R. Development between the ages of 11 and 25. In *Handbook of Educational Psychology*; Berliner, D.C., Calfee, R.C., Eds.; Routledge: New York, NY, USA, 1996; pp. 148–185.
37. Scholtzová, I. *Cesty Diskrétnej Matematiky (Kombinatoriky) na základnú školu [Routes of Discrete Mathematics (Combinatoric) to Elementary School]*; Prešovská univerzita v Prešove: Prešov, Slovakia, 2007.
38. English, L.D. Children's construction of mathematical knowledge in solving novel isomorphic problems in concrete and written form. *J. Math. Behav.* 1996, *15*, 81–112.
39. Jones, G.; Langrall, C.; Thornton, C.; Mogill, A.T. A framework for assessing and nurturing young children's thinking in probability. *Educ. Stud. Math.* 1997, *32*, 101–125.
40. Lockwood, E. A model of students' combinatorial thinking. *J. Math. Behav.* 2013, *32*, 251–265.
41. Reed, Z.; Lockwood, E. Generalizing in combinatorics through categorization. RUME 2018. In Proceedings of the 21st annual conference on research in undergraduate mathematics education, San Diego, CA, USA, 22–27 February 2018; pp. 311–325.
42. Sriraman, B. Mathematical giftedness, problem solving, and the ability to formulate generalizations: The problem-solving experiences of four gifted students. *J. Second. Gift. Educ.* 2003, *14*, 151–165.
43. Krejčová, L. *Žáci Potřebují Přemýšlet [Students Need to Think]*; Portál: Praha, Czech Republic, 2013; p. 174.
44. Häkkinen, P.; Järvelä, S.; Mäkitalo-Siegl, K.; Ahonen, A.; Näykki, P.; Valtonen, T. Preparing teacher-students for twenty-first-century learning practices (PREP21): A framework for enhancing collaborative problem-solving and strategic learning skills. *Teach. Teach. Theory Pract.* 2017, *23*, 25–41.
45. PISA 2015. Collaborative Problem Solving Framework. 2017. Available online: http://www.oecd.org/pisa/pisaproducts/Draft%20PISA%20 2015%20Collaborative%20Problem%20Solving%20Framework%20.pdf (accessed on 7 December 2020).
46. Kocak, Z.F.; Bozan, R.; Isik, Ö. The importance of group work in mathematics. *Procedia Soc. Behav. Sci.* 2009, *1*, 2363–2365.

47. Frobisher, A.; Frobisher, L. *Didaktika Matematiky I*; Raabe: Bratislava, Slovakia, 2015; p. 169s.
48. Fawcett, L.M.; Garton, A. The effect of peer collaboration on children's problem-solving ability. *Br. J. Educ. Psychol.* 2005, *75*, 157–169.
49. Erickson, S.A.; Lockwood, E. Investigating Combinatorial Provers' Reasoning about Multiplication. *Int. J. Res. Undergrad. Math. Educ.* 2020.
50. Lockwood, E.; Caughman, J.S.; Weber, K. An essay on proof, conviction, and explanation: Multiple representation systems in combinatorics. *Educ. Stud. Math.* 2020, *103*, 173–189.
51. Stahl, G. The constitution of group cognition. In *Handbook of Embodied Cognition*; Shapiro, L., Ed.; Routledge: New York, NY, USA, 2014; pp. 335–346.
52. Kirschner, P.; Sweller, J.; Kirschner, F.; Zambrano, R.J. From Cognitive Load Theory to Collaborative Cognitive Load Theory. *Int. J. Comput. Collab. Learn.* 2018, *13*, 213–233.
53. Plass, J.L.; O'Keefe, P.A.; Homer, B.D.; Case, J.; Hayward, E.O.; Stein, M.; Perlin, K. The impact of individual, competitive, and collaborative mathematics game play on learning, performance, and motivation. *J. Educ. Psychol.* 2013, *105*, 1050–1066.
54. FISME. Mathematics B-Day. 2020. Available online: https://www.uu.nl/en/education/mathematics-b-day (accessed on 17 December 2020).
55. FISME: Snakes' nest: Mathematics B-Day 2018. Available online: https://wiskundeinteams.sites.uu.nl/wp-content/uploads/sites/463/2019/03/assignment2018.pdf (accessed on 20 September 2020).
56. Norwood, R.; Poole, G.; Laidacker, M. The worm problem of Leo Moser. *Discret. Comput. Geom.* 1992, *7*, 153–162.
57. Pantziara, M.; Gagatsis, A.; Elia, I. Using diagrams as tools for the solution of non-routine mathematical problems. *Educ. Stud. Math.* 2009, *72*, 39–60.
58. Schoenfeld, A.H. Learning to think mathematically: Problem solving, metacognition, and sense making in mathematics. In *Handbook for Reasearch on Mathematics Teaching and Learning*; Grous, D., Ed.; Macmillan: New York, NY, USA, 2016.

59. R Core Team. *R: A Language and Environment for Statistical Computing*; R Foundation for Statistical Computing: Vienna, Austria, 2019.

60. Hervé, M. *RVAideMemoire: Testing and Plotting Procedures for Biostatistics*, R Package Version 0.9–70; 2020. Available online: https://cran.r-project.org/web/packages/RVAideMemoire (accessed on 15 September 2020).

61. Navarro, D.J. *Learning Statistics with R: A Tutorial for Psychology Students and Other Beginners*; Version 0.5. [Lecture notes]; School of Psychology, University of Adelaide: Adelaide, Australia, 2015.

62. Couturier, R. *'rchic: Statistical Implicative Analysis'*, R Package Version 0.25; 2018. Available online: https://github.com/rchic/Rchic/ (accessed on 12 January 2020).

63. Gras, R.; Almouloud, S.A.; Bailleul, M.; Lahrer, A.; Polo, M.; Ratsimba-Rajohn, H. L'implication statistique: Nouvelle méthode exploratoire de données, applications à la didactique. In *La Pensée Sauvage*; Pron: Paris, France, 1996.

64. Care, E.; Scoular, C.; Griffin, P. Assessment of Collaborative Problem Solving in Education Environments. *Appl. Meas. Educ.* 2016, *29*, 250–264.

65. Adegoke, B. Modelling the relationship between mathematical reasoning ability and mathematics attainment. *J. Educ. Pract.* 2013, *4*, 54–61.

66. Nunes, T.; Bryant, P.; Barros, R.; Sylva, K. The relative importance of two different mathematical abilities to mathematical achievement. *Br. J. Educ. Psychol.* 2011, *82*, 136–156.

67. Hejnová, E.; Eisenmann, P.; Cihlář, J.; Přibyl, J. Relations between Scientific Reasoning and Culture of Problem Solving. *J. Effic. Responsib. Educ. Sci.* 2018, *11*, 38–44.

68. Chimoni, M.; Pitta-Pantazi, D. Connections between algebraic thinking and reasoning processes. In Proceedings of the CERME 9, Ninth Congress of the European Society for Research in Mathematics Education (ERME), Prague, Czech Republic, 4–8 February 2015; pp. 398–404.

69. English, L.D.; Sharry, P.V. Analogical reasoning and the development of algebraic abstraction. *Educ. Stud. Math.* 1996, *30*, 135–157.

70. Martinez, V.A.; Pedemonte, B. Relationship between inductive arithmetic argumentation and deductive algebraic proof. *Educ. Stud. Math.* 2014, *86*, 125–149.
71. Lockwood, E.; Swinyard, C.; Caughman, J. Patterns, Sets of Outcomes, and Combinatorial Justification: Two Students' Reinvention of Counting Formulas. *Int. J. Res. Undergrad. Math. Educ.* 2015, *1*, 27–62.
72. Stein, M.; Burchartz, B. The Invisible Wall Project: Reasoning and Problem Solving Processes of Primary and Lower Secondary Student. *Math. Think. Learn.* 2006, *8*, 65–90.
73. McLeod, D.B. Beliefs, Attitudes, and Emotions. New Views of Affect in Mathematics Education. In *Affect and Mathematical Problem Solving: A New Perspective*; McLeod, D.B., Adams, V.M., Eds.; Springer: New York, NY, USA, 1989; pp. 245–258.
74. Susac, A.; Bubic, A.; Vrbanc, A.; Planinic, M. Development of abstract mathematical reasoning: The case of algebra. *Front. Hum. Neurosci.* 2014, *8*, 679.
75. Novotná, J. Heuristic Strategies in Solving of Mathematical Problems at School. In Proceedings of the ERIE 2016 13th International Conference Efficiency and Responsibility in Education 2016, Prague, Czech Republic, 2–3 June 2016; pp. 429–439.
76. Ellis, A.B. Connections between generalizing and justifying students' reasoning with linear relationships. *J. Res. Math. Educ.* 2007, *38*, 194–229.
77. Lou, Y.; Abrami, P.C.; Spence, J.C.; Poulsen, C.; Chambers, B.; d'Apollonia, S. Within-class grouping: A meta-analysis. *Rev. Educ. Res.* 1996, *66*, 423–458.

Chapter

COUNTERACTING DESTRUCTIVE STUDENT MISCONCEPTIONS OF MATHEMATICS

3

Uffe Thomas Jankvist [1] and Mogens Niss [2]

[1]Danish School of Education, Aarhus University, Campus Emdrup, DK-2400 Copenhagen NV, Denmark
[2]Department of Science and Environment, Roskilde University; DK-4000 Roskilde, Denmark

ABSTRACT

In this article, we ask the question of what it takes for targeted efforts to be reasonably successful in altering students' misconceptions and unproductive beliefs and ensuing myths about mathematics as a discipline and a school subject and about themselves in relation to mathematics, so as to pave the way for satisfactory learning. We attempt to answer this question through the analysis of three cases of upper secondary school students, who all struggled with mathematics-related difficulties due to myths resulting from

Citation: (APA): Jankvist, U. T., & Niss, M. (2018). Counteracting destructive student misconceptions of mathematics. Education Sciences, 8(2), 53.(17 pages).

Copyright: © Creative Commons Attribution 4.0 International License (http://creativecommons.org/licenses/by/4.0/).

misguided beliefs, erroneous proof schemes, or mistaken interpretations of the didactical contract, the three theoretical constructs we employ in the study. We describe how specially educated teachers, so-called "mathematics counsellors", taking part in a professional development program conducted by the authors, were able, firstly, to identity these students, then to diagnose more precisely the nature of their difficulties, and finally to design targeted interventions in order to assist the students in actually overcoming (parts of) their difficulties and eventually dispelling some of the myths they were influenced by. We further offer an analysis of the elements responsible for the success of these interventions. More precisely, we identify five such elements. Finally, we zoom in on the role and intricate connectedness of the three theoretical constructs mentioned above.

Keywords: misconceptions; beliefs; mathematics difficulties; interventions; maths counsellors

INTRODUCTION AND BACKGROUND

Since 2012, we have designed and been in charge of an in-service program providing particularly interested Danish upper secondary school mathematics teachers with professional development in the didactics of mathematics. The purpose has been to educate them to become so-called "maths counsellors", i.e., teachers who are able to help students with mathematics-specific learning difficulties to overcome or reduce these, by teaching the prospective counsellors to detect such students, to diagnose the nature of their difficulties, and to design interventions meant to assist the students in counteracting them [1]. Several upper secondary schools in Denmark now employ such maths counsellors to assist the increasing number of students who experience difficulties with mathematics or mathematics-related subjects. The typical maths counsellor has a master's degree in mathematics and has taught at least five years in upper secondary school mathematics programs. Typically, the teachers who enter the maths counsellors program do so on their own initiative, with, of course, the approval of their school management—as opposed to having been pointed out by the school to undertake this type of further education, which in our experience seldom functions well. Hence, the teachers who enter the program are experienced teachers, who know that the program is rather demanding but are enthusiastic about making a difference for students who experience difficulties with mathematics.

From the perspective of designing such an educational program, the important thing for us was to offer these future maths counsellors first-hand encounters with relevant research literature from the field of mathematics education and to assist them in applying this literature as an essential tool for their future detections, diagnoses, and interventions so as to make their practice research-based. To this end, the program attempts to strike a delicate balance between theory and practice, where the counsellors have to apply their newly gained knowledge, insights, and abilities towards actual students who have been identified as having mathematics-specific learning difficulties. This means that, for the duration of the program, which is three semesters, the counsellors work part-time with a good handful of students who possess difficulties related to, at least, one of the three themes of the program. These themes are: mathematical concepts and concept formation; mathematical reasoning, proofs, and proving; and mathematical models and modelling. For each of the three semesters, the counsellors are to produce a written project report, accounting for their work with the students identified (i.e., detected and invited to participate) and, more precisely, discussing their detection of these students, their diagnosis of the students' mathematics-related difficulties, and the design, implementation, and assessed effects of the intervention designed for these students by the maths counsellors. It is also part of the report to account for the specific ways the undertakings just mentioned have been informed by the research literature studied. It is in the cases reported by the maths counsellors, all of whom have been supervised by us, that we have found inspiration for the present article.

As mathematics educators, we are—and were—of course aware of the significant impact of the myths resulting from students' misconceptions and misguided beliefs about what mathematics is, both as a scientific discipline and as a school subject, on their learning of mathematics (cf. Section 2). Yet, during the last five years in our work with the maths counsellors, we have not only seen how students are under the spell of myths about mathematics and how several results from the research literature are reproduced in students' non-productive mathematics-related perceptions and views, but we have also witnessed the positive effects of the targeted efforts by the maths counsellors in our program to counteract such misguided student conceptions. The question we ask ourselves in this article, then, is what conditions are necessary and sufficient for effective counteraction to actually occur, i.e., what does it take for such targeted efforts to be reasonably successful in altering the students' misconceptions and unproductive beliefs and dispelling the myths they uphold so as to pave the way for satisfactory learning? Clearly, it is not enough to merely tell the students that their (mis)conceptions are responsible for the difficulties they experience in mathematics. They must be brought to realize this themselves on the basis of their own experience. But how to achieve this?

Below, we provide three authentic, empirical, and illustrative cases as a basis for our attempt to answer these questions. In all three cases, the efforts of the maths counsellors did in fact help the students overcome some of their difficulties, to a greater or lesser extent, by engaging them in activities that altered (or at least broadened) their conceptions of mathematics into more constructive ones regarding mathematics as a discipline and as a school subject, thus dispelling, at least partly, some of the myths that governed their view of mathematics. The three cases were structured according to, although not necessarily restricted to, one of the following three myth-generating or generated misguided conceptions that we have seen with a great many students:

- Students' conception of mathematics as a set of basically meaningless disconnected procedures that have to be memorized, and the consequences such a perception may have once the students begin to invent their own rules concerning mathematical procedures in order to create at least local meaning in what they are supposed to do.
- The interplay between students' everyday experiences and conceptions and their conceptions regarding mathematics, and in particular what happens when everyday conceptions interfere with and influence their mathematical work and reasoning in a negative manner.
- Students' conception of mathematical instruction as a decade-long training of what they perceive as essentially the same procedures and methods used over and over again, and the potential consequences such a conception may have for students' self-efficacy.

Before we move on to presenting the three cases, however, we outline a selection of theoretical constructs from mathematics education research of relevance to the cases presented, in particular pertaining to student beliefs and how non-productive specimens of these may constitute "stumbling blocks" for the students' learning of mathematics. We also briefly address methodological aspects concerning the cases reported and the data presented therein. Upon elaborating on the three cases, we finish with a slightly more general discussion of what it takes to dispel myths by counteracting destructive student misconceptions of mathematics and of how the described efforts of the maths counsellors may be considered as constructive "existence proofs" in this regard.

THEORETICAL CONSTRUCTS RELATED TO STUDENT BELIEFS

For an overall definition of beliefs, we adhere to a definition by Philipp in the second NCTM handbook on research in mathematics education. He describes beliefs as "lenses through which one looks when interpreting the world" [2] (p. 258) and further defines beliefs as:

Psychologically held understandings, premises, or propositions about the world that are thought to be true. Beliefs are more cognitive, are felt less intensely, and are harder to change than attitudes. Beliefs might be thought of as lenses that affect one's view of some aspect of the world or as dispositions toward action. Beliefs, unlike knowledge, may be held with varying degrees of conviction and are not consensual. Beliefs are more cognitive than emotions and attitudes. [2] (p. 259)

It is a widely held conviction that, in order to change students' behavior, teachers' practices, and both parties' approaches to mathematics, their beliefs first have to change, because beliefs act as filters that affect what we see, and because we have difficulty seeing what we do not already believe [3].

Mathematics education literature offers various attempts to organize notions of mathematics-related beliefs. We shall rely on the tripartition put forward by Op't Eynde, de Corte, and Verschaffel [4] with an extension due to Jankvist [5]. On the basis of an extensive literature review, Op't Eynde and colleagues propose that students' mathematics-related beliefs may be:

- **Beliefs about mathematics education**: (a) beliefs about mathematics as a subject; (b) beliefs about mathematical learning and problem solving; (c) beliefs about mathematics teaching in general
- **Beliefs about the self**: (a) self-efficacy beliefs; (b) control beliefs; (c) task-value beliefs; (d) goal-orientation beliefs
- **Beliefs about the social context**: (a) beliefs about the norms in their own class ((a1) the role and the functioning of the teacher; (a2) the role and the functioning of the students); (b) beliefs about the socio-mathematical norms in their own class [4] (p. 28)

Clearly, two of the foci of the present article, namely, students' beliefs that mathematics is a set of disconnected procedures and beliefs about the relationship between the discipline of mathematics and everyday life, fall outside this tripartition. Rather, such beliefs would fall inside what Thompson refers to as the "conception of the nature of mathematics", which covers "conscious or subconscious beliefs, concepts, meanings, rules, mental images, and preferences concerning the discipline of mathematics" [6] (p. 132). The same is true of several of the student beliefs identified by Schoenfeld:

- Mathematics problems have one and only one right answer.
- There is only one correct way to solve any mathematics problem—usually the rule the teacher has most recently demonstrated to the class.
- Ordinary students cannot be expected to understand mathematics; they expect to simply memorize it and apply what they have learned mechanically and without understanding.
- Mathematics is a solitary activity, done by individuals in isolation.
- Students who have understood the mathematics they have studied will be able to solve any assigned problem in five minutes or less.
- The mathematics learned in school has little or nothing to do with the real world.
- Formal proof is irrelevant to the processes of discovery or invention. [7] (p. 359)

Against this background, Jankvist [5] extends the tripartition by Op't Eynde and colleagues with a fourth component of beliefs about mathematics as a discipline (cf. Figure 1).

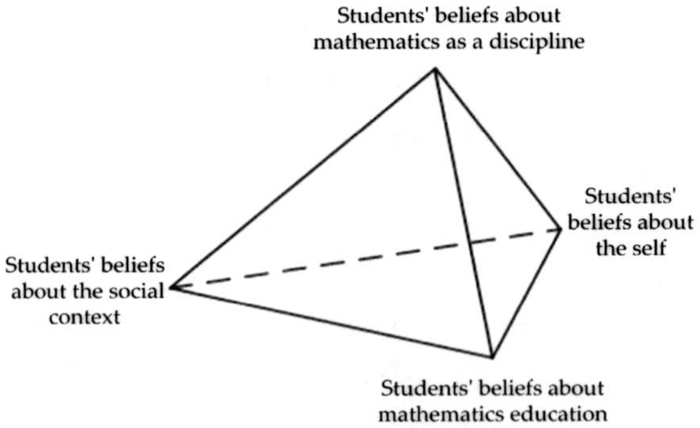

Figure 1. Students' mathematics-related beliefs as made up by a basis of their beliefs about mathematics education, the social context and the self (triangle) [4], and their beliefs about mathematics as a discipline (the point above the triangle resulting in a tetrahedron) [5].

The mathematics education literature often mentions students' beliefs about mathematics or about the nature of mathematics and, occasionally, beliefs related to aspects of the discipline of mathematics [8]. However, as shown by Jankvist, such phrases often cover very different aspects of students' mathematics-related beliefs [9]. Jankvist defines beliefs about mathematics as a discipline to include "students' beliefs about mathematics as a pure science, an applied science, a system of tools for societal practice as well as the philosophical and epistemological nature of mathematical concepts, theories, etc." [5] (p. 45). Jankvist argues for placing the component of mathematics as a discipline outside the triangle in Figure 1, as opposed to turning this into a square, since this fourth component may well be rather different from the beliefs about mathematics as a school subject (1a in the tripartition by Op't Eynde and colleagues). Still, if students are to alter non-productive beliefs about mathematics as a discipline, this can usually only happen "in the interplay between their social context (class), their mathematics education, and their self, which is to say the triangle making up the base of the tetrahedron" [5] (p. 45). Whether or not this *does* actually happen is of course a different question. Under the assumption that, when students have been exposed to mathematics education for years, they should have formed an idea about the nature of mathematics, Kloosterman asked American high school students what they would answer if an out-of-space alien turned up and asked them what mathematics is [10]. To his surprise, the students were generally incapable of providing answers and would in general begin to talk about whether they liked mathematics or not, why they needed to learn it, etc. Kloosterman concluded: "The nature of mathematics as a discipline is not an issue that United States high school students think about" [10] (p. 258) and that "students have a hard time separating the activities in a mathematics class from the nature of the discipline itself" [10] (p. 264). When the students were pressed to talk about the nature of mathematics, they mentioned "that mathematics can be used to solve a variety of problems and that it involves numbers" and they often mentioned "the procedural nature of mathematics but almost never mention deduction or proof" [10] (p. 258). This example illustrates characteristic myths about mathematics as well as the mutual influence between the components of, for example, beliefs about mathematics education and beliefs about mathematics as a discipline.

Another construct that we shall take into account is that of the didactical contract (DC). This forms part of Brousseau's theory of didactical situations and concerns students' interaction with the didactical environment (or the milieu) [11]. Brousseau says that the "student learns by adapting herself to a milieu which generates contradictions, difficulties, and disequilibria, rather as human society does. This knowledge, the result of the student's adaptation, manifests itself by new responses which provide evidence for learning" [11] (p. 30). The students' interaction with the environment can be guided by the teacher or the teacher's expectations (the situations in which this takes place are termed "didactical"), and/or be a result of the students' genuine interest in mathematics stimulated by the environment (taking place in situations referred to as "adidactical"). The teachers' expectations and classroom norms, as well as the students' expectations of the teacher's role and activities, are described as a result of a didactical contract between teacher and students. The contract is often implicit and tacit, albeit unavoidable, and is typically based on years of classroom experience on the part of both teachers and students. It guides the actors' behavior and mutual expectations, as described by the EMS Education Committee's "solid findings" article on the topic:

The DC is the set of reciprocal obligations and 'sanctions' which each partner in the didactical situation imposes or believes to have imposed with respect to the knowledge in question, explicitly or implicitly, on the other; or are imposed, or believed by each partner to have been imposed on them with respect to the knowledge in question. The DC is the result of an often implicit "negotiation" of the mode of establishing the relationships for a student or group of students, a certain educational environment, and an educational system.

[12] (p. 54)

Students'—and teachers'—mathematics-related beliefs are of course likely to affect the didactical contract, or at least the parties' perception of it. Similarly, a given didactical contract can, in itself, affect and give rise to—or suppress—certain mathematics-related beliefs in the students.

The final theoretical construct, which we shall make use of in our analyses below, is Harel's and Sowder's proof schemes: "A person's (or a community's) proof scheme consists of what constitutes ascertaining and persuading for that person (or community)" [13] (p. 809). Overall, a proof scheme belongs to one of three classes, each composed of different subclasses. The first class consist of what Harel and Sowder call external conviction proof schemes, which may be manifested as an authoritarian proof scheme, e.g., that something is held to be true because the teacher or the textbook or some other authority says so, as a ritual proof scheme, e.g., that a geometry proof in the USA must have a two-column format, or a non-referential symbolic proof scheme, e.g., that a proof must display symbols and symbolic manipulations, which do not carry any meaning for the learners. The second class consist of the empirical proof schemes, which are either inductive proof schemes, e.g., you are being convinced by one or several specific empirical examples, or by what is perceived as a "crucial" generic example, or perceptual proof schemes, e.g., when a conjecture is validated on the basis of rudimentary mental "images that consist of perceptions and a coordination of perceptions but lack the ability to transform or to anticipate the results of a transformation" [14] (p. 255). Finally, the deductive proof schemes are those with which we are familiar from the discipline of mathematics, such as direct proof, including axiomatic proofs, proof by contradiction, proofs by mathematical induction, combinatorial proofs, etc., all governed by logical deduction from a local or global set of premises.

Of course, a given student's proof scheme—or conviction scheme, as we might prefer to call it, since the notion actually concerns what it takes to convince a person—may in some way be affected by that student's beliefs. If a student perceives mathematics as a natural science in the same way as, say, physics, then that student might not only be content with, but insist on empirical examples as proof of a general mathematical statement. However, the underlying reason for such misguided beliefs and insufficient proof schemes and the resulting myths may indeed also be due to an inadequate didactical contract. Conversely, students' proof or conviction schemes will inevitably affect their beliefs about mathematics and their positions with regard to the didactical contract. Following our analyses of the three cases, we shall return to the intricate relationships between students' beliefs, proofs schemes, and perceptions of the didactical contract. First, however, we shall turn to a brief discussion of the methodological aspects surrounding the three cases.

METHODOLOGICAL ASPECTS

The core of this article is a presentation and analysis of three cases, i.e., a case study. A case study is not a presentation of singular specimens but of cases of something. For a specimen to be a case, it has to be an exemplary prime instance of a larger class of specimens, which share a number of characteristic traits and properties in certain respects, whereas they may differ in several other respects. Oftentimes, it is difficult to convincingly demonstrate that a given specimen does indeed deserve to be considered a case, since this requires scrutiny of all the specimens in the class supposedly represented by the alleged "case", at least in principle. Such a demonstration is a beyond the scope of any journal article. So, the readers usually have to take the authors' word for the exemplarity of an alleged case whilst being offered access to all the raw data on the basis of which the case has been selected.

In the present context, the cases have been selected—by us—from a large pool of some 50 students who have been chosen (out of around 700 students) as focus students by the prospective mathematics counsellors participating in our professional development program for upper secondary mathematics teachers. All these students have been described in the reports written by the teachers who worked with them in the program, and our selection rests on careful reading of all these reports. It is our best judgment that the traits and properties represented by the three case students are indeed found to manifest degrees in the entire student population, but we cannot prove it in an article such as this. Therefore, our conclusion and discussion must necessarily take a qualitative form when it comes to the generalizability of the findings we are going to present.

In each of the three cases, upon firstly having detected the students to be worked with and next having attempted to "diagnose" the individual student's mathematics-specific difficulties, the maths counsellors, under our supervision, designed interventions, inspired by or based upon the mathematics education theoretical constructs that they had studied as part of the program. These interventions took place during the middle part of one semester, i.e., they typically had a duration of three months and seldom comprised more than ten 40–80-min sessions with the students. (All quotations have been translated into English by us.)

In order to provide a bit more of a background to the three cases we are going to present in the next section, we shall briefly outline the structure of Danish upper secondary school. Danish non-vocational upper secondary education—to which the cases belong—typically takes three years, corresponding to grades 10–12 (ages 16–18/19) (even though there is, in fact, also a two-year option). The three-year upper secondary school is organized in three different streams, a general academically oriented stream, called stx, a technology- and science-oriented stream, called htx, and a commerce and business-oriented stream called hhx. All students have to study mathematics for at least one year, called C level. Many students (and soon all) must study mathematics for two years, called B level, whereas a somewhat smaller (but not small) segment of the student population study mathematics for all three years, called A level. In principle, a given level is supposed to be the same across the different streams, but, as there is a great diversity in content and approach in the three streams, they should not be taken to be identical, even though the label is the same. Students will have to take a national exam in mathematics according to the level at which they have chosen to study mathematics.

FIRST CASE: MATHEMATICS AS DISCONNECTED PROCEDURES

In the Spring of 2015, two teachers, Tina Rosendahl and Henrik Kærgaard, worked with a second-year student, with the pseudonym Christian, in the technology/science stream (called htx) of Danish upper secondary school [15]. The teachers knew Christian from a course in programming, where he had shown a clear understanding of the logical aspects involved. However, from Christian himself and from a colleague who taught him before, they learned that he was close to a failing grade in his mathematics course. This led the two prospective maths counsellors to further investigate the case of Christian. They wanted to find out what he thought of mathematics, both as a school subject and a discipline, and what his perceptions of logic and mathematical truth were. Hence, they designed a diagnostic interview in order to uncover some of his mathematics-related beliefs that might have acted as stumbling blocks and destructive myths in his learning of mathematics. The teachers found that Christian felt alright with mathematical topics which he immediately could see as having a practical application, e.g., calculation with percentages, whereas he saw no relevance of trigonometry. When performing simple arithmetical computations on specific numbers, e.g., involving addition, multiplication, and parentheses, Christian experienced no difficulties but, when presented with a simple algebraic task in a symbolic domain, he exclaimed: "I hate algebra. Because it's full of 'cheaters'" [15] (p. 147). Asked about the history of mathematics, Christian showed some knowledge concerning counting and computations as he replied "... counting with stones, later the abacus" [15] (p. 146). When asked about the hierarchy of arithmetical operations, he suggested that "... someone had invented it" and provided the following account of it: "[when] multiplying, you calculate with larger numbers to begin with. Then you want to reduce them first, so you take the simple part and pluses and minuses afterwards. Parentheses are often like… what's inside them is more important. You always 'frame' what's more important" [15] (pp. 146–147). On the basis of the interview, the two prospective maths counsellors diagnosed Christian as having a conception of mathematics as a subject "consisting of an incomprehensible swarm of weird and incoherent mnemonic rules that make no sense" [15] (p. 147). Because of this, he seemed to have convinced himself that he could not understand or do mathematics at the level required and might as well give up. In other words, he was under the spell of a destructive myth.

Taking the above as their point of departure, the teachers designed an intervention with the purpose of making Christian realize that, in fact, in mathematics you only have to memorize a very few rules, and then you are able to deduct most of the rest of what you need. The main idea for the intervention was to present Christian with a small handful of algebraic conventions and axioms, allowing him to make sense of and familiarize himself with these through a number of little exercises. This was to be followed up by some more complex tasks designed to enable him to come to grips with selected basic algebraic operations and results. The teachers hoped that Christian would then realize that these results did not drop down from the skies but followed directly from the axioms initially introduced through a process of logical reasoning. On the basis of different theoretical constructs, including Brousseau's notion of didactical and adidactical situations [11] and Hanna's notion of proofs that not only prove but also explain [16], the teachers designed a three-session intervention. In the first session (in which we were observers), Christian worked with the hierarchy of the arithmetical operations and had to multiply symbolic expressions, the results illustrating areas of quadrilaterals, e.g., $(a + b)\cdot(c + d)$ corresponding to a quadrilateral with sides $a + b$ and $c + d$. In the second session, Christian, among other things, was asked to deduce the formula for a squared binomial, $(a + b)^2 = a^2 + b^2 + 2ab$, again supported by figures showing the corresponding squares and rectangles. In the third session, he had to work with the power conventions and rules. Upon being asked to deduce that $a^0 = 1$, Christian uttered: "Oh, I've seen that on the Internet, you just have to accept that it is 1. I once asked a third year student about it, and he said that the proof was like really long" [15] (p. 177). As the math counsellors explain, once Christian was told that in less than two minutes he would have made this proof himself, he looked rather suspicious. Yet, applying the previously deduced power rules, he was indeed able to do so. At the end of the session, Christian was asked to respond to the fact that he had just deduced all the power rules and that the only mnemonic rules he had applied were some basic axioms and the hierarchy of the arithmetic operations. He was rather baffled—although obviously also proud of his accomplishments.

As a consequence of the intervention, the two teachers observed in class how Christian's attitude to the subject of mathematics changed. He now believed that it was actually possible for him to understand and learn something and he became determined to get as far as possible within the last few months of the school year. In a final exam project in mathematics that year, Christian received the grade 7 (corresponding to a C on the ECTS scale, where he used to score F or E). One year later, the maths counsellors interviewed Christian again in his final year in high school, in order to assess possible long-term effects of the intervention. He said:

I have come to view the solving of mathematics tasks in a different way. The same is true of programming. I've begun considering what possibilities I have for getting help, when there is something I can't figure out—as opposed to giving up, which is what I would have done before. I still prefer to solve the tasks by myself. When something is difficult, I have now learned that I can address the task by simplifying it—place things in a sequence and take them bit by bit, instead of looking at everything in one big hotchpotch. [...] I found out that mathematics is simpler than I thought prior to the intervention. I can now remember how the formulae [rules] are connected. Actually, I'm going to have some math as part of my chosen higher education program, where I'll be studying ICT technology. [...] I've always known that there was some kind of logic in mathematics, only I hadn't found it. Now I have. [...] After the intervention, I have gained the courage to pursue a higher education degree, where I have to apply mathematics. Before, I believed that I suffered from dyscalculia. You made me see mathematics in a completely different way. Now I know that I'm not innumerate. [17]

It seems fair to claim that this intervention actually dispelled some of the myths that used to ruin Christian's relationship with mathematics.

SECOND CASE: EVERYDAY CONCEPTIONS IN MATHEMATICS

The two teachers and prospective maths counsellors Mette B. Lomholt and Morten S. Larsen followed Student A, as they called her, closely throughout her three years in the technical branch of upper secondary school where she took mathematics and physics at the highest possible level, Lomholt and Larsen being her teachers in these subjects [18,19]. They specifically intervened towards her learning difficulties during the last semester of her second year and the first semester of her third year.

Student A believed that mathematics is completely focused on numerical computations and that mathematics tasks are prompts to rush off to do some computations as quickly as possible, a myth that, in the beginning, completely governed her relationship with mathematics. Whenever she was unsure of what to do—which was often—she would begin guessing at random or throw herself head over heels into some aimless plan. So, it was largely a matter of chance whether or not the computations she undertook had any bearing on the questions posed in the tasks. One consequence of her beliefs about mathematics was that, for ordinary students, independent thinking is neither necessary nor desirable when dealing with mathematics. She was very hostile to and afraid of tasks that either do not require any computations or place computations in contexts involving symbolic calculations. She was particularly reluctant to engage in tasks requiring reasoning in an algebraic or a geometric environment.

Based on A's response to a large variety of test questions of different kinds, the teachers first realised that her understanding and skills in algebraic contexts were very weak, in fact to such an extent that this perturbed her ability to reason mathematically. For example, when asked to consider the equation $38x + 72 = 38x$, she wrote "You cannot add 'pure' numbers and 'x-numbers'. It's like adding pears and bananas. So, you cannot reduce the equation any further" [18] (p. 78). After having made several (actually legal) transformations of another equation, she finished by writing "$1x = x \Rightarrow 0$", where it is not completely clear what the inference is supposed to be. She knew that $x \cdot x = x^2$ but not that $d \cdot d = d^2$. When examining or making statements involving formulae, she had a strong preference for working with specific numbers rather than with symbolic expressions. She was neither able to interpret, nor convinced by, conclusions obtained by manipulating formulae or expressions. When asked whether it is true that the volume of a cube is multiplied by 8 if all edge lengths are multiplied by 2, she wrote $V = s \cdot s \cdot s$ and then said, "I don't think that the new volume is eight times as large" [18] (p. 94). She maintained this response when prompted to write $V' = (2s) \cdot (2s) \cdot (2s) = 8s^3$. When asked whether it is true that if one halves the diameter of a circle then both the area and the circumference of the new circle are half those of the original circle, respectively, she was convinced that the answer is "yes", even though she was able to correctly state the formulae for the area and circumference of a circle. The teachers invited her to make drawings representing doubling the edge lengths of a square and of a cube, which she was hardly able to do. After having been assisted by the teachers in making the drawings and then asked to interpret them, she was not convinced that they show that the area and the volumes are multiplied by 4 and 8, respectively. As the teachers wrote: "She is algebraically convinced but remains uncomfortable with drawings, at least when chequered or diced patterns are involved." [18] (p. 95) In other words, A was a strong believer in the "law" that if something in a context is doubled, then everything is doubled, which is a marked instance of the over-generalization of proportionality, as previously studied [20,21].

Against this background and considering similar experiences with A in her physics class, the teachers concluded that in addition to algebraic reasoning, A's difficulties were largely to do with mathematical modelling in situations involving proportionality, scaling, ratio, rate, and speed. This was further reflected in her fear of units and unit conversion, especially in physics. So, they designed a three-week intervention program in which A was invited to work on a large variety of modelling-oriented tasks involving these notions. In the design of these intervention activities, the maths counsellors relied on Lesh's and Doerr's theoretical construct of model-eliciting activities [22]. A was asked to formulate hypotheses concerning formula-specific relationships and afterwards test these by using manipulative materials such as Lego blocks. As particularly regards the doubling of the edge lengths of squares and cubes, A assembled and disassembled the blocks numerous times, eventually convincing herself of the squaring of the area and cubing of the volume, whilst stating "This is a shaking experience" [18] (p. 144). At the end of the intervention program, A had made considerable progress, even though there was still room for further progress. She could make drawings on her own initiative, try to interpret and make sense of tasks before attempting to solve them, and no longer believed that proportionality rules the world, that mathematics consists in numerical computations, and that symbolic calculations have nothing to do with numbers or with the world outside mathematics.

As the teachers observed, their relative success in changing A's beliefs about the nature of mathematics and mathematical activity and in strengthening her mathematical competencies would not have been possible had she not been willing to invest a non-negligible amount of time and effort and to endure frustration over an extended period of time. Another key point in the story of A is that the achieved improvements of her mathematical understanding and competencies rested on the maths counsellors' targeted and sustained efforts of trying to uncover the roots of A's learning difficulties in mathematics and eventually to dispel some of the myths she was under.

THIRD CASE: LONG-STANDING TRAINING OF PROCEDURES

In the fall of 2016, three teachers, Christian Wejdemann, Karl-Kristian Bjerregaard, and Flovin Næs, worked with a second-year student in the general stream (called stx) of Danish upper secondary school [23]. Rikke (the pseudonym given by the teachers) was a conscientious student, a bit quiet in her behavior, and reluctant in showing her mathematical work to others. Her attainment level was usually assessed to around 4 or 7 (D or C on the ECTS scale). In her homework assignments as well as in the math counsellors' testing and subsequent interviewing of her, she displayed difficulties in relation to recognizing and classifying functions according to their type (e.g., linear functions, power functions, exponential functions, etc., in modelling contexts), a key component of upper secondary mathematics education in Denmark. She was offered the opportunity to take part in a four-session intervention related to aspects of mathematical modelling, where she would have to apply different types of functions as means for mathematizing real-world problem situations. At first, Rikke was reluctant to take part in this intervention, but eventually she agreed, on the condition that she could do it together with a classmate. As part of these sessions, the two students first had to pair verbal descriptions of problem situations with the correct graphical representations out of several options, and next to pair formula-based expressions of functions with graphical representations.

The following task and Rikke's problems in tackling it clearly illustrate her misconceptions in relation to identifying function types and also provide an entry point to some of her mathematics-related beliefs that were seen to act as stumbling blocks in her mathematical work. The verbal description of the problem situation was: "At 7 p.m., Axel's body temperature is the usual 37 degrees Celsius. He then becomes very ill, and in the following time period [unspecified] his body temperature rises by 1.5 degrees per hour; y denotes Axel's body temperature, and x denotes the number of hours after 7 p.m. on the night in question." The two students then had to decide whether this verbal description corresponded to linear growth, growth by a power law, or to exponential growth. Rikke's first conjecture was that the situation could be described by an increasing exponential function and that this was due to the fact that Axel's body temperature rose by 1.5 degrees per hour.

Rikke produced a (correct) table of five corresponding x- and y-values. Still, she insisted that this resulted in exponential growth. The maths counsellors then asked her to calculate more values and eventually to plot them on a piece of graph paper. Once she had completed this, she exclaimed:

- Rikke: This is pretty linear!
- Teacher: Why does this surprise you?
- Rikke: Because normally, when we have something that grows per time unit, then it is an exponential function. [23] (p. 228)

As it turns out, Rikke practiced a strategy of spotting clue words and other triggers, e.g., a phrase like "per time unit", and then based her choice of function type on this. Also, the maths counsellors discovered that she generally did not believe in her own calculated values for sample points. Rather, she attempted to choose a graphical representation based on the formula expression (if possible), often leading to an incorrect choice. During the third session, the maths counsellors interviewed Rikke about this, and in the process uncovered a central belief of hers.

- Rikke: I think it is because there are so many different ways to calculate it, and then you begin to doubt [...]. I often think about if what I do is actually correct. [...]
- Teacher: Is it the same in other subjects, i.e., that you feel this insecurity, or is it peculiar to mathematics?
- Rikke: In biotechnology, for example, we also calculate and solve problems sometimes. But there you get insecure because it is difficult, and the others think so too. So you are more in it together, finding it hard. And then we can ask the teacher. But in maths, which is a very important subject, and I have it at A level… there are other factors involved, I think.
- Teacher: But biotech is also important and at A level.
- Rikke: Yeah, I know that, but, in biotech, we are all working together on tasks that are pretty hard, and which we haven't tried before, whereas we've had mathematics since forever. Biotech is new in a completely different way, and we don't have the same background knowledge as we do in math. In biotech everything is new, and we all have to learn the same, from the same 'beginning', if you can say that.

- Teacher: Is it that you feel it is worse in math because you ought to be able to do it, since you've studied maths for so many years?
- Rikke: Yes, I think so, and then I get insecure, if there is something I can't figure out. Then I think, 'fuck, it's going to affect my grades' or something. I feel stupid.
- Teacher: In math, the pressure of expectations is greater in some way?
- Rikke: Yeah, a little I think. It seems like the expectations are a little greater, and that you should be able to do everything in math, I think.
- Teacher: Are these expectations from the teacher? The classmates? Or the textbook? Where do they stem from?
- Rikke: All of the above, kind of… Now, you are in upper secondary school and you are supposed to be able to do it.
- Teacher: A final question. In the final national written exams in mathematics, B level and A level, how many students do you think fail these?
- Rikke: Not very many.
- Teacher: Make a guess. In percentage.
- Rikke: Maybe 5%.
- Teacher: No, actually it is a lot higher. I'm not saying this to scare you, but rather to deal with your idea that 'everybody is able to do maths' and 'we've had it for such a long time, so we ought to be on top of it'. Math is hard and it is very different in upper secondary school. So, you need to view it in the same way as you view biotech. [23] (pp. 238–239)

As part of the final session of the intervention, the maths counsellors emphasized the need for Rikke to trust the methods she had been introduced to, e.g., to calculate supporting points in order to determine the type of growth at issue and to dare to experiment more as part of her mathematical problem tackling. Upon the completion of the intervention, the maths counsellors gave Rikke a test where she was invited to do a number of tasks alone in a room while being video-filmed. Although Rikke by no means was able to correctly answer all tasks presented to her, the maths counsellors did document a significant change in her approach to a task. To a larger extent, she now applied a strategy by which she tried to arrive at a formula expression informed by calculations based on the verbal task description.

Once having obtained a formula expression, she would even sometimes control it by performing calculations and see if these matched the previous ones based on the verbal descriptions. Some time after the intervention, Rikke offered her thoughts about it in an email to the maths counsellors:

Hi.

I'd like to thank you for the opportunity to participate in the intervention. It was a way for me to come to think about and view math in a different way than what I've done before. [...] in the beginning it did in fact put my self-confidence on trial, but that has changed, of course. [...] I truly feel that my motivation for raising my hand in various subjects has grown. I am now more daring, even if what I say sometimes turns out to be wrong. In principle, I have come to trust myself more than I did before. It's not that my attitude towards math has changed as such, since I've always liked maths. But the sessions with you guys made me desire to be taught more math, simply because I now feel more confident in the subject. I'm also more confident now in my own approach to solving tasks—if you see what I mean. I don't wait for my classmates to confirm that they have used the same approach as I did—now I actually trust myself, what I do, and what I know. Cooperation with my classmates in maths—I feel—has also changed to my advantage, if I may say so. The methods you provided me with in the sessions made me trust myself, without all the time needing to seek confirmation that what I'm doing is right. [...] You told me that maths has a high failure rate—this has made me aware that I'm not the only one who finds maths hard sometimes, and that I'm not alone in this.

Since I now have some tools and other ways of seeing mathematical relationships, I also have more appetite for doing cranky tasks on my own. I've learned to turn my insecurity into something positive, so I look at [mathematical] relationships from different angles and use my calculations to arrive at results that I myself believe in. An important point, I think, is not always just restricting yourself to the basics [rules, formulas, etc.], but to do experiments, make some calculations. For example, for exponential functions there is a certain formula, a certain way the graph has to look, and some requirements that have to be fulfilled. Previously—as I believe you also noticed—I restricted myself to considering a mathematical relationship solely based on the information given [in the task] and the basics and used that to decide whether a function is exponential, linear, and so on. This didn't always work so well for me [...]. You taught me that with different tools and methods you can easily determine the relationship, without initially being able to see it on the basis fof the given information. [...] I can definitely use these experiences and my newly acquired tools and methods from the intervention in my everyday maths classes. And I want to say that I'm quite positively surprised at what I've actually gained from all of this.

I hope you can use my answer for something.

Rikke

[23] (pp. 256–257)

At least two of Rikke's myths about mathematics were dispelled by this intervention, namely, that, when you have received teaching of mathematics for 10 years, you should have grasped it by then, and that the ingredients required to deal with a mathematics tasks are based on surface features.

ANALYSIS OF THE THREE STUDENTS' BELIEFS

Christian was a student who had come to believe that mathematics is a subject that is out of reach in terms of understanding, at least for him. His own encounters with the subject had failed, also according to his previous teachers, and he accepted this and in fact agreed. He acknowledged that it is potentially possible for others to learn and do mathematics, but unfortunately not for him. He believed to be operating under a didactical contract saying that most of mathematics needs to be memorized and learnt by rote, because the subject makes no sense from a logical point of view. In particular, things related to algebra and geometry appeared almost as "abracadabra" to him. Still, he did accept that some aspects of mathematics can be of practical importance.

As the reader will notice, some of Christian's views coincide with Schoenfeld's list of students' counterproductive beliefs about mathematics [7]. Christian's "beliefs about mathematics as a discipline" were misguided, and this affected this "beliefs about mathematics education"—hence in the tetrahedron of Figure 1, we are dealing with the edge between these two vertices—because he did not see the subject of mathematics as concerning reasoning and logic, but rather as an arbitrary collection of mnemonic rules. This, however, had consequences for his negative "beliefs about the self". Seen in this light, it may not be so strange that Christian performed poorly in mathematics. Also, considering that he did not suffer from more general cognitive disabilities, since he did well in his programming class, it seems fair to infer that his difficulties in relation to mathematics in large part stemmed from his counterproductive beliefs and the myths that created them.

The maths counsellors chose to invest efforts in making it clear to him that mathematics is a discipline of reasoning rather than of rote learning. They decided to focus on a topic about which Christian uttered fear and contempt, namely algebra, and did so with the hope that if he could come to grips with the fundamentals of algebra, this might have a spillover effect into other topic areas. Because algebra apparently had caused Christian so much trouble, they decided to begin with the most elementary symbol conventions, rules, and manipulations of symbolic expressions and formulae. They established a local axiomatization of algebra, which could and would form the basis for Christian's mathematical work as part of the intervention. They supplemented the algebraic manipulations, e.g., of the formula for a squared binomial, with drawings in order to—hopefully—provide a meaning to and plausibility of the axiomatic rules. In so doing, they hoped to convince Christian that he could learn to obtain mathematical conclusions, that he actually could do mathematics.

The intervention was a success—as mentioned, we witnessed part of it. Christian experienced that he was actually able to do mathematics, not just memorize it, but perform reasoning and carry out deductions of his own. Moreover, the things he had to do made sense to him. The two maths counsellors hit the nail on its head with their design of the intervention. Christian got to realize and experience that people—people like himself—can in fact be brought to understand mathematics. He experienced that one can act mathematically and activate mathematical processes—and, not least, he experienced a set of rules (axioms), the ones on which the intervention was based, that he could make sense of and relate to. This not only changed his attitude towards mathematics and his behavior in relation to the subject, but also boosted his self-confidence in general.

Concerning Student A, it is difficult to say whether her beliefs and the myths she were under primarily concerned "mathematics as a discipline" or "mathematics education", although we lean towards the latter. This was not least due to the fact that she considered mathematics to be a subject dealing mainly with computations. In fact, she expressed almost fear—or "math anxiety" [24]—when having to perform any type of mathematical reasoning or algebraic manipulation. From a didactical contract perspective, Student A knew that she was required to do this as part of mathematics, but she was very reluctant to embark on such undertakings.

These were the symptoms that the teachers noticed at first and the reason why they offered Student A to participate in the interventions in the first place. During the course of the interventions, the teachers uncovered some of her more deeply rooted mathematical stumbling blocks, namely, her conceptual difficulties related to proportionality, scaling, ratio, and volume, especially in relation to tasks involving mathematical modelling. A dominant aspect of A's difficulties was her reliance on an empirical proof scheme, a proof scheme which, on the one hand, was one source of her troubles but, on the other hand, turned out also to be a source of inspiration in the attempts to reconcile this proof scheme with the deductive proof scheme. Through the design of a number of model-eliciting activities, the teachers sought to provide meaning to Student A in relation to the concepts involved. She was to gain experience—also of a concrete and empirical nature—of what happens to proportions and ratios when the constituents, including units, change. As part of the intervention, it became clear that what was actually essential was to draw on her belief that mathematical convictions must be empirically founded, if she was not to simply doubt her own calculations. We admit that this seems rather extreme and we have not encountered a similar case, neither before nor after Student A.

Rikke was a student who was embarrassed by the fact that she did not do as well in mathematics as she, according to her own beliefs, thought she ought to, since she had been learning mathematics for 10 years already. Furthermore, she believed that she was one of only a very few students who was performing somewhat poorly. These beliefs mainly correspond to the edge of the tetrahedron connecting the vertices "beliefs about mathematics education" and "beliefs about the self", although there are also links to the vertex "beliefs about the social context". Rikke was concerned about her academic appearance and worked hard to avoid exposing her difficulties to other students and to her teachers. She was also concerned—in fact even nervous—about her grades in mathematics, which was probably one reason why she was rather reluctant to participate in the intervention in the first place.

Unlike the other two cases, we do not see clear links to "beliefs about mathematics as a discipline". An interesting aspect of Rikke's case was her perception of her biotechnology class. Since this subject was new to all students, it was okay to find it hard and difficult—in contrast to mathematics. This seems to constitute a rather peculiar version of the didactical contract, one that relates to the entire school system as opposed to just her current mathematics class. From a slightly different perspective, Rikke could be considered to show elements of "math anxiety" [24]. Her teacher's approach to this was to inform her that the number of students who actually fail mathematics in upper secondary school is much higher than she imagined, indicating to her that not only she, but many students at this level actually do find mathematics difficult and hard.

Rikke was selected for intervention because of her marked lack of ability to choose the correct type of function in tasks about mathematical modelling. On closer inspection, the teachers found that her choice of functions relied on certain "trigger words". On the one hand, this was of course due to Rikke's insufficient understanding of different representations of functions, including her limited ability to make conversions between these [25]. On the other hand, it might also partly be explained by her perception of the didactical contract as requiring her to be able to immediately "guess" the correct function type on the basis of some superficial features of the task description. She simply believed that, in mathematics, there is no need to think and reflect about one's choice. From a proof scheme perspective, Rikke was markedly dependent on an external conviction scheme of the authoritarian type, which was also evident from the fact that she constantly sought confirmation from both teacher and classmates in everything she did in mathematics. As part of the intervention, the mathematics counsellors attempted to move her towards a more empirical conviction scheme, demonstrating to her why and how it is okay to "experiment'" to some extent, as part of carrying out mathematical processes, and that she herself could in fact do this—which, as her letter illustrates, she actually came to realize. Although the trees did not grow to the sky, Rikke did become more capable of switching between different representations of functions and used this to correctly select functions from the types at her disposal. Another outcome of the intervention was that Rikke began to take the initiative in mathematics classes, both in relation to her own problem-solving and modelling and when collaborating with other students. Finally, we should like to call attention to the fact that Rikke's letter to the teachers actually shows a very high degree of self-reflection and an explicit acknowledgement of myths having been dispelled—which again underlines the impact of the intervention.

DISCUSSION OF THE EFFICACY OF THE INTERVENTIONS

What, then, are the elements responsible for the success of the interventions described? As we see it, these are mainly five.

Firstly, satisfactory boundary conditions, an organizational framework, and exterior practical circumstances must be in place. The (upper secondary) schools and their management have to support and facilitate the initiative. They must acknowledge that many students need this kind of assistance and they must agree to cater for education to teachers that enables them to actually do it. This implies that teachers must be provided with a research-based insight that allows them to carry out both diagnoses and interventions in addition to identifying the students in need of assistance. Not least, sufficient time must be allotted to teachers and students for undertaking the tasks in the interventions.

Secondly, well-founded diagnostic work must be done by the teachers in order to uncover the roots of the students' misguided conceptions of mathematics, i.e., which are the misconceptions of the didactical contract or the unproductive beliefs, etc., that foster destructive myths and should be "blamed" for the students' mathematics-related difficulties. To increase the likelihood of successful and productive diagnostic work, this must be grounded in pertinent mathematics education research. If the diagnoses are only superficial and merely point to the symptoms of the misconceptions rather than to their roots, then the following interventions are likely to be much less effective than they could be.

Thirdly, clever interventions have to be designed, i.e., interventions which, to a sufficient degree, capture and address significant elements of the problems diagnosed with the specific student(s). Interventions, too, have to be research-based as much as possible but they cannot be so to the same extent as the preceding diagnosis. This is because research tells us much more about mathematical learning and learning difficulties than about successful teaching, which is much more dependent on the setting, context, and conditions within which it takes place and, above all, on the students being taught. So, teachers have to be very inventive in combining research findings with their own intervention designs targeting the specific students with whom they work. Surely, there is an element of luck involved in this enterprise.

Fourthly, not only the teachers, but also the students have to be prepared to invest the necessary amounts of effort and time. The interventions are only likely to work if the students cooperate and are willing to engage in the designed activities in the manner intended by the teachers. In other words, students and teachers have to come to an agreement on a didactical contract for the intervention. In order for this to happen, a significant degree of mutual trust between the students and the teachers have to be established.

Fifthly and finally, if the four elements just mentioned are present, including the element of luck in relation to the intervention design, it is not unlikely that the intervention may have a number of positive spillover effects, especially in the form of generalizability of the experiences. For instance, in the case of Christian, even though the intervention concerned basic algebra, the increased self-confidence that he acquired alongside his improved understanding of algebra positively affected his entire life with the subject of mathematics. Similarly, a spillover from Rikke's intervention was that she, too, began to trust herself in the social context of doing mathematics.

The three cases in unison illustrate that even though the five elements listed above are undoubtedly somewhat demanding, they are not insurmountable. In terms of time invested, the interventions for both Christian and Rikke lasted less than three months. For Student A, the intervention was a lengthier one, as the two teachers met with her more or less regularly over a period of one year. Still, the cases illustrate quite clearly that, although these three students' misconceptions and unproductive beliefs were the products of 10 years or more of mathematics teaching and learning, something effective could actually be done about them in a markedly smaller timespan, both in absolute and in relative terms. However, we have to admit that we do not know much about the long-term duration of the documented changes. Yet, the changes lasted long enough to be significant throughout students' remaining years in upper secondary school, including their final exams.

The theoretical constructs that we have applied in our analyses of the case study of the three students are those of students' mathematics-related beliefs, proof schemes, the didactical contract. Although these three constructs are usually not perceived as directly related to each other, they are indeed rather closely connected in the sense that misguided beliefs may cause misconceptions of the didactical contract and vice versa and that erroneous or rudimentary proof schemes may stem from a mistakenly construed didactical contract. An obvious question to ask is, of course, why the didactical contract goes awry? We can think of two main reasons for this. Firstly, the mathematics teacher may possess, him- or herself, misguided mathematics-related beliefs, e.g., in relation to mathematics as a discipline. Secondly, even though the teacher actually knows how to construe the didactical contract in a reasonable, sensible, and productive fashion, the possibilities, e.g., of a practical or financial nature, for actually implementing it are not present to the degree needed. Which of these two reasons are to blame for the misguided and destructive conceptions of Christian, Student A, and Rikke is not to say, especially since their misconceptions are the results of 10 or more years of mathematics education, with different teachers, at different schools, etc. Still, as singular instances and taken together, the cases of these three students illustrate the connectedness of the theoretical constructs of beliefs, proof schemes, and didactical contract. In fact, a separate study of the interconnectedness of these three constructs would be of both theoretical and practical importance. Unfortunately, this goes beyond the scope of the present article but does deserve to be taken up at a later point in time.

CONCLUSIONS

The question we set out to answer in this article was: what does it takes for targeted efforts to be reasonably successful in altering students' misconceptions and unproductive beliefs and dispelling myths so as to pave the way for satisfactory learning? We have tried to answer this question by way of a study of three cases in which it proved possible to design and implement such targeted efforts with a reasonable degree of success. We have further tried to identify the factors that were responsible for the achievements obtained, obtaining the five elements presented in the previous section. A common trait of these elements is that it is necessary to pay serious attention to the substantive aspects of the teaching and learning of mathematics, rather than being content with its superficial features.

It is clear that the efforts described did not rest on implementing a fixed, one-size-fits-all template of detection, diagnosis, and intervention to all students in all situations. Even though mathematical learning and mathematical learning difficulties do indeed have some universal constituents, which have been uncovered by research, they also manifest themselves very differently in different students. The specific combination and realization of misconceptions and unproductive—or even destructive—mathematics beliefs and myths held by a student is typically particular to that student. This has to be reflected in the interventions put in place to help a student overcome his or her difficulties, and some degree of "a lucky hand" is necessary for a successful outcome.

ACKNOWLEDGMENTS

We want to thank the mathematics counsellors educated in our program at Roskilde University, not only the seven teachers who have provided us with the three cases discussed in the present article, but all the 72 who have completed the program so far, for giving us access to unique first-hand experiences with students as well as with cases and data like the ones presented here.

AUTHOR CONTRIBUTIONS

This article has been written jointly and integratedly by the two authors.

REFERENCES

1. Jankvist, U.T.; Niss, M. A framework for designing a research-based 'math counsellor' teacher programme. *Educ. Stud. Math.* **2015**, *90*, 259–284.
2. Philipp, R.A. Mathematics teachers' beliefs and affect. In *Second Handbook of Research on Mathematics Teaching and Learning*; Lester, F.K., Jr., Ed.; Information Age Publishing: Charlotte, NC, USA, 2007; pp. 257–315, ISBN-13: 978-1-59311-587-6.
3. Parajes, M.F. Teachers' beliefs and educational research: Cleaning up a messy construct. *Rev. Educ. Res.* **1992**, *62*, 307–332.
4. Op't Eynde, P.; de Corte, E.; Verschaffel, L. Framing students' mathematics-related beliefs. In *Beliefs: A Hidden Variable in Mathematics Education?* Leder, G.C., Pehkonen, E., Törner,, G., Eds.; Kluwer Academic Publishers: Dordrecht, The Netherlands, 2002; pp. 13–37. ISBN 1-4020-1058-3.
5. Jankvist, U.T. Changing students' images of "mathematics as a discipline". *J. Math. Behav.* **2015**, *38*, 41–56.
6. Thompson, A.G. Teachers' beliefs and conceptions: A synthesis of the research. In *Handbook of Research on Mathematics Teaching and Learning*; Grouws, D.A., Ed.; Macmillian Publishing Company: New York, NY, USA, 1992; pp. 127–146. ISBN 0-02-922381-4.
7. Schoenfeld, A.H. Learning from instruction. In *Handbook of Research on Mathematics Teaching and Learning*; Grouws, D.A., Ed.; Macmillian Publishing Company: New York, NY, USA, 1992; pp. 334–370. ISBN 0-02-922381-4.
8. Spangler, D.A. Assessing students' beliefs about mathematics. *Arith. Teacher* **1992**, *40*, 148–152.
9. Jankvist, U.T. Using History as a 'Goal' in Mathematics Education. Ph.D. Thesis, Roskilde University, Roskilde, Denmark, August 2009. Published as no. 464 in *Tekster fra IMFUFA*. Available online: **http://milne.ruc.dk/imfufatekster/pdf/464.pdf** (accessed on 18 April 2018).
10. Kloosterman, P. Beliefs about mathematics and mathematics learning in the secondary school: Measurement and implications for motivation. In *Beliefs: A Hidden Variable in Mathematics Education?* Leder, G.C., Pehkonen, E., Törner, G., Eds.; Kluwer Academic Publishers: Dordrecht, The Netherlands, 2002; pp. 247–269. ISBN 1-4020-1058-3.

11. Brousseau, G. *Theory of Didactical Situations*; Kluwer Academic Publishers: Dordrecht, The Netherlands, 1997; ISBN 0-7923-4526-6.
12. Education Committee of the European Mathematical Society. What are the reciprocal expectations between teacher and students? Solid findings in mathematics education on didactical contract. *Newsl. Eur. Math. Soc.* **2012**, *84*, 53–55.
13. Harel, G.; Sowder, L. Toward comprehensive perspectives on the learning and teaching of proof. In *Second Handbook of Research on Mathematics Teaching and Learning*; Lester, F.K., Jr., Ed.; Information Age Publishing: Charlotte, NC, USA, 2007; pp. 805–842, ISBN-13: 978-1-59311-589-0.
14. Harel, G.; Sowder, L. Students' proof schemes: Results from an exploratory study. In *Research in College Mathematics Education III*; Schoenfeld, A.H., Kaput, J., Dubinsky, E., Eds.; AMS: Providence, RI, USA, 1998; pp. 234–283. ISBN 978-0-8218-0882-5.
15. Rosendahl, T.; Kærgaard, H.N.L. *Udvikling af Matematikforståelse—Hos Elever Med Snublesten*; Diploma Report; Roskilde University: Roskilde, Denmark, January 2016.
16. Hanna, G. Some pedagogical aspects of proof. *Interchange* **1990**, *21*, 6–13.
17. Rosendahl, T.; Kærgaard, H.N.L. Udvikling af matematikforståelse med elevens egen logik. In *Matematikvejledning i Gymnasiet*; Niss, M., Jankvist, U.T., Eds.; Frydenlund: Frederiksberg, Denmark, 2018.
18. Lomholt, M.B.; Larsen, M.S. *Matematikvejledning i htx—Forsøg Med, og Refleksioner over, Identifikation, Diagnosticering og Intervention over for Elever Med Særlige Vanskeligheder i Matematik*; Diploma Report; Roskilde University: Roskilde, Denmark, January 2014.
19. Lomholt, M.B.; Larsen, M.S. A's historie. In *Fra snublesten til Byggesten—Matematikdidaktiske Muligheder*; Niss, M., Jankvist, U.T., Eds.; Frydenlund: Frederiksberg, Denmark; pp. 98–121. 2016; ISBN 978-87-7118-609-3.
20. Van Dooren, W.; De Bock, D.; Janssen, D.; Verschaffel, L. The linear imperative: An inventory and conceptual analysis of students' overuse of linearity. *J. Res. Math. Educ.* **2008**, *39*, 311–342.
21. De Bock, D.; Van Dooren, W.; Verschaffel, L. Students' over-use of linearity: An exploration in physics. *Res. Sci. Educ.* **2011**, *41*, 389–412.

22. Lesh, R.; Doerr, H.M. (Eds.) *Beyond Constructivism, Models and Modeling Perspectives on Mathematics Problem Solving, Learning, and Teaching*; Lawrence Erlbaum Associates, Publishers: Mahwah, NJ, USA, 2003; ISBN 0-8058-3822-8.
23. Wejdemann, C.; Bjerregaard, K.-K.; Næs, F.T.N. *Matematikvejlederuddannelsen—Afsluttende Rapport*; Diploma Report; Roskilde University: Roskilde, Denmark, January 2017.
24. Fennema, E.; Sherman, J.A. Fennema-Sherman mathematics attitudes scales: Instruments designed to measure attitudes toward the learning of mathematics by females and males. *J. Res. Math. Educ.* **1976**, *7*, 324–326.
25. Duval, R. A cognitive analysis of problems of comprehension in a learning of mathematics. *Educ. Stud. Math.* **2006**, *61*, 103–131.

Chapter

ADVERSITY QUOTIENT AND RESILIENCE IN MATHEMATICAL PROOF PROBLEM-SOLVING ABILITY

4

Fauziah Hakim[1], Murtafiah[2]

[1,2]Program Studi Pendidikan Matematika, Universitas Sulawesi Barat
[1,2]Jl. Baurung, Banggae Timur, Kabupaten Majene, Sulawesi Barat

ABSTRACT

Various mathematical abilities require a high level of struggle to be achieved, one of which is the ability to solve mathematical proof problems. Several factors are also associated with this ability, including adversity quotient and resilience. This research aimed to look at the effect of adversity quotient and resilience on the mathematical proof problem-solving ability. This research used a quantitative approach with a correlational method. The sample in this study were 31 students randomly selected from all students of the Mathematics Education Study Program at Universitas Sulawesi Barat who programmed Abstract Algebra course in the academic year of 2019/2020,

Citation: (APA): Hakim, F., & Murtafiah, M. (2020). Adversity Quotient and Resilience in Mathematical Proof Problem-Solving Ability. MaPan: Jurnal matematika dan Pembelajaran, 8(1), 87-102. (16 pages).

Copyright: © Creative Commons Attribution 4.0 International License (http://creativecommons.org/licenses/by/4.0/).

amounting to 71 students. The analysis technique used is multiple linear regressions. The results of hypothesis testing indicate that adversity quotient and resilience affect the mathematical proof problem-solving ability positively both individually and simultaneously.

Keywords: Adversity Quotient, Resilience, Problem-Solving, Mathematical Proof

INTRODUCTION

Not only at the elementary and secondary level, mathematical problem-solving is also a problem at the undergraduate level, even for mathematics class students. Many studies conducted at the elementary school level show that students' mathematical problem-solving abilities are still low. The results of the observation of Indarwati, Wahyudi, & Ratu (2014) showed that many fifth grade students were unable to work on problem-solving problems in calculating operations, as well as the results of trials conducted by Novianti & Kurniadi (2016) that the majority of students had difficulty working on problem-solving, especially in the form of word problems. Furthermore, it was revealed that the low ability of these students was because at the elementary school level, students were accustomed to being given the fastest formula in solving a problem, not using mathematical concepts (Rostika & Junita, 2017; Mustafia & Widodo, 2018). According to Mulyati (2011), the low mathematical problem-solving ability of elementary school students is an indication of the low mathematical problem-solving ability of students at the secondary level. As for the undergraduate level, various previous studies related to mathematical problem-solving at the undergraduate level (Ningsih, 2017; Octaria & Sari, 2017; Amalia & Widodo, 2018; Gurat, 2018) showed that students' mathematical problem-solving ability is still low. In fact, mathematical problem-solving is a goal and the result of the teaching and learning process, which is very important. There is no mathematics without thinking, and there is no thought process without problems (Aljaberi, 2015). Mathematical problemsolving has long been seen as an important aspect of mathematics, teaching mathematics, and learning mathematics (Liljedahl, Santos-Trigo, Malaspina, & Bruder, 2016). Problem-solving plays an important role in mathematics education and most mathematics learning is the result of the problem-solving process (Ersoy, 2016). According to Arvanitakis & Hornsby (2016), there are some general skills that universities try to develop at the student level. The term "Citizen Scholar" is introduced to describe graduates who have key skills in the form of critical thinking skills, communication skills, problem-solving, research, and collaboration that will help students adapt to the changing needs of society.

In many subjects, the problem of mathematical proof is one of the mathematical problems often faced by undergraduate students. Ozdemir & Ovez (2012) suggested that in preparing formal mathematical proofs, there were still many serious difficulties faced by prospective mathematics teacher students. Mathematical evidence provides guarantees for mathematical knowledge and is an important activity in doing and understanding mathematics. Proof is a tool in learning mathematics. Proof has an important role in many courses in the Department of Mathematics. According to Hanna & Barbeau (2010), mathematical propositions can be said to be true if they have been proven. Stefanowicz (2014) also revealed that the essence of mathematics learning is proof. Lesseig, Hine, & Boardman, (2019) who argue "the centrality of proof of mathematics is indisputable." Some opinions also suggested the importance of mathematical proof (Balacheff, 2010; Basturk, 2010; CadwalladerOlsker, 2011; Rocha, 2019). However, Mukmin & Fa'ani (2020) found the fact that in solving mathematical proof problems, there were still so many misconceptions faced by students that resulted in the low mathematical proof ability of students. Some preliminary studies that have been conducted also showed the low mathematical proof ability of undergraduate students (Abdussakir, 2014; Hasan, 2016; Muliawati, 2018; Nurrahmah & Karim, 2018; Herutomo, 2019). Mathematical proof requires a high level of ability that requires great effort to be able to get it (Hasan, 2016). The difficulty faced by students in understanding mathematical proofs depends on how the students process information and then solve the problem. In addition, the success of students in solving mathematical proof problems also depends on whether they will surrender or conquer the difficulties or continue to struggle to conquer the difficulties encountered.

One of the factors that influence hard work in solving problems is adversity quotient. Many people have high IQ and EQ, but they fail. This is due to the low adversity quotient (AQ) (Stoltz, 2004). In psychological studies, adversity quotient (AQ) is a person's ability to survive in the face of difficulties and efforts to resolve difficulties (Hastuti & Sari, 2017). If related to the learning process, then AQ is needed for learners in dealing with problems in learning (Matore, Khairani, & Razak, 2015). Learners who have a good AQ, can survive in the face of various difficulties in learning mathematics. Therefore, it is necessary to study AQ in learning mathematics (Parvathy & Praseeda, 2014). Various opinions regarding AQ have also been expressed by several previous researchers (Rukmana, Hasbi, & Paloloang, 2016; Mardika & Insani, 2017; Qin, Zhou, & Tanu, 2019).

There are 3 categories in AQ, namely quitter (low AQ), camper (medium AQ), and climber (high AQ). Climbers choose to survive and struggle to face the problems, challenges, and obstacles that will continue to hit, campers have the willingness to try to deal with problems, challenges, and obstacles but stop because they feel they are no longer able, while the quitters lack the ability to accept challenges in life. AQ consists of four dimensions used to measure and understand AQ and in determining the level of adversity quotient, control, origin and ownership, reach, and endurance termed CO2RE (Stoltz, 2004).

Another factor related to hard work in solving problems is resilience. Rojas (2015) stated resilience as the ability to face challenges that will be seen when someone is facing a difficult experience and knows how to deal with or adapt to it. In relation to the world of education, especially problem-solving, according to Maharani & Bernard (2018), students are afraid and avoid challenging problem-solving because of difficulties in working on problemsolving problems. This fear causes anxiety for students and they need to have a strong, persistent, and confident attitude called resilience. Reivich & Shatte (2002) suggested that seven abilities make up resilience, namely (1) emotional control, (2) ability to control impulses, (3) optimism, (4) ability to analyze, (5) ability to empathize, (6) self-efficacy, and (7) ability to achieve what is desired.

Adversity quotient and resilience, which are factors related to hard work to face challenges, make research to see the effect of these two variables on the ability to solve mathematical proof problems very interesting to do. This research is very important to be carried out as a first step to improve students' mathematical proof proving ability. It takes data related to what factors affect the ability of these students. Several related studies have been conducted by other researchers related to this research. Rukmana, Hasbi, & Paloloang (2016) found that there is a significant positive relationship between adversity quotient with student mathematics learning outcomes. This is supported by the results of Mardika & Insani's research (2017), it was found that problem-solving ability has a relationship with adversity quotient. Regarding resilience factors, (Zanthy, 2018) found that resilience has a positive contribution to students' academic abilities in Mathematics Statistics. Some researchers have conducted various studies related to these three variables, but separately. This study will examine at the same time, the relationship between adversity quotient and resilience with the mathematical proof problem-solving ability. Research hypotheses to be tested are (1) there is an effect of adversity quotient on the ability to solve mathematical proof problems, (2) there is an effect of resilience on the ability to solve mathematical proof problems, and (3) there is an effect of adversity quotient and resilience simultaneously on the ability to solve mathematical proof problem.

RESEARCH METHOD

This research was a correlational study with a quantitative approach that aimed to determine the effect of adversity quotient and resilience on the ability to solve mathematical proof problems. The samples in this study were 31 students chosen randomly from all students of the Mathematics Education Study Program at Universitas Sulawesi Barat who programed Abstract Algebra courses in the academic year 2019/2020, amounting to 71 students. Researchers collected research data during the COVID-19 pandemic, so it was not possible to make the entire population a sample in this research. This size of the research sample is consistent with several theories regarding sample size. Roscoe (1975) revealed that the sample size used was at least 10 times the number of factors if there were many factors in the research. Gay & Diehl (1992) argue that at least 30 sample units are used if research is correlational research.

The non-test instrument in this research was the Quick Take ™ Adversity Response Profile (ARP), a test by Stoltz (2004), which has been tested by more than 7,500 people from all over the world with various careers, ages, races, and cultures. After repeated and continuous tests, this instrument proved to be very reliable. This instrument was used to obtain an adversity quotient score. The second non-test instrument was Resilience Factor Inventory (RFI), a test that was successfully developed by Reivich & Shatte (2002) and has been used extensively throughout the world. This instrument was used to obtain resilience scores. As for the test instrument used was a mathematical proof problemsolving test with Group Theory materials.

The data analysis technique used in this research was multiple linear regression analysis with the assumption test consisting of tests of normality, multicollinearity, autocorrelation, and heteroscedasticity. Then used partial regression coefficient test (t-test) to see the effect of the independent variables (adversity quotient and resilience) on the dependent variable (mathematical proof problem-solving ability), regression coefficient test together (F test) to see the effect of independent variables on the dependent variable simultaneously, multiple correlation analysis (R) to determine the type and category of influence between the independent variables on the dependent variable, and the analysis of determination (adjusted R2) to determine the percentage contribution of the influence of the independent variable simultaneously on the dependent variable. The analysis was carried out by a two-party test using a significance level of 5%. Analyzes were performed using IBM SPSS Statistics 21.

RESULTS AND DISCUSSION

The theory presented in the previous section revealed that AQ can be classified into 3 categories, and based on research data, the classification results are obtained as in figure 1 below.

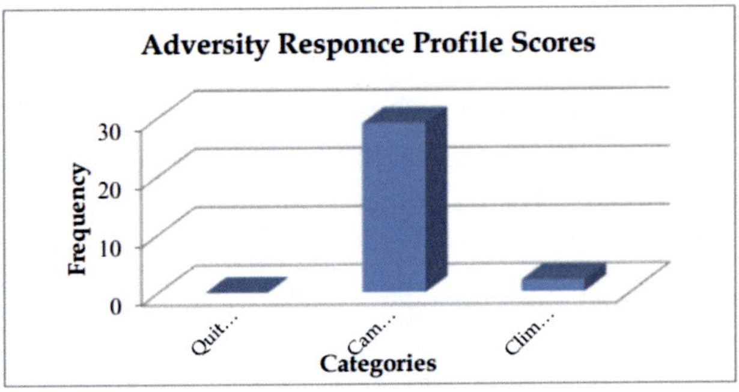

Figure 1. AQ Categories.

The resilience data obtained are categorized and presented in figure 2 below.

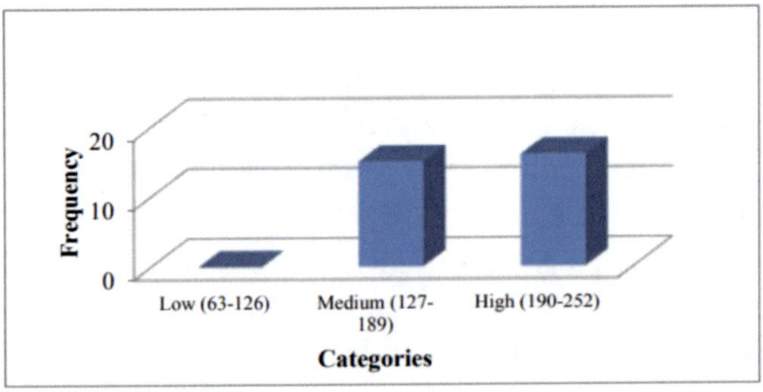

Figure 2. Resilience Categories.

The mathematical proof problem-solving ability data are categorized as in figure 3 below.

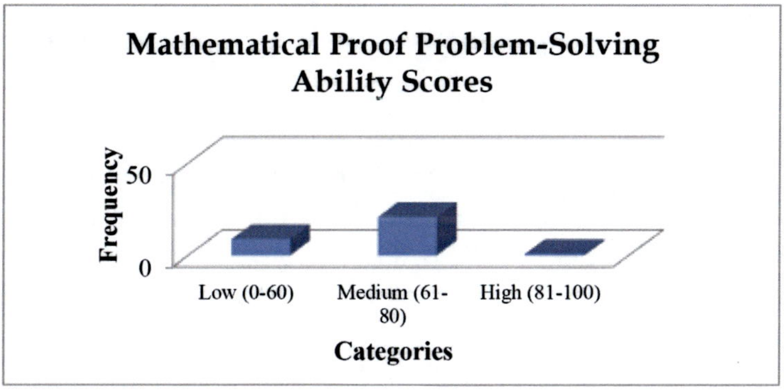

Figure 3. Mathematical Proof Problem-Solving Ability Categories.

The first assumption test, the normality test, gave the results of adversity quotient, resilience, and mathematical problem-solving ability variables data in table 1.

Table 1. Summary of Normality Tests Results

	Shapiro-Wilk		
	Statistic	df	Sig.
ARP_Score	.970	31	.528
RFI_Score	.957	31	.236
PMPM_Score	.967	31	.434

Based on the data in table 1, the significance for the ARP score (adversity quotient score) is 0.528, the significance for the RFI score (resilience score) was 0.236, and the significance for the mathematical proof-solving ability problem score is 0.434. Because all three variables have a significance value of more than 0.05, it can be concluded that all three data (adversity quotient, resilience, and the ability to solve mathematical proofs) come from normally distributed populations. For the second assumption test, multicollinearity test between independent variables, the results are summarized in the following table 2.

Table 2. Summary of Multicollinearity Test

Model		Collinearity Statistics	
		Tolerance	VIF
1	(Constant)		
	ARP_Score	.978	1.022
	RFI_Score	.978	1.022

Based on the data in table 2, it is known the VIF variable adversity quotient (1,022) and VIF resilience variable (1,022). Because the VIF value for all variables is less than 10, it can be concluded that there is no multicollinearity disorder. For the autocorrelation test results, the third assumption test, the results are summarized in the following table 3.

Table 3. Summary of Autocorrelation Test

Model	R	R Square	Adjusted R Square	Std. Error of the Estimate	Durbin-Watson
1	.524[a]	.275	.223	8.258	1.727

Based on the data in table 3, the Durbin-Watson (DW) value obtained is 1,727. The Durbin Lower (DL) and Durbin Upper (DU) values obtained from the Durbin-Watson Table are 1.29685 and 1.57011, respectively. Because DW> DU, there is no positive autocorrelation and because (4 - DW) > DU, there is no negative autocorrelation. While the heteroscedasticity test results are summarized in the following figure 4.

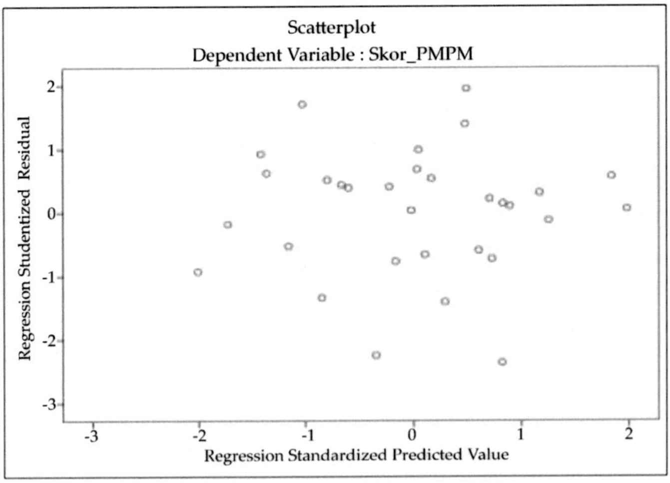

Figure 4. Result of Heteroscedasticity Test.

Heteroscedasticity test results show that the data do not form a specific pattern and are spread above and below the number 0. So it can be concluded that there are no symptoms of heteroscedasticity. Thus all assumption tests for multiple linear regression tests are met.

The equation of multiple linear regression obtained is

$$Y' = -47{,}471 + 0{,}303X_1 + 0{,}387X_2$$

with:

Y' = Mathematical Proof Problem-Solving Ability Variable

X_1 = Adversity Quotient Variable

X_2 = Resilience Variable

Constants and equation coefficients in (1) are obtained from the following table 4.

Tabel 4. Summary of Coefficients

	Model	Unstandardized Coefficients		Standardized Coefficients
		B	Std. Error	Beta
1	(Constant)	-47.471	35.636	
	ARP_Score	.303	.123	.400
	RFI_Score	.387	.157	.402

The results of the partial regression coefficient test (t-test) to see the effect of individual independent variables (adversity quotient and resilience) on the dependent variable (mathematical proof of problem-solving ability) are summarized in the following table 5.

Table 5. Result of Partial Regression Coefficient Test (t-test)

	Model	T	Sig.
1	(Constant)	-1.332	.194
	Skor_ARP	2.459	.020
	Skor_RFI	2.472	.020

In table 5, it can be seen that the significance value for the adversity quotient variable is 0.020. Because the significance is less than 0.05, it can be concluded that at a significance level of 5%, there is a significant effect of adversity quotient on the ability to solve mathematical proofs. The significance of the resilience variable is 0.020. Since the significance is less than 0.05, it can be concluded that at a significance level of 5%, there is a significant effect of resilience on the ability to solve mathematical proofs.

Next, to test the effect of simultaneous independent variables (adversity quotient and resilience) on the dependent variable (the ability to solve mathematical proof problems), a regression coefficient test is conducted together (F test) with the results summarized in table 6.

Table 6. Result of Regression Coefficient Test Simultaneously (F test)

	Model	Sum of Squares	df	Mean Square	F	Sig.
1	Regression	722.610	2	361.305	5.298	.011[b]
	Residual	1909.584	28	68.199		
	Total	2632.194	30			

In table 6, it appears that the significance value is 0.011. Because the significance is less than 0.05, it can be concluded that at a significance level of 5%, there is a significant effect of adversity quotient and resilience simultaneously on the ability to solve mathematical proofs.

Furthermore, to determine the types and categories of influence of the independent variables on the dependent variable simultaneously multiple correlation analysis (R) is used and to determine the percentage of the contribution of the independent variable simultaneously on the dependent variable the determination analysis (adjusted R2) is used with the results summarized in table 7.

Table 7. Result of Multiple Correlation Analysis (R) dan Determination Analysis (adjusted R2)

Model	R	R Square	Adjusted R Square	Std. Error of the Estimate
1	.524[a]	.275	.223	8.258

Based on table 7, the value of R is 0.524. This shows that the effect of adversity quotient and resilience simultaneously on the ability to solve mathematical proof problems is a positive influence and is in the medium category. The adjusted R2 value obtained is 0.223. This shows that the percentage contribution of the effect of variable adversity quotient and resilience simultaneously on the ability to solve mathematical proof problems by 22.3%.

The results obtained in the multiple linear regression test showed that there was an effect of adversity quotient and resilience on the ability to solve mathematical proofs, both individual and simultaneous influences. Furthermore, the coefficients of the two variables in the multiple linear regression equation are positive so that it can be concluded that adversity quotient has a positive effect on the ability to solve mathematical proofs and resilience has a positive effect on the ability to solve mathematical proofs. It can also be concluded that adversity quotient and resilience simultaneously have a positive effect on the ability to solve mathematical proofs and the effect is in the moderate category.

Different efforts and attitudes by students when solving mathematical proof problems have a positive effect on adversity quotient on the ability to solve mathematical proof problems. Student climbers, campers, and quitters have different efforts and attitudes in solving mathematical proofs. Climber students will make a concerted effort to solve the problem, camper students have an effort to solve the problem but will leave it when it finds the problem getting complicated, while the quitter really will not make an effort to solve the problem because they think they will not be able to solve the problem. Qin, Zhou, & Tanu (2019) revealed that adversity quotient is the greatest strength of a learner to solve existing problems. In addition, several research findings also support this finding. Oliveros (2014) revealed that adversity quotient determines the level of achievement that students ultimately achieve in challenging tasks such as solving mathematical problems. Mardika & Insani (2017) also concluded that adversity quotient has a relationship with problemsolving abilities. Malik, Mulyono, & Mariani (2019) argue that the success of problem-solving learning is not only influenced by IQ and EQ, but also by AQ.

In addition to adversity quotient, resilience also has a positive effect on the ability to solve mathematical proofs. According to Murphey, Barry, & Vaughn (2013), someone who has a different level of resilience, will have different characteristics. One of the characteristics of someone who has high resilience is to have good thinking skills (intelligence). Dilla, Hidayat, & Rohaeti (2018) concluded in the results of their study that there is a positive influence between resilience and gender differences in the achievement of students' mathematical creative thinking abilities. As for thinking skills affect the ability to solve mathematical proof problems. Suhendris' (2018) research results show that there is an influence, although not significant, resilience and mathematical creativity on the ability to solve mathematical problems. Rojas (2015) suggests that by growing resilience, people can develop lifelong skills such as communication and problem-solving skills, the ability to make realistic plans, and be able to follow up on those plans. The influence of resilience on problemsolving abilities was also stated by Supardi (2014), who revealed that resilience is an attitude to recognize a problem and its causes.

Mathematics education is a discipline that is closely related to various disciplines, one of which is psychology. Nevertheless, the relationship between mathematics and psychology education is still debated (Obersteiner, Reiss, & Heinze, 2018). Star & Rittle-Johnson (2016) put forward the fact that we are currently in the midst of a new phase in the relationship between the disciplines of mathematics education and psychology. Thus, research on psychology needs to be done in the policy and teaching of mathematics education. The results obtained in this study also indicated psychological aspects that have a relationship with the mathematical problem-solving in specific and towards mathematics education in general. Psychological aspects must be a special concern of educators in the world of mathematics education because through various related studies also find facts related to psychological aspects and various mathematical abilities.

CONCLUSION

Adversity quotient and resilience affect the ability to solve mathematical proof problems, both the effect of individual adversity quotient and resilience on the ability to solve mathematical proof problems, as well as simultaneous influence. It was further found that the effect of adversity quotient and resilience on the ability to solve mathematical proof problems individually was a positive influence and the effect of adversity quotient and resilience on the ability to solve mathematical proofs simultaneously was a positive influence and was in the medium category.

REFERENCES

1. Abdussakir, A. (2014). Proses berpikir mahasiswa dalam menyusun bukti matematis dengan strategi semantik. Journal of Science, Mathematics, and Vocational Education, 2(3), 132-140. http://doi.org/10.17977/jps. v2i3.4506.
2. Aljaberi, N. M. (2015). University students' learning styles and their ability to solve mathematical problems. International Journal of Business and Social Science, 6(4), 152-165. Retrieved from http://ijbssnet.com/journal /index/3132.
3. Amalia, S. R., & Widodo, A. N. A. (2018). Analisis kemampuan pemecahan masalah mahasiswa melalui model PBL berbasis etnomatematika ditinjau dari kepribadian topologi hippocrates dan galenus tipe cholearis dan phlegmantis. Aksioma: Jurnal Matematika dan Pendidikan Matematika, 9(1), 1-10 http://doi.org/10.26877/aks. v9i1.2467.
4. Arvanitakis, J., & Hornsby, D. (2016). Universities, the citizen scholar and the future of higher education. Springer.
5. Balacheff, N. (2010). Bridging knowing and proving in mathematics: a didactical perspective. In Explanation and proof in mathematics. Springer, Boston, MA. 115-135. https://doi.org/10.1007/978-1-4419-0576-5_9.
6. Basturk, S. (2010). First-year secondary school mathematics students' conceptions of mathematical proofs and proving. Educational Studies, 36(3), 283-298. https://doi.org/10.1080/03055690903424964.
7. CadwalladerOlsker, T. (2011). What do we mean by mathematical proof?. Journal of Humanistic Mathematics, 1(1), 33-60. https://doi.org/10.5642/ jhummath.201101.04.
8. Dilla, S. C., Hidayat, W., & Rohaeti, E. E. (2018). Faktor gender dan resiliensi dalam pencapaian kemampuan berpikir kreatif matematis siswa SMA. Journal of Medives: Journal of Mathematics Education IKIP Veteran Semarang 2(1), 129-136. https://doi.org/10.31331/medives.v2i1.553.
9. Ersoy, E. (2016). Problem solving and its teaching in mathematics. The Online Journal of New Horizons in Education, 6(2), 79-87. Retrieved from https:// www.tojned.net/journals/tojned/articles/v06i02/v06i02-11.pdf.

10. Gay, L. R. dan Diehl, PL (1992), Research methods for business and management, New York: MacMillan Publishing Company.
11. Gurat, M. G. (2018). Mathematical problem-solving strategies among student teachers. Journal on Efficiency and Responsibility in Education and Science, 11(3), 53-64. https://doi.org/10.7160/eriesj.2018.110302.
12. Hanna, G., & Barbeau, E. (2010). Proofs as bearers of mathematical knowledge. in Explanation and Proof in Mathematics: Philosophical and Educational Perspectives. 85-99. https://doi.org/10.1007/978-1-4419-0576-5.
13. Hasan, B. (2016). Proses berpikir mahasiswa dalam mengkonstruksi bukti menggunakan induksi matematika berdasarkan teori pemerosesan informasi. APOTEMA: Jurnal Program Studi Pendidikan Matematika, 2(1), 33-40. Retrieved from http://www.jurnal.stkippgri-bkl.ac.id/index.php /APM/article/view/304.
14. Hastuti, T. D., & Sari, S. (2018). Student profile with high adversity quotient in math learning. JPhCS, 983(1), http://doi.org/10.1088/1742-6596/983/ 1/012131.
15. Herutomo, R. A. (2019). Kesalahan mahasiswa dalam pembuktian matematik. Jurnal Didaktik Matematika, 6(1), 53-67. https://doi.org/10.24815/jdm. v6i1.13262.
16. Indarwati, D., Wahyudi, W., & Ratu, N. (2014). Peningkatan kemampuan pemecahan masalah matematika melalui penerapan problem based learning untuk siswa kelas V SD. Satya Widya, 30(1), 17-27. https:// doi.org/10.24246/j.sw.2014.v30.i1.p17-27.
17. Lesseig, K., Hine, G., Na, G. S., & Boardman, K. (2019). Perceptions on proof and the teaching of proof: A comparison across preservice secondary teachers in Australia, USA and Korea. Mathematics Education Research Journal, 31(4), 393-418. https://doi.org/10.1007/s13394-019-00260-7.
18. Liljedahl, P., Santos-Trigo, M., Malaspina, U., & Bruder, R. (2016). Problem solving in mathematics education. In Problem Solving in Mathematics Education. Springer, Cham. 1-39. https://doi.org/10.1007/978-3-319-40730-2.
19. Maharani, S., & Bernard, M. (2018). Analisis hubungan resiliensi matematik terhadap kemampuan pemecahan masalah siswa pada materi lingkaran. JPMI (Jurnal Pembelajaran Matematika Inovatif), 1(5), 819-826. http://doi.org/10.22460/jpmi.v1i5.p819-826.

20. Malik, I., Mulyono, M., & Mariani, S. (2019). Ability in mathematics problem solving based on adversity quotient. Jurnal Profesi Keguruan, 5(1), 90-95. https://doi.org/10.7290/jpk.v5i1.18878.
21. Mardika, F., & Insani, S. U. (2017). Adversity quotient and students' problem solving skill in mathematics. 4th ICRIEMS Proceedings. Published by The Faculty of Mathematics and Natural Sciences Yogyakarta State University.
22. Matore, M. E. E. M., Khairani, A. Z., & Razak, N. A. (2015). The influence of aq on the academic achievement among malaysian polytechnic students. International Education Studies, 8(6), 69-74. http://doi.org/10.5539/ies. v8n6p69.
23. Mukmin, M. I., & Fa'ani, A. M. (2020). Identification of students' misconceptions in proving onto and one-to-one function in abstract algebra using certainty response index. International Journal on Teaching and Learning Mathematics, 2(1), 1. https://doi.org/10.18860/ijtlm. v2i1.8582.
24. Muliawati, N. E. (2018). Proses berpikir mahasiswa dalam memecahkan masalah pembuktian grup berdasarkan langkah polya. JP2M (Jurnal Pendidikan dan Pembelajaran Matematika), 4(2), 32-42. http://doi.org/10. 29100/jp2m.v4i2.954.
25. Mulyati, T. (2011). Kemampuan pemecahan masalah matematis siswa sekolah dasar. Jurnal Pendidikan Dasar. 3(2). 35-46. https://doi.org/10.17509/eh. v3i2.2807.
26. Murphey, D., Barry, M., & Vaughn, B. (2013). Positive mental health: resilience. Adolescent Health Highlight, 3, 1-6. Retrieved from https://www. childtrends.org/publications/positive-mental-health-resilience.
27. Mustafia, I. D., & Widodo, S. A. (2018). Problem solving skill: effectiveness on think pair share with comic. International Journal on Teaching and Learning Mathematics, 1(2), 76-83. http://doi.org/10.18860/ijtlm.v1i2. 7181.
28. Ningsih, S. C. (2017). Meningkatkan kemampuan pemecahan masalah mahasiswa pendidikan matematika UPY pada mata kuliah teori bilangan melalui model pembelajaran creative problem solving (CPS). Jurnal Mercumatika. 1(2). 132-138. https://doi.org/10.26486/mercumatika. v1i2.212.
29. Novianti, E., & Kurniadi, Y. (2016). Pendekatan Eksploratif untuk meningkatkan kemampuan pemecahan masalah matematis dan

motivasi belajar siswa. Jurnal Pena Ilmiah, 1(1), 401-410. https://doi.org/10.23819/ pi.v1i1.3046.

30. Nurrahmah, A., & Karim, A. (2018). Analisis kemampuan pembuktian matematis pada matakuliah teori bilangan. Jurnal eDuMath, 4(2), 21-29. https://doi.org/10.26638/je.753.2064.
31. Obersteiner, A., Reiss, K., & Heinze, A. (2018). Psychological theories in mathematics education. Journal für Mathematik-Didaktik, 39, 1-6. https:// doi.org/10.1007/s13138-018-0134-3.
32. Octaria, D., & Sari, E. F. P. (2017). Kemampuan pemecahan masalah matematis melalui problem based learning. Jurnal Dosen Universitas PGRI Palembang, 1(9), 1-14. https://jurnal.univpgri-palembang.ac.id/index. php/prosiding/article/view/1688.
33. Oliveros, J. C. (2014). Adversity quotient and problem-solving skills in advanced algebra. JPAIR Multidisciplinary Research, 17(1). http://doi.org /10.7719/jpair.v17il.282.
34. Ozdemir, E., & Ovez, F. T. D. (2012). A research on proof perceptions and attitudes towards proof and proving: some implications for elementary mathematics prospective teachers. Procedia-Social and Behavioral Sciences, 46, 2121-2125. https://doi.org/10.1016/j.sbspro.2012.05.439.
35. Parvathy, U., & Praseeda, M. (2014). Relationship between adversity quotient and academic problems among student teachers. Journal of HumanitiesandSocialScience,19(11),23-26.Retrievedfromhttps://pdfs.semanticscholar.org/3dfb/f7c8ff92cf1e84fc8c79503cf8fbd4721a79.pdf.
36. Qin, L., Zhou, Y., & Tanu, W. T. (2019). The Analysis of mathematics adversity quotient of left behind junior high school students in rural areas. Open Journal of Social Sciences, 7(10), 331. https://doi.org/10.4236/ jss.2019.710028.
37. Reivich, K., & Shatte, A. (2002). The resilience factor: 7 essential skills for overcoming life's inevitable obstacles. Broadway books.
38. Rocha, H. (2019). Mathematical proof: from mathematics to school mathematics. Philosophical Transactions of the Royal Society A, 377 (2140), 20180045. http://doi.org/10.1098/rsta.2018.0045.
39. Rojas, L. F. (2015). Factors affecting academic resilience in middle school students: A case study. Gist: Education and Learning Research Journal, (11), 63-78. https://doi.org/10.26817/16925777.286.

40. Roscoe, J. T. (1975). Fundamental research statistics for the behavioral sciences (Internasional Series in Decision Processes). Holt, Rinehart & Winston.
41. Rostika, D., & Junita, H. (2017). Peningkatan kemampuan pemecahan masalah siswa SD dalam pembelajaran matematika dengan model diskursus multy representation (DMR). EduHumaniora| Jurnal Pendidikan Dasar Kampus Cibiru, 9(1), 35-46. https://doi.org/10.17509/eh.v9i1.6176.
42. Rukmana, I., Hasbi, M., & Paloloang, B. (2016). Hubungan adversity quotient dengan hasil belajar matematika siswa kelas XI SMA Negeri Model Terpadu Madani Palu. Jurnal Elektronik Pendidikan Matematika Tadulako, 3(3). Retrieved from http://jurnal.untad.ac.id/jurnal/index.php/JEPMT/article/view/7220.
43. Star, J. & Rittle-Johnson, B. (2016). Toward an educational psychology of mathematics education. In L. Corno & E. Anderman (Eds.) Handbook of Educational Psychology. Third edition. American Psychological Association. New York, United States: Routledge.
44. Stefanowicz, A., Kyle, J., & Grove, M. (2014). Proofs and mathematical reasoning. University of Birmingham.
45. Stoltz, P. G. (2004). Adversity quotient: mengubah hambatan menjadi peluang. Jakarta: Gramedia Widiasarana Indonesia.
46. Suhendri, H. (2018). The role of resilience (adversity intelligence) and creativity in mathematics learning. Unnes Journal of Mathematics Education, 7(2), 90-94. https://doi.org/10.15294/ujme.v7i2.25367.
47. Supardi, U. S. (2014). Pengembangan Instrumen Pengukuran Ketahanmalangan (Kecerdasan Adversitas) Siswa. In Prosiding Seminar Nasional Evaluasi Pendidikan Program Pacsasarjana UNJ. 978-608. Jakarta.
48. Zanthy, L. S. (2018). Kontribusi resiliensi matematis terhadap kemampuan akademik mahasiswa pada mata kuliah statistika matematika. Mosharafa: Jurnal Pendidikan Matematika, 7(1), 85-94. https://doi.org/10.31980/ mosharafa.v7i1.344.

Chapter 5

PROFILE OF STUDENTS' ERRORS IN MATHEMATICAL PROOF PROCESS VIEWED FROM ADVERSITY QUOTIENT (AQ)

Arta Ekayanti[1], Hikma Khilda Nasyiithoh[2]

[1]Faculty of Education and Teacher Training, Universitas Muhammadiyah Ponorogo, Indonesia

[2]Faculty of Education and Teacher Training, Institut Agama Islam Negeri Ponorogo, Indonesia

ABSTRACT

Mathematical proof is an important aspect in mathematics, especially in analysis. An error in the mathematical proof construction process often occurs. This study aims to analyze the students' errors in producing proof. Each of the categories of students' Adversity Quotient (AQ) is identified related to the type of students' error. The type of students' errors used according to Newmann's Error Analysis. This study used a qualitative approach. This study was conducted to 25 students who were taking real

Citation: (APA): Ekayanti, A., & Nasyiithoh, H. K. (2018). Profile of Students' Errors in Mathematical Proof Process Viewed from Adversity Quotient (AQ). Tadris: Jurnal Keguruan dan Ilmu Tarbiyah, 3(2), 155-166. (12 pages).

Copyright: © Creative Commons Attribution-ShareAlike 4.0 International License (http://creativecommons.org/licenses/by-sa/4.0/).

analysis course. Documentation, test, and interview were used to gather the data. Analyzing the students' test result and then interviewing them for each AQ category were done for the analysis process. The results show that there are 48% climber students, 52% camper students, and no one is identified as a quitter student. Climber students tend to make some proving error such as transformation error, process skill error, and encoding error while camper students make the comprehension error, transformation error, process skill error, and encoding error when they are producing proof.

INTRODUCTION

Analysis is one branch of mathematics. This is stated in one of the courses called real analysis. It should be noted that mathematics is not merely numbers. This is in line with Hernadis' (Hernadi, 2016), an opinion which says that so far the views on mathematics were still within the scope of the calculation activities relating to variables and numbers. However, it should be noted that important activities in the study of mathematics are mathematical proof the truth or facts applied and communicated in mathematics. Therefore, Yi Yin Ko and Eric Knuth say that mathematical proof is one of the basic abilities for advanced mathematical thinking (Ko & Knuth, 2009). Besides that, Knuth also says that mathematical proof plays a role in systematizing statements into axiomatic systems (Sucipto & Mauliddin, 2016). Mathematical proof includes thinking about new concepts, focusing on important aspects, using relevant prior knowledge, defining new things (if needed), and compiling valid argument (Hidayat, 2017; C. K. Sari, Waluyo, Ainur, & Darmaningsih, 2018). This must be based on a deductive mindset so that students are able to understand the mathematical proof process (Ekayanti, 2017). There is often a misunderstanding in solving mathematical proof problems, including the use of empirical arguments in the process of mathematical proof (Stavrou, 2014). This is not an easy job, and it can be seen from the many errors made by students in completing mathematical proof cases.

Some errors in mathematical proof occur because students have not fully understood the true nature of mathematical proof. Students still often do mathematical proof using examples. Of course, this is invalid for the mathematical proof process. Besides that, the argument used is illogical. There are times when the mathematical proof made does not conclude. This can occur because of failures or errors in the first few stages are failing to reach the next stage (D. P. Sari, Darhim, & Rosjanuardi, 2018; Wijaya, Heuvel-panhuizen, Doorman, & Robitzch, 2014). The problem, critical thinking skills are needed so that students can plan and execute it effectively and accurately (Sukoriyanto, Nusantara, Subanji, & Chandra, 2016). This also applies in the case of mathematical proof where the students are required to have tenacity and resilience in facing existing difficulties.

Tenacity and resilience in facing challenges or difficulties are called Adversity Quotient (Stoltz, 2000). Stoltz divides three types of AQ, namely quitters, campers and climbers. The quitters tend to lack the willingness to accept the challenges that exist in their lives. The campers already have the willingness to try facing the challenges and problems, but this type of individual thinks that the effort is enough. The climbers tend to survive and struggle in facing problems, challenges, and obstacles (Yanti, Koestoro, & Sutiarso, 2018).

Considering that, the real analysis course is more dominated by mathematical proof, including in the rules of proof derived from formal definitions, as well as the theorems or lemmas associated previously (Ah, 2016). This is considered a difficult thing for students. Because of these difficulties, AQ is needed in learning mathematics (Guntur Suhandoyo, 2016). Therefore, this research was carried out in real analysis course to know more about the types of errors made by the students in learning mathematics, especially mathematical proof in terms of Adversity Quotient.

THEORETICAL SUPPORT

Hernadi says that mathematical proof is a method of communicating a mathematical truth to others who also understand the language of mathematics (Hernadi, 2016). A proof is a series of logical arguments that explain the truth of a statement or proposition. (Stefanowich, 2014) states that proof is a series of logical statements, where one statement influences the other statement, of course, there must be a valid explanation of the truth of the statement. Logically, in this case, it is intended that each step in the mathematical proof must be based on previous steps or other facts with guaranteed truth.

Anne Newmann classified types of errors into five types, including reading errors, comprehension errors, transformation errors, process skill errors, and encoding errors (Bagus Nur Iman, Toto Nusantara, 2016). Students are said to make a reading error if they experience errors in reading and understanding the command of the questions and errors in recognizing the symbols on the question. Comprehension error occurs when the students did not know what is known and asked from the question. Transformation errors occur if students experience errors in determining problem-solving strategies. Students experience a process skill error if they make algebraic operational errors and are wrong in carrying out completion procedures. While encoding errors occur when the students are able to determine the solution to the problem, but they are unable to write the procedure and form the answer correctly.

Intelligence is one of the psychological factors that influence learning achievement (Leonard, 2017). There are several types of intelligence including Adversity Quotient. Adversity Quotient (AQ) is a person's ability to struggle with and overcome obstacles, difficulties, or problems that exist and will turn them into opportunities for success (Stoltz, 2000). Understanding the importance of AQ in achieving success will encourage the students to always struggle in the learning process even though they must face various obstacles and difficulties (Rukmana & Paloloang, 2016). AQ possessed by each individual in facing and overcoming difficulties is different. The level of ability possessed will have an impact on the ability to go through life and be able to provide great benefits for success (Nurhayati, 2015). Stoltz illustrates that life is like climbing a mountain. Therefore, Stoltz divides AQ into three types, namely Quitters (groups of individuals who stop) are groups of individuals who lack the willingness to accept the challenges that exist in their lives. The quitter will be more likely to reject challenges or problems (Hidayat, Herdiman, Aripin, Yuliani, & Maya, 2018; Christina Kartika Sari, Sutopo, & Aryuna, 2016). In the world of education, students who belong to the quitter type are students who are easy to give up and despair in facing the problems. Campers (groups of individuals who camp) are groups of individuals who already have the will to try to deal with challenges and problems but then they feel that it is enough. These individual groups prefer safe situations or prefer to be in a comfort zone. Students who belong to the campers usually type already struggle, but one factor could make them give up and eventually lost the challenge. Climbers are groups of individuals who tend to survive and struggle in facing problems, challenges, and obstacles. Students who belong to the climber type are learners who always sought and unyielding (Wardiana, 2014; Yani, Ikhsan, & Marwan, 2016). Students of the climber type tend to have the desire to get better (Indra Kurniawan, Kusmayadi & Sujadi, 2015).

Someone with high AQ will be encouraged to get the best results by actively acting, always taking advantage of the opportunities that exist, and having the willingness to learn independently (Novilita & Suharnan, 2013). Yanti and Syazali suggest that the high and low AQ can be measured using an indicator which consists of four dimensions including Control, Origin, Reach and Endurance (Yanti & Syazali, 2013), as shown in Table 1. The AQ score can be counted using the formula C + O_2+ R + E = AQ (Stoltz, 2000).

Table 1. The Indicators of Adversity Quotient

	Indicators (AQ Dimension: CO_2RE)	Description
C	Control; the level of control toward the events lead to problems	Students' self-control when sensing a problem
O_2	Origin and Ownership	O_r : The ownership of the origin of problems O_w : The ownership toward the problems
R	Reach; how far the problem could reach other aspects of live	The students' ownership of how far the problem could reach other aspects of live
E	Endurance	Students' perception of how long will the problems going on

METHOD

This study uses the qualitative approach with descriptive research type. This research was conducted at the Mathematics Education Study Program. The research subjects were the students who took Real Analysis courses in the second semester of the 2017/2018 Academic Year with a total of 25 students. Sampling technique used was purposive. The data collecting techniques were documentation, tests, and interviews. The students first fill out a questionnaire of Adversity Quotient to later group them into three categories namely climbers, campers, and quitters. From the questionnaire, the AQ score was obtained.

Furthermore, the categorization of AQ in this study refers to the determination of the interval category (Azwar, 2002), based on the theoretical mean (μ) and standard deviation (σ). The categorization criteria can be seen in Table 2 below. Where X states, the AQ score obtained.

Table 2. Categorization of AQ

Criteria	Category
$\mu + 1{,}0\sigma \leq X$	High
$\mu - 1{,}0\sigma \leq X < \mu + 1{,}0\sigma$	Medium
$X < \mu - 1{,}0\sigma$	Low

After analysis of the AQ, questionnaire had been conducted, and the results were obtained as presented in Table 3.

Table 3. Results of Adversity Quotient Questionnaire

Dimension	Number of Subjects	Score				Mean		Standard Deviation	
		t-Min	t-Max	e-Min	e-Max	Theoretical	Empirical	Theoretical	Empirical
C	25	8	32	20	28	20	23.32	4	1.95
O_2	25	11	44	26	41	27.5	32.84	5.5	3.80
R	25	12	48	28	41	30	35.80	6	3.11
E	25	9	36	21	35	22.5	26.28	4.5	3.17
AQ	25	40	160	101	140	100	118.24	20	9.04

Furthermore, from the data in Table 3, the theoretical mean and standard deviations were then used to determine the AQ categorization criteria in this research. The categorization criteria are in Table 4.

Table 4. AQ Categorization

Criteria	Category
$120 \le X$	High
$80 \le X < 120$	Medium
$X < 80$	Low

For the category of the Adversity Quotient, the highest category is assumed to be the Climbers category, and the medium category is assumed to be the Campers category, while the lowest category is assumed to be the Quitters category. Then the students were given a test of mathematical proof, the results of the tests are analyzed as a determination for the next process, namely interviews. From each category selected the work results of students with the type of error that represents other students and then selected as a subject who will be confirmed the results of their work through interviews. As for the analysis of the results of interviews conducted by going through several stages, namely data reduction, data presentation, and final conclusion.

RESULT AND DISCUSSION

Based on the data obtained, grouping students is based on Adversity Quotient by referring to Table 5.

Table 5. Student Grouping Results

Category	Number of Students	Percentage
High	12	48%
Medium	13	52%
Low	0	0%

This study did not find any students with the quitter Adversity Quotient category. The result is taken from the campers and climbers category. The test questions given were three mathematical proof questions. The first problem is as follows:

Given the function $f: \mathbb{R} \to \mathbb{R}$ defined by

$$f(x) := \begin{cases} x & if\ x\ rational \\ 0 & if\ x\ irrational \end{cases}$$

Prove that $\lim_{x \to 0} f(x)$ exist, but $\lim_{x \to c} f(x)$ do not exist if $c \neq 0$.

The answer from the climber- type students can be seen in the following Figure 1.

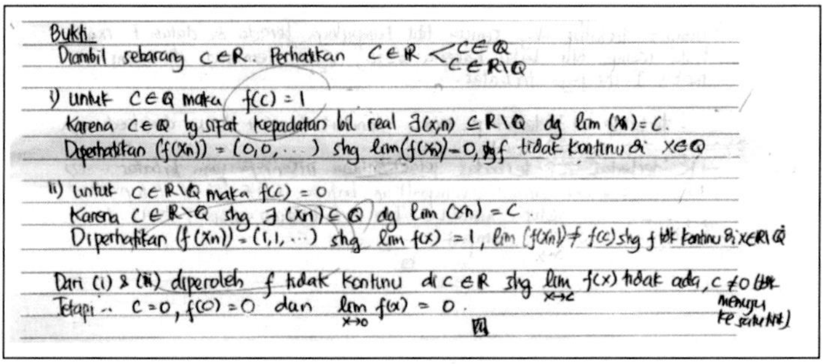

Figure 1. The Results of Climbers Type Students' Work for Question Number 1.

Based on Figure 1, it appears that students proved that the function did not have a limit by connecting it to a continuous function. Students thought that if a function is not continuous, the function has no limit. This is a wrong understanding. Furthermore, when further analyzed was conducted, it appears that when $c \in \mathbb{Q}$ obtained $f(c) = 1$. Of course, this is not true based on the function definition given in the question. After these things were confirmed to the students concerned, it turns out that students were still referring to the example discussed in the previous lecture. In addition, students still had the wrong understanding regarding the limit of functions and continuous functions. Therefore, students experienced errors in determining strategies to solve these problems. While in the process, there were still a number of incorrect steps. The results of this analysis can be seen in the following Table 6.

Table 6. Results of Analysis of Type Student Work Climber for Number 1

Types of Errors	Analysis Results
Reading Error	Students do not experience problems in reading errors. Students understand the problem given in question number 1.
Comprehension Error	Students know and understand what information given by question number 1 and what must be proven. It is seen that students are able to write the definition of functions given in mathematical language.
Transformation Error	Students make mistakes in this type. It is seen that the strategy used by students is to show that a continuous function has no limit. Of course, this is in contrast to the facts.
Process Skill Error	Students still make mistakes in carrying out some verification steps. It can be seen $c \in \mathbb{R} \backslash \mathbb{Q}$ is written $f(c) = 0$. course this is not in accordance with the definition given.
Encoding Error	Students have not been able to determine the resolution of this problem correctly.

Thus, it can be seen that students experience a tendency for transformation error and process skill error. Next in Figure 2, the results of Camper type students for question number 1. The results of this work indicated that there was a mismatch between the answers and the questions. The students were required to prove that the limit for $x \to 0$ exists, while the limit for $x \to c$ and $c \neq 0$ do not exist. However, it can be seen that students show f continuous in x=0 and not continuous in $x \neq 0$. After being confirmed through interviews, it turns out that students were fixated on the sample questions that were discussed at the lecture. Students understood when they were asked to prove that the limit exists, but did not know what can be used from the information given by the question. So that students had difficulty in determining the next step for the verification process.

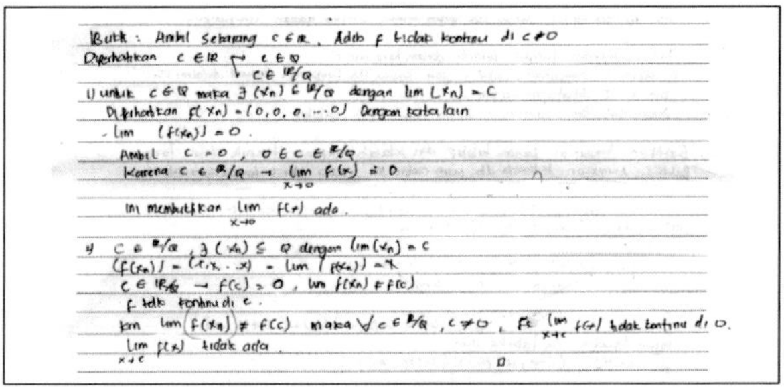

Figure 2. The Results of Camper-type Students' Answer for Question Number 1.

The results of the analysis are in the following Table 7.

Table 7. The Analysis Results of Camper-type Students' Answer for Question Number 1

Type of Error	Analysis Results
Reading Error	If you see the results of student work above, it seems that there is an error in the reading process. Because there is a mismatch between questions and answers. However, after being confirmed through interviews, it turned out that students were aware of that. So, students know that the answer given is not by the question.
Comprehension Error	Students provide such answers because they only know a little from the information. The rest of the students did not know what could be used from the information provided by the question.
Transformation Error	Students did not know what strategies to use to solve problems in this question.
Process Skill Error	Students do not carry out verification procedures correctly.
Encoding Error	Students have not been able to determine the resolution of this problem correctly.

Thus, on the question, it can be seen that the student tend to do comprehension errors, transformation errors, and process error skills. For the second question, it is still in the form of mathematical proof. The second question is as follows:

For example, let $f: A \to \mathbb{R}$ be continuous on R and let x_n sequences in A is convergent. Prove the $\lim(f(x_n)) = f(\lim(x_n))$.

Following is the work results from the climber-type students for the second question.

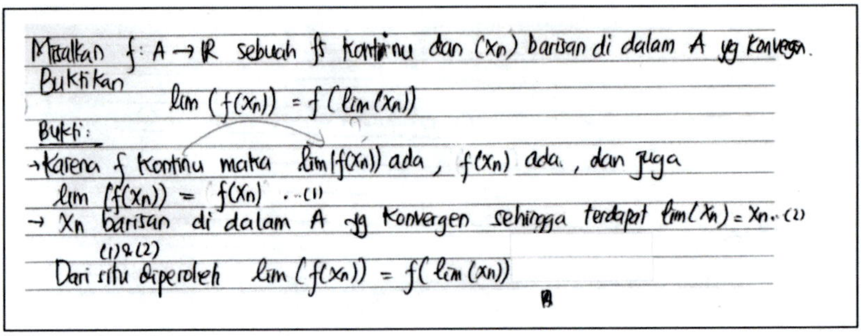
Misalkan $f: A \to \mathbb{R}$ sebuah fs kontinu dan (x_n) barisan di dalam A yg konvergen.
Buktikan $\lim(f(x_n)) = f(\lim(x_n))$
Bukti:
→ Karena f kontinu maka $\lim(f(x_n))$ ada, $f(x_n)$ ada, dan juga
$\lim(f(x_n)) = f(x_n)$ ...(1)
→ x_n barisan di dalam A yg konvergen sehingga terdapat $\lim(x_n) = x_n$ ...(2)
(1)&(2)
Dari situ diperoleh $\lim(f(x_n)) = f(\lim(x_n))$
∎

Figure 3. The Results of Climber-type Students' work for question number 2.

From Figure 3, it appears that students proved this case by connecting the concepts of continuous functions and limit lines. In the first stage, students took advantage of the concept that when a function is continuous, the limit is exist, the function value is exist, and the limit value is the same as the function value. However, students did not provide a justification regarding the line of (x_n) used. Next, students used the concept of converging sequence for the next process. However, it appears that students wrote down $\lim(x_n) = x_n$ since (x_n) was a convergent sequence. Then the conclusion was that the limit value was equal to the value of its function. After being confirmed to the students concerned, information was obtained that the students used line (x_n) on continuous concepts so that they could be linked to the information given, namely sequences (x_n) convergent. Furthermore, when the students wrote $\lim(x_n) = x_n$ in hopes that they could be connected to the concept of continuous function. From the results of this confirmation, it can be seen that the students used the correct strategy, but at the time of execution, it seems that students used inappropriate methods. Thus, it can be seen that in this problem students did not make a transformation error, but a process skill error. Furthermore, students had led to solving the problem, but the form of the answer given was still incorrect. It can be concluded that this thing is included in encoding errors. The results are presented in Table 8.

Table 8. The Results of analysis of the Climbertype Students' Work for Question Number 2

Type of Error	Analysis Results
Reading Error	Students do not experience problems related to reading.
Comprehension Error	Students understand the purpose of the problem, and it seems that students use all the information provided by the problem.
Transformation Error	Students have had a strategic idea to prove this case, namely by connecting the limit of the line and the continuous function. This is done by utilizing the properties that apply to the limit of functions and continuous functions.
Process Skill Error	In the process, it appears that students are still writing inappropriate procedures. This can be seen from the statement $\lim(x_n) = x_n$. Of course, this statement raises questions, but there is no justification for this statement.
Encoding Error	Student answers have led to solving the problem, but the form of the answer given is not correct. Because there are some steps that are not clear and there is no justification.

Furthermore, the following (Figure 4) is the work result of the camper-type students for question number 2. After further observing the results of student work in Figure 4, students intend to prove this case by using formal definitions of continuous functions and limit functions.

However, students do not provide a definition of continuous functions but a definition of limit functions. It seems that students have not been able to correctly identify what is informed by the problem and what can be utilized from the question information. It seems that students experience error comprehension.

Furthermore, in the process, the definition of convergent sequence does not appear in the results of students' work. There appears to be a statement

$\forall \varepsilon > 0, \exists \delta > 0 \ni |x_n - \lim(x_n)| < \delta \rightarrow |f(x_n) - f(\lim(x_n))| < \varepsilon$ caused by the convergence of lines , but there should be an explanation before writing the statement above because if so, the causal relationship above is not suitable.

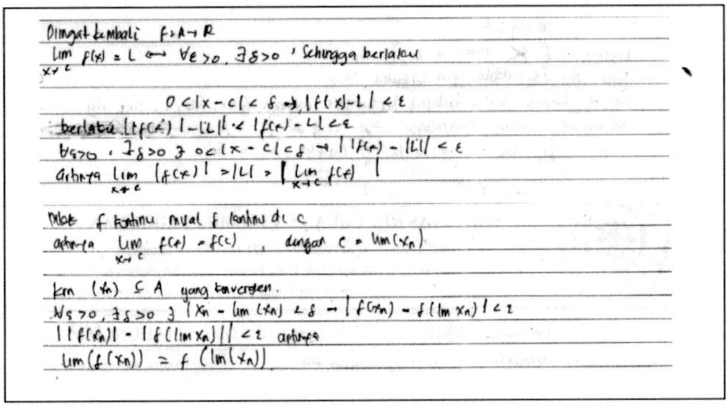

Figure 4. Results of Camper-type Students' Work for Question Number 2.

After being confirmed with the students concerned, it turns out that students are still confused about how to use the concept of converging sequence. Therefore, students direct the answer to statement 1. It appears that in this problem, Camper-type students are similar to Climber-type students in the sense that they have the right problem-solving strategies, but made mistakes in carrying out the strategy.

Table 9. The Results of the Analysis of the Camper-type Students' Work for Question Number 2

Type of Error	Analysis Results
Reading Error	Students do not experience problems related to reading.
Comprehension Error	Students cannot use the information provided by the problem. The actual concept that needs to be reviewed in this problem is the definition of continuous functions. However, students instead provide a definition of limit functions.
Transformation Error	Students have the right strategy, namely by utilizing formal definitions. Furthermore, students try to associate the concept of continuous function with a line limit.

Process Skill Error	In the process, the student only mentions once the line limit is locked and there is no justification at all regarding the line limit.
Encoding Error	The results of the students' work have led to the completion of the desired final form. However, the verification procedure provided is still incorrect.

Thus, for question number 2, the Camper-type students have a tendency toward the comprehension errors, process skills error, and encoding errors. The third question is still in the form of mathematical proof. The third question is as follows:

Prove that the set of limit points of a set is closed.

The following is the result of students' work for the third question.

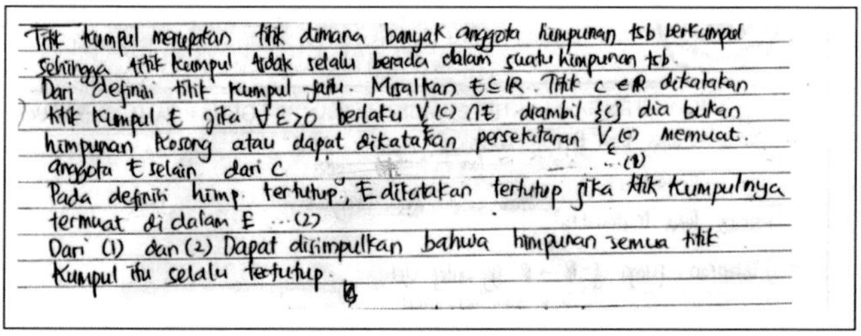
Titik kumpul merupakan titik dimana banyak anggota himpunan tsb berkumpul sehingga titik kumpul tidak selalu berada dalam suatu himpunan tsb.
Dari definisi titik kumpul jadi. Misalkan $E \subseteq \mathbb{R}$. Titik $c \in \mathbb{R}$ dikatakan titik kumpul E jika $\forall \varepsilon > 0$ berlaku $V_\varepsilon(c) \cap E$ diambil $\{c\}$ dia bukan himpunan kosong atau dapat dikatakan persekitaran $V_\varepsilon(c)$ memuat anggota E selain dari c ... (1)
Pada definisi himp. tertutup, E dikatakan tertutup jika titik kumpulnya termuat di dalam E ... (2)
Dari (1) dan (2) Dapat disimpulkan bahwa himpunan semua titik kumpul itu selalu tertutup.

Figure 5. Results of Climber-type Students' Work for Question Number 3.

The result of the student work above shows that the strategy used to prove the case was the definition of a closed set. However, the reason or explanation given was not so strong to conclude. After being confirmed to the students, they were still confused to provide mathematical proof of justification. As a result, students provided compelling conclusions. In this case, it can be seen that the students already had mathematical proof of ideas or strategies, but the same as in solving the previous questions, they still had problems with the execution of the strategy.

Table 10. Results of Analysis of the Climber-type Students' Work for Question Number 3

Types of Errors	Analysis Results
Reading Error	Similar to other cases, students do not experience problems related to reading.
Comprehension Error	Students have understood information that can be used from the questions given. It is seen that students can provide definitions of gathering points and definitions of closed sets.
Transformation Error	Students have the right strategy, namely by using the definition of closed set.
Process Skill Error	When executing an existing strategy, students are still lacking in giving justification at each step.
Encoding Error	The results of this students' work have led to completion and have the desired final form. But the justification is still lacking, and it can be said that the justification is still not strong enough to conclude this proof.

Thus, it can be seen that the climber-type students have a tendency to make mistakes in the process skills error and encoding errors (Indra Kurniawan, Kusmayadi & Sujadi, 2015). The following is the work of the camper-type students for question number 3.

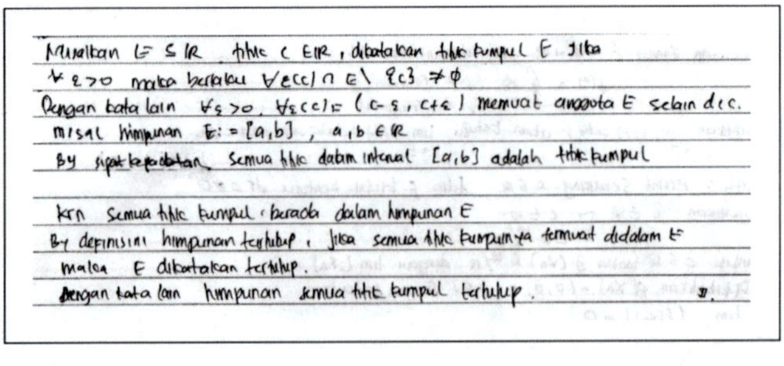
Misalkan $E \subseteq \mathbb{R}$ titik $c \in \mathbb{R}$, dikatakan titik kumpul E jika
$\forall \varepsilon > 0$ maka berlaku $V_\varepsilon(c) \cap E \setminus \{c\} \neq \emptyset$
Dengan kata lain $\forall \varepsilon > 0$, $V_\varepsilon(c) = (c-\varepsilon, c+\varepsilon)$ memuat anggota E selain dari c.
misal himpunan $E := [a,b]$, $a, b \in \mathbb{R}$
By sifat kepadatan semua titik dalam interval $[a,b]$ adalah titik kumpul
krn semua titik kumpul berada dalam himpunan E
By definisi himpunan tertutup, jika semua titik kumpulnya termuat didalam E
maka E dikatakan tertutup.
dengan kata lain himpunan semua titik kumpul tertutup.

Figure 6. The results of the Camper-type Students' Work for Question Number 3.

The result of question number 3 is generally similar to the answer of the climber-type students. The students use the definition of closed sets to prove this case. However, students encountered problems in how to justify this verification. So, students don't have problems in determining strategies, and problems arise when executing the strategy. This shows that the error that tends to occur for question number 3 is the process skill error and encoding error.

CONCLUSION

There are several types of errors that students tend to do in solving mathematical cases in the form of mathematical proof. For climber-type students, some types of errors that they tend to do in doing mathematical proof are transformation errors, process skill errors, and encoding errors. The camper type students tend to do comprehension errors, transformation errors, process skills error, and encoding errors. In comprehension error, it can be seen that in compiling proof, the students understand the intention of the problem but do not know what information can be taken. For transformation error, it can be seen from the misunderstanding between the concept of continuous functions and limit functions. As for the process skill error, it can be seen from students' errors in writing mathematical proof, and there are still steps that are not accompanied by justification, or even steps that are not written correctly. The encoding error can be seen from the evidentiary steps that have been written down, have not been compiled with the correct flow, and there is still a lack of mathematical proof justification for drawing conclusions.

REFERENCES

1. Ah, N. I. (2016). Pengembangan Bahan Ajar Mata Kuliah Analisis Real Berbasis Pembuktian Pada Semester V UNMUH JEMBER. Jurnal Gammath, I Nomor 2, 26–32.
2. Azwar, S. (2002). Penyusun Skala Psikologi. Yogyakarta: Pustaka Pelajar Offset.
3. Bagus Nur Iman, Toto Nusantara, D. (2016). Diagnosis Kesalahan Siswa Kelas XI SMA Negeri Candipuro pada Materi Penerapan Turunan Fungsi. In Seminar Nasional Matematika dan Pembelajarannya (pp. 146–153). Malang: Universitas Negeri Malang.
4. Ekayanti, A. (2017). Diagnosis Kesalahan Mahasiswa Dalam Proses Pembuktian Berdasarkan Newmann Error Analysis. Mosharafa: Jurnal Pendidikan Matematika, 6(1), 105–116.
5. Guntur Suhandoyo, P. W. (2016). Profil Kemampuan Berpikir Kreatif Siswa dalam Menyelesaikan Soal Higher Order Thinking Ditinjau dari Adversity Quotient (AQ). MathEdunesa: Jurnal Ilmiah Pendidikan Matematika, 3(5), 156–165.
6. Hernadi, J. (2016). Fondasi Matematika Dasar Berfikir Logis dan Kritis dalam Matematika. Ponorogo: UMPO Press.
7. Hidayat, W. (2017). Adversity Quotient dan Penalaran Kreatif Matematis Siswa SMA dalam Pembelajaran Argument Driven. KALAMATIKA, 2(1), 15–28.
8. Hidayat, W., Herdiman, I., Aripin, U., Yuliani, A., & Maya, R. (2018). Adversity Quotient (AQ) dan Penalaran Kreatif Matematis Mahasiswa Calon Guru. Jurnal Elemen, 4(2), 230–242. https://doi.org/10.29408/jel.v4i2.701
9. Indra Kurniawan, Kusmayadi, T. A., & Sujadi, I. (2015). Proses Berpikir Kreatif Siswa Climber dalam Pemecahan Masalah pada Materi Peluang (Penelitian Pada Siswa Kelas XI IPA Semester II SMA Negeri I Polanharjo Tahun Ajaran 2014 / 2015), 3(6), 599–612.
10. Ko, Y. Y., & Knuth, E. (2009). Undergraduate mathematics majors' writing performance producing proofs and counterexamples about continuous functions. Journal of Mathematical Behavior, 28(1), 68–77. https://doi.org/10.1016/j.jmathb.2009.04.005
11. Leonard, L. (2017). Pengaruh Adversity Quotient (AQ) dan Kemampuan Berpikir Kritis Terhadap Prestasi Belajar Matematika, 28(April 2014).

12. Novilita, & Suharnan. (2013). Konsep Diri Adversity Quotient dan Kemandirian Belajar Siswa. Jurnal Psikologi, 8(1), 619–632.
13. Nurhayati. (2015). Pengaruh Adversity Quotient dan Motivasi Berprestasi Terhadap Prestasi Belajar Matematika. Jurnal Formatif, 3(1), 72–77.
14. Rukmana, I., & Paloloang, B. (2016). Hubungan Adversity Quotient dengan Hasil Belajar Matematika Siswa Kelas XI SMA Negeri Model Terpadu Madani Palu, 3(3), 325–333.
15. Sari, C. K., Sutopo, S., & Aryuna, D. R. (2016). The Profile of Students' Thinking in Solving Mathematics Problems Based on Adversity Quotient. Journal of Research and Advances in Mathematics Education, 1, 36–48.
16. Sari, C. K., Waluyo, M., Ainur, C. M., & Darmaningsih, E. N. (2018). Logical errors on proving theorem. Journal of Physics: Conference Series, 948(1). https://doi.org/10.1088/1742-6596/948/1/012059
17. Sari, D. P., Darhim, D., & Rosjanuardi, R. (2018). Errors of Students Learning with React Strategy in Solving The Problems of Mathematical. Journal on Mathematics Education, 9(1), 121–128.
18. Stavrou, S. G. (2014). Common Error and Misconceptions in Mathematical Proving by Education Undergraduates. Issues in the Undergraduate Mathematics Preparation of School Teachers, 1(March), 1–8.
19. Stefanowich, A. (2014). Proofs and Mathematical Reasoning. University of Birmingham.
20. Stoltz. (2000). Mengubah Hambatan Menjadi Peluang. Jakarta: Grasindo.
21. Sucipto, L., & Mauliddin. (2016). Analisis Kesulitan Mahasiswa dalam Memahami Konsep Bilangan Real. Beta Jurnal Tadris Matematika, 9(2).
22. Sukoriyanto, S., Nusantara, T., Subanji, S., & Chandra, T. D. (2016). Students' Errors in Solving the Permutation and Combination Problems Based on Problem Solving Steps of Polya. International Education Studies, 9(2), 11. https://doi.org/10.5539/ies.v9n2p11
23. Wardiana, W. dan S. Z. (2014). Hubungan Antara Adversity Quotient (AQ) dan Minat Belajar dengan Prestasi Belajar Matematika pada Siswa Kelas V SD di Kelurahan Pedugan. Jurnal Mimbar PGSD Universitas Pendidikan Ganesha, 2(1).

24. Wijaya, A., Heuvel-panhuizen, M. Van Den, Doorman, M., & Robitzch, A. (2014). Difficulties in solving context-based PISA mathematics tasks : An analysis of students ' errors. The Mathematics Enthusiast, 11(3), 555–584. https://doi.org/10.1139/t02-118
25. Yani, M., Ikhsan, M., & Marwan, M. (2016). Proses Berpikir Siswa Sekolah Menengah Pertama dalam Memecahkan Masalah Matematika Berdasarkan Langkah-langkah Polya. Jurnal Pendidikan Matematika, 10(1), 43–58.
26. Yanti, A. P., Koestoro, B., & Sutiarso, S. (2018). The Students' Creative Thinking Process based on Wallas Theory in Solving Mathematical Problems viewed from Adversity Quotient /Type Climbers. Al Jabar: Jurnal Pendidikan Matematika, 9(1), 51–61.
27. Yanti, A. P., & Syazali, M. (2013). Analisis Proses Berpikir Siswa dalam Memecahkan Masalah Matematika Berdasarkan Langkah-Langkah Bransford dan Stein Ditinjau dari Adversity Quotient. Al-Jabar: Jurnal Pendidikan Matematika, 7(3), 64.

Chapter 6

INTRODUCING A MEASURE OF PERCEIVED SELF-EFFICACY FOR PROOF (PSEP): EVIDENCE OF VALIDITY

Benjamin Shongwe[(1)], Vimolan Mudaly[(2)]

(1) Department of Mathematics Education, University of KwaZulu-Natal, South Africa

(2) Department of Mathematics Education, University of KwaZulu-Natal, South Africa

ABSTRACT

It is widely recognized that students encounter difficulties with proof across all grades and beyond, yet standardized instruments related specifically to students' perceived self-efficacy for mathematical proof have not been readily available. The purpose of this study was to develop and investigate preliminary validity evidence for a new instrument for measuring self-efficacy for mathematical proof that can be of importance to the field. The new Perceived Self-Efficacy for Proof (PSEP) questionnaire is a self-administered, 8-item questionnaire that quantifies experimentation,

Citation: (APA): Shongwe, B., & Mudaly, V. (2021). Introducing a measure of perceived self-efficacy for proof (PSEP): Evidence of validity. JRAMathEdu (Journal of Research and Advances in Mathematics Education), 6(3), 260-276.(17 pages).

Copyright: © Creative Commons Attribution 4.0 International License (http://creativecommons.org/licenses/by/4.0/).

conjecturing, inductive reasoning, justification, and validation. To validate the PSEP, two studies with 260 eleventh grade students—recruited from three Dinaledi schools in EThekwini metropolitan area, South Africa—were conducted. In Study 1 (n=128), face and content validity were evaluated, and an exploratory factor analysis (EFA) was performed. In Study 2 (n=132), a confirmatory factor analysis (CFA) was conducted and external validity was investigated. In both samples, the PSEP was found to possess good internal consistency reliability with relatively high factor loadings on a single component. Although the findings in this report represent preliminary validation evidence, it can be concluded that the PSEP is a valid, reliable and sensitive measure of 11th grade students' perceptions of their ability to construct a proof and may serve as a meaningful outcome in mathematical proof research and classroom proof education.

Keywords: Calibration, factor analyses, instrument development, self-efficacy for proof

INTRODUCTION

It is widely recognized that students with higher levels of self-efficacy—also known as perceived ability, which denotes here the belief in one's abilities to succeed in a task—are more effortful, attempt more cognitively challenging problems, and persist longer in the face of difficulties in tasks (Oriol, Amutio, Mendoza, Da Costa, & Miranda, 2016). In addition, self-efficacy is a significant predictor of students' academic achievement (Daher, Gierdien, & Anabousy, 2021). This relationship is not new. For instance, Pajares and Kranzler (1995) investigated the influence of self-efficacy and mental ability of a sample of high school students engaging in mathematical problem solving tasks. They found that selfefficacy influenced both students' observed performance in the problem solving tasks and their mental ability with equal strength. However, traditional research on students' performances in proof has concentrated primarily on cognitive factors and less on affect (Lau, Fang, Cheng, & Kwong, 2019). Yet, "[p]urely cognitive" behavior is extremely rare, and what is often taken for pure cognition is actually shaped – if not distorted – by a variety of factors (Schoenfeld, 1983, p. 330). Thus, an understanding of this construct (i.e., perceived self-efficacy for proof) could be of utmost importance to the field, particularly in drawing attention to the sources of students' difficulty with mathematical proof (hereinafter proof), a difficulty widely recognized as universal. However, existing validity studies of instruments designed to measure students' perceived ability to construct a proof remain unexplored despite the domain of Euclidean geometry demanding significantly different factors.

Broadly speaking, perceived self-efficacy for proof refers to a student's mechanism based on expectation to accomplish proof-related tasks. The purpose of the present study was threefold: distinguishing perceived self-efficacy from related constructs; developing an instrument designed to measure it; and investigating tentative validity evidence for it. The Perceived Self-Efficacy for Proof (PSEP) questionnaire is a brand new, self-administered, 8- item questionnaire operationalized by items describing the activities in which students engage in a meaningful proving process which characterizes the practices of research mathematicians: experimentation, conjecturing, seeking counterexamples, justification, and validation. To explore evidence of validity for the new measure of self-efficacy for proof, two studies were conducted.

Extensive research has been devoted to understanding the important place that the concept of self-efficacy occupies in the learning and instruction area. The most used and frequently cited instrument designed to measure high school students' self-efficacy is Betz and Hackett's (1993) Mathematics Self-Efficacy Scale. In addition, the researchers who have investigated students' performance in mathematics report similar effects of selfefficacy and the cognitive component.

In the past decade, large-scale international studies attempted to measures students' self-efficacy toward mathematics. For instance, Wilkins (2004) analyzed the Trends in International Mathematics and Science Study (TIMSS) data of several countries to understand the relationship between students' self-belief concepts in mathematics and respective achievement. He made contradictory findings; there were positive relationships at the student level, yet negative ones at the country level. A similar study undertaken by Shen and Tam (2008) found that whereas students in Korea and Singapore showed the highest achievement in TIMSS, they had lower self-efficacy than those in the Unites States (US).

Conceptual Framing: Self-efficacy

Conceptually, self-efficacy for proof is situated within Bandura's (1986) social cognitive perspectives of self-beliefs. The conceptual domain of self-beliefs constructs includes confidence, self-efficacy, self-concept, and anxiety (Morony, Kleitman, & Lee, 2012). These four constructs are substantively distinct. However, there is no consensus concerning their definitions and specificity in part due to the overlap between them, various perspectives from which these constructs have been investigated, and their operationalization and measurement (Bong & Skaalvik, 2003; Ferla, Valcke, & Cai, 2009; Hughes, Galbraith, & White, 2011). Whereas confidence statements are described as utterances intended to affect the belief of others about one's ability to provide correct answers to a task (Charness, Rustichini, & Van de Ven, 2018), self-efficacy examination "includes both an affirmation of a capability level and the strength of that belief" (Bandura, 1997, p. 282). The lack of consensus notwithstanding, it is possible to delineate an explicit.

According to Stankov, Lee, Luo, and Hogan (2012), confidence is an individual's state of being certain about successful performance in a task. Efklides (2011) sees confidence as a task-specific metacognitive experience in which an individual provides particular responses to indicate the level of certainty about the correctness of responses to a mathematical problem. For instance, an item such as "Mathematics is harder for me than for many of my classmates" in the TIMSS Students' Self-Confidence in Learning Mathematics index suggest a comparison between the student and his or her peers. Thus, measures of confidence are confined to making judgment about the correctness of an answer to a problem.

Now, let us turn to the distinguishing elements of the concept of self-efficacy. Central to the construct of self-efficacy is the distinction between perceived capability and observed accomplishment, respectively a non-cognitive component (judgment of capability) and an evaluative component (judgment of accomplishment). One other distinctive feature of self-efficacy is that no comparison of individuals with others takes place (Pajares, 2006). From these perspectives, it is reasonable to conclude that whereas confidence is viewed as a product, self-efficacy is taken as a process (learning and performing).

Calibration of Self-efficacy

Calibration is the degree to which an individual student's judgment of his or her capability reflects the actual (observed) accomplishment in an objective task. Researchers have also reported that high achieving students have high self-efficacy and are relatively accurately calibrated in that they have more accurate self-perceptions (Zimmerman, 2009). In contrast, the notion of miscalibration (either over- or under-confident) arises when an individual's self-efficacy does not mirror the observed accomplishment (Alexander, 2013; Pajares, 2006). According to Chiu and Klassen (2010), this phenomenon (i.e., miscalibration) is associated with students in countries that are poor, culturally hierarchical, less more tolerant of uncertainty, or less flexible regarding gender roles; attributes broadly describing Felbrich, Kaiser, and Schmotz (2012) notion of collectivist cultures. Triandis' (1989) assertion that African cultures are collectivist was particularly important for this study. This assertion is important in light of Lawson' (2015) claim that in collectivist cultures, individuals need to adhere to group expectations which tend to be less tolerant of deviation from the norm.

An appreciation of miscalibration requires a reflective thought by the student. Baron, Gürçay, and Metz (2017) use the term "reflective thought" to denote moment in which one chooses "to be careful at the expense of speed" (p. 108). Thus, reflective thought relates to an individual's moment in which he or she ascribes meaning to the task at hand. Further, in the reflective thinking process, the basis of students' knowledge and beliefs are reviewed and evaluated (Kim & Silver, 2016). Thus, reflection is a process that can be seen as a means through which students learn mathematics meaningfully, including construction of proofs in ways typical of the practices of theoretical mathematicians. As Stylianides and Stylianides (2018) point out, these ways include seeking patterns, conjecturing, seeking counterexamples, proving, and justifying.

Significance of Researching Students' Self-efficacy for Proof

As already mentioned, despite the persistent and pervasive nature of students' difficulty with proof, limited attention is given to developing measures of self-efficacy for proof in the current literature. An assessment of students' self-efficacy for proof warrants attention as it could be of utmost importance for a variety of reasons. Understanding students' self-efficacy for proof could draw attention to the sources of students' difficulty with the concept of proof. Students' self-efficacy for proof may be a beneficial construct to employ when investigating correlations of students' difficulty with the concept of proof. Task-specificity assessments can provide prediction indexes and insights not available from general assessments of influence of self-efficacy on observed performance (Seegers & Boekaerts, 1996).

Conceptualizing and measuring students' self-efficacy for proof may also be useful for understanding their overall achievement in mathematics. Given the central role which self-efficacy plays in the exercise of personal agency by its strong impact on thought, affect, motivation, and action, calibration may also be one way students metacognitively observe their academic progress (Bandura, 2014). Metacognitive monitoring entails the evaluation of self-efficacy relative to task demands to assess whether the standards being pursued are attainable or beyond a student's reach. However, as Oriol et al. (2016) point out, in human endeavors, failures do not necessarily lower aspirations. Final but not least, miscalibration could be unraveled. The next section provides a brief background on the theoretical model of self-efficacy for proof.

RESEARCH METHODS

Current Research

The present study consisted of two phases (i.e., Study 1 for scale development and Study 2 for testing the reliability of the scale with an independent sample) to give credence to the reliability of the scale (Field, 2017). In Study 1, the definition of students' selfefficacy for proof was used to generate an initial set of items. These items were based on Dreyfus and Hadas' (1987) principles of proof understanding whose outline is outside the scope of the present study. It was hypothesized that the items on the scale would indeed load onto a single factor and thus enable an examination of self-efficacy for proof as a unidimensional factor.

In Study 2, additional evidence of validity was the factor structure obtained by conducting a confirmatory factor analysis (CFA) and examining external validity. It was hypothesized that the single-factor solution would provide good fit to the data and that the students' PSEP scale would be empirically distinct from related constructs. In addition to providing the methodology of this study, the next section documents the development of the measurement instrument and its psychometric properties in line with the principle that prior to being utilized, an instrument must be rigorously assessed for its scores to be labeled valid and reliable (reproducible).

Study 1: Participants and Procedure

The PSEP measurement instrument was administered to 128 grade 11 students conveniently recruited from three Dinaledi schools in EThekwini metropolitan area, South Africa. Dinaledi schools followed a mandatory mathematics curriculum that includes algebra, trigonometry, analytical geometry, and Euclidean geometry in addition to financial mathematics, statistics and probability. The poor performance of previously disadvantaged students in mathematics and physical science has been of national concern and a priority for South African government, as evidenced in the establishment of the Dinaledi School Project in 2001 (Department of Basic Education [DBE], 2009).

Of the 128 students participating in the study, 56 were male, 67 were females, and 5 did not report their gender. The ages of the participants ranged from 15 to 22 (M = 16.26, SD = 1.44). These data were collected during February of 2017. In addition, the participants selected multiple ethnic/racial categories in which 76 identified themselves as Black, 41 as Asian, 4 as Colored, and 4 as "Other." Three participants chose not to report ethnicity. Exploratory factor analysis (EFA) was performed using data obtained from this sample.

Study 2: Participants and Procedures

Participants were 132 students who were in the second half of the larger sample recruited to determine whether the items and their factor were the same across two independent samples (Brown, 2014). In this sample, there were 56 males, 71 were females, and 5 did not report their ages. The ages of the participants ranged from 14 to 22 (M = 16.96, SD = 1.04). The racial/ethnic composition of the sample was Black (52.3%), Asian (41.7%), Colored (3.8%), and "Other" (2.3%). Three participants did not indicate their race or ethnicity. The procedure for Study 2 mirrored that of Study 1.

Prior to describing item development process and factor analyses methods in detail, it is worth reminding the reader that the purpose of the present study was to (1) develop a measure of students' self-efficacy for proof and (2) determine the psychometric properties of this instrument. The measure was developed after a thorough review of the proof literature, self-efficacy instruments, expert judgments, and cognitive interviews with respondents. The next sections describe these studies, in turn. Specifically, the sections demonstrate a "hybrid" approach in which an exploratory factor analysis (EFA) is performed to help generate a new theory and following up on its results using a CFA with separate data to test the theory (Hair, Black, Babin, & Anderson, 2019).

Study 1: Item Development and Content and Face Validation

Based on the preliminary definition of students' self-efficacy for proof, an initial pool of 13 items was developed. Content validity was evaluated by soliciting feedback from five experts (two faculty colleagues, two doctoral students, and one postdoctoral researcher) with knowledge of the concepts of self-efficacy and mathematical proof. However, one doctoral student invited to comment on content validity of the instrument did not respond. One of the experts has specific expertise in scale development and another has published extensively on the subject matter (i.e., proof).

In the examination of content validity, some items which were deemed (a priori) as very similar were included so that the experts might help in determine a more appropriate way to phrase them. The experts also evaluated face validity, a component of content validity pertaining to whether the test "looks valid" to the participants (Creswell, 2018). Following Adelson and McCoach's (2011) and McCoach, Gable, and Madura's (2013) description of the best practices for the development of new, valid, and reliable scales, the experts rated three aspects of content validity: agreement if each item measured selfefficacy for proof or a different construct; certainty for categorizing the items as selfefficacy for proof (1 = not very sure, 2 = pretty sure, 3 = very sure); and the extent to which an item is relevant for measuring self-efficacy for proof (1 = low/no relevance, 2 = mostly relevant, 3 = highly relevant).

Next, cognitive interviews with five respondents whose characteristics were similar to the target population were conducted to refine and assess item interpretation. Specifically, further face validation took place through conducting two rounds of cognitive interviewing with these respondents. Questions were carefully considered for gender neutrality, clarity, brevity, interpretability, and subsequent translations into other languages (Creswell, 2018).

RESULTS AND DISCUSSION

Results

An analysis of experts' feedback provided both qualitative and quantitative data. From a qualitative perspective, they commented on individual items and provided feedback in relation to the appropriateness of the operational definition, construct coverage, item phrasing (clear and unambiguous), and appropriate level of the scale for use with high schoolers. The definition was revised to acknowledge two critical aspects.

First, it was to acknowledge that students' self-efficacy for proof is indeed a matter of their own perceptions rather than an objective fact. Second, the definition was revised to acknowledge that students also lack an appreciation of the meaning of the concept of proof in mathematics. Specifically, the revision was prompted by respondents' inability to distinguish between the common use of the term "proof" and its technical meaning in the mathematics discipline. As a consequence, self-efficacy for proof was redefined as a construct that describes student's naïve or informed and perceived competence in successfully accomplishing a proof construction task; hence the questionnaire is hereinafter referred to a Perceived Self-Efficacy for Proof (PSEP). On the one hand, "naïve" means viewing proof as merely a means to verify the truth of a proposition using few cases. On the other hand, "informed" denotes viewing proof as serving not only the function verification but also explanation, communication, discovery, and systematization (de Villiers, 1990).

Although one item and the response categories were somewhat modified by the experts, the theoretical concepts essentially remained the same. Only two rounds are reported here because relatively few new insights emerged from the third round (Willis, 2005). Based on the feedback responses from cognitive interviews, one item was dropped for lack of clarity. In addition, modification to grammar and the word choice were also made.

From a quantitative perspective, the percentage of experts who agreed that the items operationalized the construct of perceived self-efficacy for proof ranged from 85.7% to 100.0%. Items with over 80% agreement (9 items) were retained for further consideration. Mean certainty and mean relevance ranged from 2.50 to 3.00 and from 2.14 to 3.00, respectively. The agreement rate for the item ("Seek counterexamples") that had under 85.7% was, however, retained because it was considered a key feature in proof construction. One more item ("Verify the truth of a proposition") was deleted due to its low relevance rating (2.14) and multiple expert comments about its redundancy ($M_{certainty} = 2.49$, $M_{relevance} = 2.55$). Based on the feedback from the expert at scale development, midpoint anchor label "Uncertain" was removed to better assess miscalibration; it was replaced with "Neither yes or no" so as to maintain a seven-point Likert scale. The final version of PSEP contained 8 items along with 3 demographic questions (i.e., age, ethnicity or race, and gender). The advantage of using a multiple-item PSEP scale lies in its ability to measure internal consistency reliability and thus reflect the "cohesiveness" of different items within a unique factor (Creswell, 2018).

Exploratory Factor Analysis 1

The questionnaire was presented on the same paper as the proof construction task in which participants were expected to experiment with few triangles (plane), conjecture, seek counterexamples, construct, and validate a proof (however, on different pages). The participants were asked to self-report on the PSEP questionnaire whose stem was, "Do you think you have the ability to do the following task?" They responded on a seven-point Likert response scale: 1 = strong no, 2 = no, 3 = somewhat no, 4 = neither yes or no, 5 = somewhat yes, 6 = yes, and 7 = strong yes. At the start of the administration of the survey, participants reported on demographic data. Then, directions on the instrument informed participants that they were to express their perceived ability to engage in the list of different activities related to proof construction. On average, the completion of the survey took 15 minutes after which they were directed to turn to the next page which contained the proof construction task.

As depicted in Figure 1, participants were provided with a few triangles with which observe and seek a pattern, generalize, conjecture, seek counterexamples, prove, and validate the correctness of the proof. The figure also shows possible activities (observation, generalization, and conjecturing). The proving task took on average 25 minutes to complete. Participants completed the survey and the task after normal school hours.

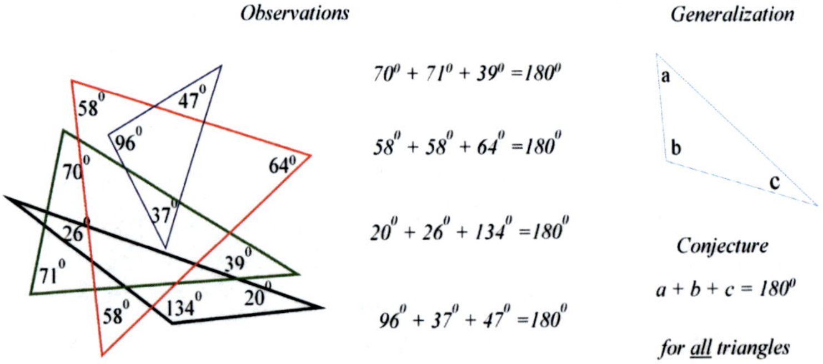

Figure 1. Observation of few cases resulting in a generalization about angles on a plane triangle.

Exploratory Factor Analysis 2

Two separate SPSS extraction methods used were principal axis factoring (PAF) and principal components analysis (PCA). In an attempt to identify the factors that underlie the structure of the PSEP, PAF as an extraction method was initially performed because it is the "classic factor analytic approach" (Pett, Lackey, & Sullivan, 2003, p. 103). Item 5 ("Give up in the face of difficulties") in Table 1 was reverse coded so that all items were in the same direction and consisted of positive correlations. In addition to the fact that the determinant score was above the rule which indicated an absence of multicollinearity, the Barlett's test of sphericity (Chi-square = 413.42, df = 28, $p < .001$) which rejected the null hypothesis that the correlation matrix is an identity matrix, and the Kaiser-Meyer-Olkin (KMO) measure of sampling adequacy which was closer to 1 (.838) suggested the data were suitable for EFA (Field, 2017).

In summarizing the variables into fewer components, parallel analysis (PA), which Hayton, Allen, and Scarpello (2004) describe as "one of, if not the, most accurate method for determining the number of factors to retain" (p. 197) was considered together with the Kaiser-Guttman rule of retaining eigenvalues greater than 1.0, and the "elbow joint" of the scree plot. The observed eigenvalues exceeded the comparison (i.e., "parallel") eigenvalues as derived from randomly permutated observed data having the same sample size and items as the original dataset thus supporting the single-factor structure observed in scree plot. Thus, the eight items loaded onto one factor to create a construct than can explain the interrelationships among these items.

Factor Analysis and Internal Consistency Reliability Results

The results of EFA identified the factor structure and other items for fine tuning. A PAF extraction provided pattern and structure coefficients as well as factor correlations. Items with pattern coefficients of at least .40 were retained (McCoach et al., 2013; Pett et al., 2003). The pattern matrix of the PAF solution with oblique rotation, structure matrix, and communalities are presented in Table 1.

Given that the factor explained 65.90% of the common variance, it is reasonable to suggest that the PSEP scale consisted of items that operationally defined the construct of perceived self-efficacy for proof. The resulting internal consistency of the items within the PSEP scale (i.e., Cronbach's coefficient alpha) in this sample, which represents the proportion of variability attributable to the true score (Pett et al., 2003), was strong ($\alpha = .92$). Cronbach's coefficient alpha is defined as the degree to which the set of items in the scale co-vary relative to their sum score (Raykov & Marcoulides, 2011). Although an alpha coefficient of .70 is regarded as an acceptable threshold for reliability, .80 and .95 are preferred for the psychometric quality of scales (Hair et al., 2019). The first two columns in Table 1 present oblimin-rotated pattern and structure matrices following PAF extraction of the single factor. In the last column, means and standard deviations for these items are presented. Factor loadings (i.e., the strength of the relationship that each item has with the factor) were significantly high.

Table 1. The PSEP scale items and their coefficients

Item		Coefficients			
		Pattern	Structure	(h^2)	*M* (*SD*)
1.	Engage in experimentation to seek patterns	.48	.49	.44	4.70 (0.94)
2.	Make a conjecture	.88	.79	.63	4.65 (0.73)
3.	Verify if the conjecture using few cases	.63	.67	.45	4.73 (1.06)
4.	Seek counterexamples	.63	.73	.54	4.45 (1.00)
5.	Give up in the face of difficulties*	.74	.74	.58	4.37 (0.99)
6.	Use previously proven statements	.73	.58	.59	4.00 (1.05)
7.	Formally write out each step of the proof	.86	.65	.79	4.22 (0.94)
8.	Examine the proof for accuracy	.52	.47	.46	3.63 (1.09)

Note: Pattern coefficient values less than .20 were suppressed; h^2 = communalities of the measured items using PCA. Oblimin rotation was used. *Reverse-coded item scores after the item reverse coding

Interpretation of PSEP Factor

The PSEP measures students' perceived self-efficacy for proof. High scale scores reflect a perception that a student holds in relation to accomplishment in a proof construction task and low scores reflect a perception that a student holds in relation to accomplishment in a proof construction task. Having reverse-coded item 5 of the PSEP scale, item scores were averaged to obtain a scale score ranging from 0 to 7. Overall, scores reflected participants' varying degrees in which they perceived themselves to have the ability to accomplish a proof construction task (M = 3.90, SD = .76).

The total scores in the PSEP scale ranged from 0 to 48; the sample contained participants whose perception that they could hardly construct a proof was strong and those who felt strongly that they could successfully construct a proof. Also, the proportion of variance in the items accounted for by the factor was strong for all the items loading on it. Further, a fairly good variability for responses on each individual item in PSEP was observed. As Hair et al. (2019) propose, given that the dataset had several high factor loading scores (> .80), then the sample size was reasonably sufficient.

The total variance explained by the self-efficacy for proof construct was at the threshold (49.99%) which suggested a reasonably good factor solution. In social science research, unlike in the natural sciences where information is often more precise, extracted factors usually explain only 50% to 60%

(Hair et al., 2019). Put another way, the single extracted component (factor) explained nearly 50% of the variability in the eight items on SEP. Having developed this theory using EFA, the next section uses CFA to hypothesize an a priori model of the underlying factor structure and examines if this model fits the data adequately (Bandalos, 1996).

Confirmatory Factor Analysis: Dimensionality Investigation

A CFA was performed on the second half of the sample to test dimensionality (i.e., to assess the fit of the data to the theoretical structure). This approach was adopted to maintain independence of the EFA and CFA samples. However, prior to estimating how well the model fit the hypothesized structure, the P-P plot was examined to determine the shape of the distribution of data; that is, whether the data approximated a normal distribution. P-P plots compare the cumulative probability of observed data with a theoretically ideal normal distribution (Field, 2017). As displayed in Figure 2, in addition to absence of outliers, the data values in the P-P plot lay along the diagonal line. This suggested that the data approximately resembled a normal distribution; thus, performing and interpreting the linear regression model was appropriate for the data.

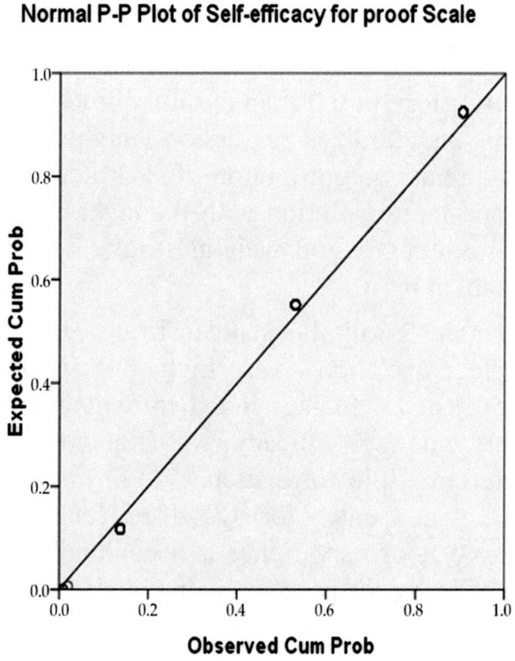

Figure2. Standardized normal P-P Plot for the PSEP data.

The CFA was performed with the use of SPSS analysis of moment structures (AMOS) because there were no missing data. The CFA model consisted of a single latent variable, representing the singe factor derived from the results of EFA (i.e., perceived self-efficacy for proof). The test for the hypothesized unidimensional structure obtained from a satisfactory initial factor analysis in Study 1 followed systematic fit assessment procedures in line with recommendations by Creswell (2018). The fit statistics that were selected and evaluated in this study were Chi-square test of exact fit, the comparative fit index (CFI), the Tucker-Lewis Index (TLI), and the root mean square error of approximation (RMSEA).

CFA Model and Internal Consistency Reliability

The nonsignificant Chi-square value suggested a good fitting model. In assessing the extent to which the hypothesized model fit the actual data, Hair et al's. (2019) standards were adopted: evidence of reasonable fit included obtaining a nonsignificant χ^2 value, a normed χ^2 (i.e., χ^2/df) value below 5, comparative fit index (CFI) and incremental fit index (IFI) values of .90 or higher, and root mean square error of approximation (RMSEA) value of .08 or less. Although these statistics indicate how well the model fits the data, the strength of the structural paths in the model is determined by the squared multiple correlations (SMC) (Hair et al., 2019).

Squared multiple correlations in Table 2 suggested the absence of singularity (i.e., SMC close to 0.00) and multicollinearity (i.e., SMC close to 1.00). In addition, standardized regression weights were determined to compare directly the relative contribution of each item on the factor (Hair et al., 2019). The one-factor solution with the eight observed PSEP items comprised standardized regression weights (factor loadings) which were statistically significant at $p< .01$.

As shown in Table 2, all the standardized regression weights and the squared multiple correlations were high. For instance, standardized regression weight of item 1 ("Engage in experimentation to seek patterns") on the factor it represents (self-efficacy) was higher than the threshold of .70. Also, the squared multiple correlation (.71) of this item suggested that 71% of its variance is accounted for by self-efficacy for proof construct while the remaining 29% of its variance is accounted for by measurement error. In short, self-efficacy for proof explained half (50%) of the variation in the item with the other half being error variance. Taken together, self-efficacy for proof is theoretically meaningful and contributes appreciably to account for the variance in the dataset. Put another way, these results suggested that self-efficacy for proof is indeed a construct that defines a distinct cluster of interrelated items. As a consequence, all the eight items in EFA were retained.

The Chi-square for the model suggested a good model fit: 2 (17) = 27.108, $p > .05$. The other alternative fit indices also indicated a good fit to the data: normed χ^2 = 1,60, CFI = .98, TLI = .96, and, RMSEA = .07. Put another way, the results suggested a reliable model for future forecasts. Similar to Study 1, the internal reliability consistency was high ($\alpha = .91$). Similar to Study 1, participants in Study 2 varied in terms of the extent to which they were certain of their proof construction ability (M = 4.32, SD = 1.24). In addition to descriptive statistics for each item, Table 3 shows that there was relatively also good variability observed for each item in the PSEP scale.

Table 2. Standardised regression weights (factor loadings) and squared multiple correlations (communalities) for the one-factor CFA model (n = 132)

Item	Factor loadings	Squared multiple correlations	
1. Engage in experimentation to seek patterns	.74	.20	4.52 (1.37)
2. Make a conjecture	.75	.56	4.05 (1.43)
3. Verify if the conjecture is true using few cases	.70	.48	4.32 (0.96)
4. Seek counterexamples	.71	.51	4.30 (1.65)
5. Persevere in the face of difficulties*	.74	.55	4.44 (0.69)
6. Use previously proven statements	.67	.38	4.15 (1.35)
7. Formally write out each step of the proof	.68	.46	4.78 (1.29)
8. Examine the proof for accuracy	.79	.26	4.03 (1.17)

External Validity

Further evidence of validity was provided by assessing construct validity through examining the correlations between PSEP and other measures. Confirmatory factor analysis has gained the most prominence as an alternative model to simultaneously evaluate both discriminant and convergent validity to probe construct validity (Boateng, Neilands, Frongillo, Melgar-Quiñonez, & Young, 2018) of the conceptualization of students' self-efficacy. The purpose of investigating convergent validity is to estimate the relationship between the self-efficacy for proof construct and a similar construct in which stronger correlation coefficients would suggest support for convergent validity (Creswell, 2018). In contrast, the purpose of discriminant validity is to estimate the relationship between scale scores and distinct constructs in which weaker correlation coefficients would suggest support for discriminant validity (Creswell, 2018). Specifically, convergent validity was examined by correlating PSEP scores with the four-item mathematics rigor scale. Discriminant validity was assessed by correlating PSEP scores with the measure of perceived confidence. The next section describes the measures used to assess external validity of the PSEP scale.

Mathematics Rigor

Although TIMSS is hypothesized as low-stakes tests—that is, tests without consequences for the student but potentially high-stakes for schools and countries which may be judged deficient if scores are poor (Jackson, Khavenson, & Chirkina, 2020)—its measure of mathematics rigor provided an excellent opportunity to test students' perceptions. A student who rarely, if ever, thought about the rigor of mathematical problems would have a low score on the instrument. The participants responded to 4 items on the scale (e.g., "I need to invest greater effort in mathematics" and "I learn things quickly in mathematics (or science)" based on a 5-point Likert scale (1 = strongly disagree, 2 = disagree, 3 = neither agree nor disagree, 4 = agree, 5 = strongly agree). The item, "There is no need to study harder in mathematics", was reverse-coded so that all items were in the same direction. The total score for each participant ranged from 4 to 16, and this score was then divided by 4 to obtain an average score for rigor. TIMSS reported a Cronbach's α coefficient of .94. A similarly excellent internal consistency reliability α coefficient of .81 was obtained in this study.

An assessment of this construct in participants was deemed important following Shen and Tam's (2008) report that students in low-performing countries, for example, South Africa and Morocco, are likely to say that they perceive mathematics (including proof) as being easy and that they learn the subject matter quickly. An examination of correlations coefficients of PSEP scores and those of the four-item measure of perceived mathematics rigor provided additional evidence of convergent validity. Put another way, strong coefficients between perceived self-efficacy and rigor were evidence in support of convergent validity.

Confidence

Given that Stankov et al. (2012) found that confidence alone explained the major part of variance on the self-beliefs constructs including self-efficacy, it was used as a measure to determine discriminant validity. Students' confidence is typically assessed by asking participants to indicate, on a percentage scale, how confident they are that their justprovided responses to the proving task are correct (Morony et al., 2012). According to Bandura (1997), asking participants to express their confidence in solving a mathematical problem (in this case, constructing a proof) similar to that presented in a cognitive test or examination serves to increase prediction of academic outcomes. He points out that the expression of confidence can provide prediction indices and insights not available from broader assessments of self-efficacy.

Participants rated the strength of their confidence to successfully accomplish the activity on the PSEP scale using a five-point Likert scale ranging from 1 (not confident at all) to 5 (completely confident). The total score for each participant ranged from 8 to 32. This score was then divided by 8 to obtain an average self-efficacy for proof score. An excellent internal consistency reliability α coefficient of .92 was obtained.

Table 3 models the interrelationships among items by representing them with three (latent) variables. An examination of correlations coefficients of PSEP scores and those of the confidence measure provided additional evidence of discriminant validity (i.e., perceived self-efficacy for proof is empirically distinct from confidence). The relationship between perceived self-efficacy for proof and confidence was weak, an indicator of discriminant validity. Put differently, the results indicated that although these two constructs were related, they were conceptually distinct. In contrast, the high and statistically significant correlation coefficient between PSEP and mathematics rigor demonstrated that self-efficacy for proof and rigor were related constructs, an indicator or convergent validity. Correlations between PSEP and each of the three external measures are provided in Table 3.

Table 3. Correlations among measures for Study 2

Measure	1	2	3
SEP	–		
Rigor	.79**	–	
Confidence	.20**	.28**	–

Note: Correlation is significant at the 0.01 level (2-tailed).

The correlation matrix of all the scales shows interrelationships in a pattern that supports their construct validity. On the one hand, ratings were significantly correlated within each item across the participants, evidencing convergent validity. On the other hand, ratings were not significantly correlated with each other, providing evidence for discriminant validity. Thus, it can be claimed that this PSEP measurement scale has construct validity as demonstrated by the results of both convergent and discriminant validity. The results were consistent with the hypothesis that there exist weak correlations between PSEP and the measure of confidence. The results were also consistent with the hypothesis that there exists a strong relationship between the PSEP and the measure of rigor. Considered together, these findings provided tentative evidence for convergent and divergent validity. Further, these results suggested that PSEP is a unidimensional measure comprising multiple items that capture the underlying domain of students' perceived self efficacy for proof.

Discussion

Findings from the present study represent an important first step in conceptualizing and measuring high school students' perceived self-efficacy for proof. In Study 1, a rigorous content-evaluation process was performed to establish evidence of content validity for the PSEP scale. The exploratory factor analysis of the scale yielded a single-factor structure on which all items loaded highly. Also, an excellent internal consistency reliability index was obtained. Similar to Study 1, a new independent sample in Study 2 yielded scores that demonstrated high internal consistency reliability and good model fit (i.e., the items on the scale can be used to predict or explain students' self-efficacy for proof). Further, this sample demonstrated convergent validity with the measure of rigor and discriminant validity with the measure of confidence. The CFA results show that students' self-efficacy for proof can be understood in terms of eight variables.

Scores on PSEP for both Study 1 and Study 2 samples demonstrated good variability. Self-efficacy for proof is a valuable metacognitive monitoring process in a learning environment because it helps students to correct distorted self-efficacy. Self-efficacy for proof also provides useful diagnostic information for teachers and schools by comparing the confidence score to accuracy (i.e., the percentage of correct answers in a task) to examine confidence judgments (Moore & Healy, 2008).

Additional findings were made concerning miscalibration. The finding that participants tended to overestimate their self-efficacy for proof relative to actual achievement was similar to Chiua and Klassen's (2010) findings. They found that miscalibration was a feature of students in countries that were poor, less egalitarian, or less flexible regarding gender roles. Vancouver and Kendall (2006) have shown that overestimation of one's self-efficacy is characteristic of students in a collectivist society, which Hofstede (1986; 2011) defines as the tendency within a culture toward gregariousness and group orientation (e.g., South Africa; Triandis, 1989). Overestimation may result in less preparation, less help-seeking, and ultimately poor achievement

(Kim & Silver, 2016). According to Triandis (1989), urban samples tend to be individualistic, and traditional-rural samples tend to gravitate toward collectivism within the same culture. The samples in this study were urban thus suggesting that urban societies tend to embrace individualism rather than collectivism. Thus, the results related to miscalibration are inconsistent with Triandis' (1989) findings. One explanation for this contradictory result is that, although African cultures are considered collectivist, the sample consisted of participants from other ethnic groups whose culture could be different.

Limitation

The results reported in this study must be treated with caution. The process of validating the PSEP scale was complicated by the absence of appropriate known-to-be reliable measures of self-efficacy for proof. The use of other measures is regarded as the best option available. Although the design of this study incorporated an attempt to reduce the threats to external validity of the results by using two studies with different participants from the population of eleventh graders because random selection could not be done, some caution on inference of generalizability must be made. Specifically, some students may already have been harboring intentions to change to the other mathematics whose content excludes proofs (e.g., mathematical literacy) due to career choices, parental disapproval, peer pressure, and so on. Because of these factors, it is difficult to generalize the results of this study to other contexts.

The fact that only Dinaledi high schools were sampled is considered a limitation. Students in these schools may have demonstrated evidence of self-efficacy for proof that may be unique. Therefore, caution must be exercised in generalizing the findings of this study to the general high school population because conclusions regarding a global view of self-efficacy in proof-related tasks were not made. However, given the sparse but needed research on perceived self-efficacy for proof, it is critically important to understand competence beliefs in schools with a focus on mathematics and physical sciences in light of the budget spending allocated to these schools. Also, direct input from the participants in the questionnaire should not be underestimated; such input can broaden our understanding of the difficulties that students encounter when learning proof. It is through this understanding that will guide us to new methods that make the learning and teaching of proof meaningful. In addition, based on the operative definition of self-efficacy for proof, the statements in the scale represent substantively valid measures.

CONCLUSION

In conclusion, previous studies have focused on self-efficacy for mathematics in general. The present study has considered self-efficacy specifically for proof. The purpose of this study was to develop and psychometrically validate an instrument specifically designed to measure students' perceived self-efficacy for proof in high schoolers and thus add to the literature on understanding students' difficulty with proof which exists across national and cultural boundaries. The PSEP instrument was developed and validated with two samples utilizing EFA and CFA.

Measuring students' self-efficacy for proof is important because proof construction is considered a vital practice for understanding mathematics and how its knowledge is constructed. For instance, measuring students' self-efficacy for proof may help in the identification and assistance of those with high self-self-efficacy scores but underachieve (i.e., miscalibration). This study offers insight into the challenges and skills current school leaders must face to leverage technology to improve student learning. Embarking on an instrument development design was intended to provide the mathematics education field with a tool to make sound generalizations about calibration.

On the basis of tentative validity evidence in the present study, PSEP scale has demonstrated to be a metacognitive tool to measure and diagnose students' distorted selfefficacy for proof perceptions. Thus, measuring students' perceived self-efficacy for proof is particularly important for the improvement and deepening of our understanding of the mathematical domain of proof which persistently continues to be an area of great difficulty for many students. The findings of this study would be further strengthened through replication with reliable measures of self-efficacy for proof. Future research efforts may be directed at calibration (using bias scores) to compare subgroups such as gender or age, examination of demonstrative geometry resources (e.g., dynamic geometry software), and ethnic differences. Policy decisions can be guided by research efforts focusing on such comparisons.

ACKNOWLEDGMENT

The authors deeply appreciate the participation of learners in this study. The study was supported in part by the University Capacity Development Programme (UCDP) grant at the University of KwaZulu-Natal (UKZN). Any opinions, findings, and conclusions or recommendations expressed in this paper are those of the authors and do not necessarily reflect the views of the UCDP.

REFERENCES

1. Adelson, J. L., & McCoach, D. B. (2011). Development and psychometric properties of the Math and Me Survey: Measuring third through sixth graders' attitudes toward mathematics. Measurement and Evaluation in Counseling and Development, 44(4), 225-247. http://doi.org/10.1177/0748175611418522
2. Alexander, P. A. (2013). Calibration: What is it and why it matters? An introduction to the special issue on calibrating calibration. Learning and Instruction, 24, 1-3. http://doi.org/10.1016/j.learninstruc.2012.10.003
3. Bandalos, B. (1996). Confirmatory factor analysis. In J. Stevens (Ed.), Applied multivariate statistics for the social sciences (3rd ed., pp. 389-420). Mahwah, NJ: Lawrence Erlbaum Associates.
4. Bandura, A. (1986). Social foundations of thought and action: A social cognitive theory. Englewood Cliffs, NJ: Prentice Hall.
5. Bandura, A. (1997). Self-efficacy: The exercise of control. New York: WH Freeman.
6. Bandura, A. (2014). Social cognitive theory of moral thought action. In W. M. Kurtines, & J. L. Gewirtz (Eds.), Handbook of moral behavior and development (pp. 69-128). Hillsdale, NJ: Psychology Press.
7. Baron, J., Gürçay, B., & Metz, S. E. (2017). Reflective thought and actively open-minded thinking. In M. E. Toplak, & J. Weller (Eds.), Individual differences in judgment and decision making from a developmental context (pp. 107-126). New York, NY: Routledge.
8. Betz, N. E., & Hackett, G. (1993). Mathematics self-efficacy scale. (Mental Measurements Yearbook 14, No. 14081939). Abstract retrieved from Mental Measurements Yearbook.
9. Boateng, G. O., Neilands, T. B., Frongillo, E. A., Melgar-Quiñonez, H. R., & Young, S. L. (2018). Best practices for developing and validating scales for health, social, and behavioral research: A primer. Frontiers in Public Health, 6:149. http://doi.org/10.3389/fpubh.2018.00149
10. Bong, M., & Skaalvik, E. M. (2003). Academic self-concept and self-efficacy: How different are they really? Educational Psychology Review, 15, 1-40.
11. Brown, T. (2014). Confirmatory factor analysis for applied research. New York, NY: Guilford Press.
12. Charness, G., Rustichini, A., & Van de Ven, J. (2018). Self-confidence and strategic behavior. Experimental Economics, 21(1), 72-98.

13. Chiua, M. M., & Klassen, R. M. (2010). Relations of mathematics self-concept and its calibration with mathematics achievement: Cultural differences among fifteen-year-olds in 34 countries. Learning and Instruction, 20, 2-17.
14. Creswell, J. W. (2018). Educational research: Planning, conducting, and evaluating quantitative and qualitative research (6th ed.). Boston, MA: Pearson.
15. Daher, W., Gierdien, F., & Anabousy, A. (2021). Self-efficacy in creativity and curiosity as predicting creative emotions. Journal of Research and Advances in Mathematics Education, 6(2), 86-99. http://doi.org/10.23917/jramathedu.v6i2.12667
16. de Villiers, M. D. (1990). The role and function of proof in mathematics. Pythagoras, 24, 17-24.
17. Department of Basic Education [DBE]. (2009). The Dinaledi Schools Project: Report from a strategic engagement between the national department of education and business on increasing support for mathematics and science in education in schools. Pretoria: Department of Basic Education.
18. Dreyfus, T., & Hadas, N. (1987). Euclid may stay—and even be taught. In M. M. Lindquist, & A. P. Shulte (Eds.), Learning and teaching geometry, K-12 (pp. 47-58). Reston, VA: National Council of Teachers of Mathematics.
19. Efklides, A. (2011). Interactions of metacognition with motivation and affect in self-regulated learning: The MASRL model. Educational Psychologist, 46, 6-25.
20. Felbrich, A., Kaiser, G., & Schmotz, C. (2012). The cultural dimension of beliefs: An investigation of future primary teachers' epistemological beliefs concerning the nature of mathematics in 15 countries. ZDM Mathematics Education, 44, 355-366.
21. Ferla, J., Valcke, M., & Cai, Y. (2009). Academic self-efficacy and academic self-concept: Reconsidering structural relationships. Learning and Individual Differences, 19, 499-505.
22. Field, A. (2017). Discovering statistics using IBM SPSS statistics (5th ed.). London: Sage.
23. Hair, J. F., Black, W. C., Babin, B. J., & Anderson, R. E. (2019). Multivariate data analysis (8th ed.). New Jersey, NJ: Pearson.

24. Hayton, J. C., Allen, D. G., & Scarpello, V. (2004). Factor retention decisions in exploratory factor analysis: A tutorial on parallel analysis. Organizational Research Methods, 7, 191-205.
25. Hofstede, G. (1986). Cultural differences in teaching and learning. International Journal of Intercultural Relations, 10, 301-320.
26. Hofstede, G. (2011). Dimensionalizing cultures: The Hofstede model in context. Online Readings in Psychology and Culture, 2(1), 26 pages.
27. Hughes, A., Galbraith, D., & White, D. (2011). Perceived competence: A common core for self-efficacy and self-concept? Journal of Personality Assessment, 93, 278-289. http://doi.org/10.1080/00223891.2011.559390
28. Jackson, M., Khavenson, T., & Chirkina, T. (2020). Raising the stakes: Inequality and testing in the Rtussian education system. Social Forces, 98(4), 1613-1635. http://doi.org/10.1093/sf/soz113
29. Kim, Y., & Silver, R. E. (2016). Provoking reflective thinking in post observation conversations. Journal of Teacher Education, 67(3), 203-219. http://doi.org/10.1177/0022487116637120
30. Lau, Y., Fang, L., Cheng, L. J., & Kwong, H. K. (2019). Volunteer motivation, social problem solving, self-efficacy, and mental health: A structural equation model approach. Educational Psychology, 39(1), 112-132.
31. Lawson, D. M. (2015). Family violence: Explanations and evidence-based clinical practice. New York: Wiley.
32. McCoach, D. B., Gable, R. K., & Madura, J. P. (2013). Instrument development in the affective domain. School and corporate applications (3rd ed.). New York, NY: Springer.
33. Moore, D. A., & Healy, P. J. (2008). The trouble with overconfidence. Psychological Review, 115(2), 502-517.
34. Morony, S., Kleitman, S., & Lee, Y. P. (2012). Predicting achievement: Confidence vs self-efficacy, anxiety, and self-concept in Confucian and European countries. International Journal of Educational Research, 58, 79-96.
35. Oriol, X., Amutio, A., Mendoza, M., Da Costa, S., & Miranda, R. (2016). Emotional creativity as predictor of intrinsic motivation and academic engagement in university students: The mediating role of positive emotions. Frontiers in Psychology, 7:1243. http://doi.org/10.3389/fpsyg.2016.01243

36. Pajares, M. F. (2006). Self-efficacy during childhood and adolescence. In M. F. Pajares, & T. Urban (Eds.), Adolescence and education: Self-efficacy and adolescence (Vol. 5, pp. 339-367). Greenwich, Connecticut: Information and Age.
37. Pajares, M. F., & Kranzler, J. (1995). Self-efficacy beliefs and general mental ability in mathematical problem-solving. Contemporary Educational Psychology, 20, 426-443.
38. Pett, M. A., Lackey, N. R., & Sullivan, J. J. (2003). Making sense of factor analysis: The use of factor analysis for instrument development in health care research. Thousand Oaks, CA: Sage.
39. Raykov, T., & Marcoulides, G. A. (2011). Introduction to psychometric theory. New York, NY: Routledge.
40. Schoenfeld, A. H. (1983). Beyond the purely cognitive: Beliefs systems, social cognitions, and metacognitions as driving forces in intellectual performance. Cognitive Science, 7, 329-363.
41. Seegers, G., & Boekaerts, M. (1996). Gender-related differences in self-referenced cognitions in relation to mathematics. Journal for Research in Mathematics Education, 27(2), 215-240.
42. Shen, C., & Tam, H.-P. (2008). The paradoxical relationship between student achievement and self-perception: A crossnational analysis based on three waves of TIMSS data. Educational Research and Evaluation, 14(1), 87-100.
43. Stankov, L., Lee, J., Luo, W., & Hogan, D. J. (2012). Confidence: A better predictor of academic achievement than self-efficacy, self-concept and anxiety? Learning and Individual Differences, 22, 747-758.
44. Stylianides, A. J., & Stylianides, G. J. (2018). Addressing key and persistent problems of students' learning: The case of proof. In A. J. Stylianides, & G. Harel (Eds.), Advances in mathematics education research on proof and proving: An international perspective (pp. 99-113). Cham, Switzerland: Springer.
45. Triandis, H. C. (1989). The self and social behaviour in differing cultural contexts. Psychological Review, 96, 506-520.
46. Vancouver, J. B., & Kendall, L. N. (2006). When self-efficacy negatively relates to motivation and performance in a learning context. Journal of applied psychology, 91(5), 1146-1153. http://doi.org/10.1037/0021-9010.91.5.1146

47. Wilkins, J. L. (2004). Mathematics and science self-concept: An international investigation. The Journal of Experimental Education, 72(4), 331-346.
48. Willis, G. B. (2005). Cognitive interviewing: A tool for improving questionnaire design. Thousand Oaks, CA: Sage.
49. Zimmerman, B. J. (2009). Self-efficacy and educational development. In A. Bandura (Ed.), Self-efficacy in changing societies (pp. 202-231). Cambridge, UK: Cambridge University Press.

Chapter 7

DEDUCTIVE OR INDUCTIVE? PROSPECTIVE TEACHERS' PREFERENCE OF PROOF METHOD ON AN INTERMEDIATE PROOF TASK

Tatag Yuli Eko Siswono, Sugi Hartono, Ahmad Wachidul Kohar

Universitas Negeri Surabaya, Gedung C8 FMIPA Unesa Ketintang, Surabaya, Indonesia

ABSTRACT

The emerging of formal mathematical proof is an essential component in advanced undergraduate mathematics courses. Several colleges have transformed mathematics courses by facilitating undergraduate students to understand formal mathematical language and axiomatic structure. Nevertheless, college students face difficulties when they transition to proof construction in mathematics courses. Therefore, this descriptive-explorative study explores prospective teachers' mathematical proof in the

Citation: (APA): Siswono, T. Y. E., Hartono, S., & Kohar, A. W. (2020). Deductive or Inductive? Prospective Teachers' Preference of Proof Method on an Intermediate Proof Task. Journal on Mathematics Education, 11(3), 417-438.(22 pages).

Copyright: © Creative Commons Attribution-ShareAlike 4.0 International License (http://creativecommons.org/licenses/by-sa/4.0/).

second semester of their studies. There were 240 pre-service mathematics teachers at a state university in Surabaya, Indonesia, determined using the conventional method. Their responses were analyzed using a combination of Miyazaki and Moore methods. This method classified reasoning types (i.e., deductive and inductive) and types of difficulties experienced during the proving. The results conveyed that 62.5% of prospective teachers tended to prefer deductive reasoning, while the rest used inductive reasoning. Only 15.83% of the responses were identified as correct answers, while the other answers included errors on a proof construction. Another result portrayed that most prospective teachers (27.5%) experienced difficulties in using definitions for constructing proofs. This study suggested that the analytical framework of the Miyazaki-Moore method can be employed as a tool to help teachers identify students' proof reasoning types and difficulties in constructing the mathematical proof.

Keywords: Deductive-inductive reasoning; proving difficulties; mathematical proof; prospective teachers

The construction of formal mathematical proof is an important component of advanced mathematics courses for undergraduate degree (Shaker & Berger, 2016). In recent years, some universities have transformed mathematics courses by introducing the transition of proof or introduction to mathematical reasoning courses (Selden & Selden, 2007; Smith, 2006), which facilitates college students to understand formal mathematical language and axiomatic structure. However, Clark and Lovric (2008) explored challenges faced by college students as they make the transition to proof construction in mathematics courses. This transition requires college students to change their types of reasoning, for instance, shifting the informal language to formal one, reasoning from mathematical definition, understanding and applying the theorem, and making connections between mathematical objects (Clark & Lovric, 2008).

In addition, college students, including prospective teachers, are also demanded to conceive several skills: a) recognizing reasoning and proof as fundamental aspects of mathematics, b) making and investigating allegations of mathematical conjectures, c) developing and evaluating mathematical arguments and proofs, and d) selecting and using different types of reasoning and methods of proof (National Council of Teachers of Mathematics [NCTM], 2000). Blanton, Stylianou, and David, (2003) agreed that college students need to develop required proving skills to construct a proof. In this case, teachers' knowledge about proof must be given to students because that can help the students strengthen the concept and skill of proof (Carrillo, et al, 2018; Stylianides, 2007). Such skills, more particularly, are also necessary for prospective teachers due to the teacher's need for perceiving a deep understanding of nature and the role of proof for conducting instructional practices (Jones, 1997). Moreover, the math teachers' rationales beyond teaching proof and proving in schools are due to the fact that students have experienced similar reasoning to the mathematicians, such as learning a body of mathematical knowledge and gaining insight about why assertions are true. They also can teach students logical thinking, communication, and problem-solving skills in mathematics.

Although proving is an important part of advanced mathematics, many studies indicated that students often have difficulties in constructing a proof (Moore, 1994; Selden, Benkhalti, & Selden, 2014; Selden & Selden, 2007). Epp (2003) reported that a 'poor' mathematical proof process is caused by the lack of proofwriting attempts. In addition, Moore (1994) carried out an observation of some students' transition to college in which most of them stated that they only memorized the proof since they did not understand proof and how to write it. Furthermore, Edwards and Ward (2004) said that the students could not use mathematical definitions or construct the relation between every day and mathematical languages.

In connection with examining student's mathematical proof, Miyazaki and Moore methods might have inspired many researchers for analyzing student's proof with particular objectives. For example, Kögce, Aydin and Yildiz (2010) adopted Miyazaki's (2000) classification of proof to investigate high school students' level of proof based on types of reasoning. Furthermore, Ozdemir and Ovez (2012) looked for the relationship between prospective teachers' perception proof types proposed by Almeida (2000) and their proving processes related to the experienced types of difficulties (Moore, 1994). In relation with students' common error and misconceptions in mathematical proving associated with the use of Moore's error category of proof, Stavrou (2014) found that the students did not necessarily understand the content of relevant definitions or how to apply them in writing proofs. Another study found that students got difficulties in creating definitions that conformed their concept images or accepted definitions of basic concepts (Dickerson & Pitman, 2016). While those studies concern on a single objective on how students' proof is assessed, the present study highlights that individuals' performance regarding mathematical proof can be explained from at least two aspects namely types of reasoning they involve (Miyazaki's method) and types of difficulties they experience during the proving process (Moore's method). Hence, the researchers argue that obtaining data on both aspects simultaneously leads a broader and more likely fruitful knowledge on how individuals, prospective teachers rather, deal with mathematical proof. Therefore, the present study aims to explore the prospective teachers' proof regarding their types of reasoning and difficulties in a mathematical proof.

Proof is an important aspect in mathematics because it is the main component in understanding mathematics (Kögce et al., 2010) and mathematical thinking (Hanna et al., 2009). Consequently, learning mathematics by mastering mathematical proof along with how to construct it becomes a strategic view (Balacheff, 2010). De Villiers (1990) and Knuth (2002a) stated that the role of mathematical proof is to verify the correctness of a result or truth of a statement, to communicate mathematical knowledge, and to apply an axiomatic system. Its purpose helps investigate the trueness or falseness of an argument regardless the cases and conditions (Baki, 2008) and shows the relevance of the justifications (Lee, 2002).

There are two universally recognized proving methods namely deduction and induction (Kögce et al., 2010; Miyazaki, 2000). Deduction method of proof involves several methods encompassing direct proof, proof by contraposition, and proof by contradiction (Baki, 2008; Moralı et al., 2006). Deduction method in mathematics begins with a general statement or hypothesis and examines the possibilities to reach a specific logical conclusion (Morris, 2002). Induction method is generally used by 8th grade students or secondary school students because they have already learned to prove numerical or geometrical proposition (Miyazaki, 2000). These two methods are based on the types of reasoning used by someone in carrying out a proving process, of which each respectively refers to deductive reasoning and inductive reasoning. Deductive reasoning is unique because it is a process of deducing conclusions from known information (premise) based on formal logic rules, where the conclusions must come from information provided and do not need to validate them with experiments (Ayalon & Even, 2008). Whereas, Christou and Papageorgiou (2007) conveyed that inductive reasoning is a reasoning process from specific premises or observations to reach a general conclusion or an overall rule. Of those two, deductive reasoning, which is used in a deductive proof, is considered the preferred tool in many mathematical communities to verify mathematical statements and demonstrate universality. Therefore, Ayalon and Even (2008) argued that deductive reasoning is often used as a synonym for mathematical thinking.

Knowledge of mathematical proof is considered as one essential component of subject matters (Shulman, 1986) in which mathematics teachers must acquire. Jones (1997) argued that teachers will have an extremely secured subject knowledge base of mathematical proof if they teach it accurately and confidently.

Most researchers who have examined teachers' knowledge of proof have centered on teachers' acceptance of empirical versus deductive arguments as valid proofs. Knuth (2002b) investigated sixteen in-service secondary school mathematics teachers' knowledge about what constituted a proof. Martin and Harel (1989) assessed the notions of proof performed by 101 pre-service elementary school teachers by giving their participants some statements accompanied by predetermined arguments and asking them to rate for revealing the validity. Both Knuth (2002b) and Martin and Harel (1989) concluded that most teachers correctly identified a valid argument and also wrongly accepted invalid arguments as proofs. Some pre-service elementary teachers accepted empirical arguments as proofs (Martin & Harel, 1989; Morselli, 2006; Simon & Blume, 1996).

Teacher's knowledge about proof is indeed not limited to understanding how to construct valid proofs. More broadly, it is related to the knowledge of both content and pedagogical aspects of proof. Steele and Rogers (2012) propose the so-called 'Mathematical knowledge for teaching proof' that can be considered as a meaningful framework to assess teacher's knowledge about proof. Steele and Roger (2012) also mention components of proof knowledge comprising knowledge of defining proof, identifying proofs and non-proofs, creating mathematical proofs, and understanding the roles of proof in mathematics. The first three components give main features on the content knowledge of mathematical proof, while the last component gives attention about the work of teaching proof since it is related to, for instance, checking or confirming students' thinking on the truth of a known idea, unpacking students' thinking and reasoning beyond their decision why a statement is true, confirming students' conjectures, and developing students' new mathematical ideas.

This study focuses on one of the teacher's knowledge, particularly on studying prospective teachers' proof identification and creating mathematical proofs through recognizing them across representations in case of two types of reasoning namely deductive and inductive reasoning and types of difficulties during a proving process.

METHOD

This study used descriptive-explorative research design to explore prospective teachers' types of reasoning and difficulties in carrying out a mathematical proving process. There were 240 prospective teachers who studied at Department of Mathematics of a state university in Surabaya, Indonesia, as the research participants. They were in the first semester so that the present study used convenience sampling method (Miles & Huberman, 1994). The participants consisted of 216 female and 24 male students (aged around 18 years old in average). The data were collected in three years (from 2013 to 2015) by providing initial tests to the participants, as many as 80 prospective teachers per academic year. There were 2 classes per academic year, in which each class consisted of 40 prospective teachers.

Even though Moore's (1994) research used interview to see errors in mathematical proof, the present study tended to examine more on deductive and inductive proof without employing interview like what Miyazaki (2000) did. The data were collected using a simple task of constructing one mathematical proof "Prove that the sum of two odd numbers is an even number". Actually, the task type could be more than one, such as the sum of two even numbers is even, the sum of odd and even numbers is odd, or the subtraction variation from two even, odd, or even-odd numbers. However, the main point of this study was a proof method whether using deductive common symbols or certain numbers that tended to be inductive. This, therefore, only required one sufficient problem determined to represent the use of deductive and inductive reasoning. This simple task consisted of a question that was similar to the task used by Özer and Arõkan (2002) in constructing proof in accordance with the types of reasoning. Instead of only types of reasoning, the present study's analysis also concerned on exploring prospective teachers' proof based on the types of difficulties. Moreover, this task was selected since it was often found in Indonesian secondary school curriculum, which had been frequently learned by prospective teachers who studied early mathematical proof. The process of data collection was carried out during the beginning of number theory and elementary algebra courses in their study. Each prospective teacher had been given 15 minutes to complete the task. Afterwards, each prospective teacher's responses were assessed to investigate the prospective teacher's comprehension of their deductive and inductive knowledge in constructing a proof along with their difficulties in constructing their proofs.

Table 1. Moore's types of difficulties in performing mathematical proof (Moore, 1994)

Type of difficulties	Discussion
D1	Prospective teachers did not know the definitions. That is, they were unable to state the definitions.
D2	Prospective teachers had a little intuitive understanding of the concepts.
D3	Prospective teachers' concept images were inadequate for doing the proofs.
D4	Prospective teachers were unable, or unwilling, to generate and use their own examples.
D5	Prospective teachers did not know how to use definitions to obtain the overall structure of proofs.
D6	Prospective teachers were unable to understand and use mathematical language and notation.
D7	Prospective teachers did not know how to start making a proof.

The data of participants' responses about proof correctness were analyzed by using Miyazaki's (2000) classification for types of reasoning in mathematical proof and Moore's (1994) classification for types of errors in a mathematical proof. Table 1 and Table 2 show the scheme of proof analysis used to analyze prospective teachers' proof.

Table 2. Miyazaki's types of reasoning in performing mathematical proof (Miyazaki, 2000)

Representation	Contents	
	Inductive reasoning	Deductive reasoning
Functional language used for demonstration	Proof D	Proof A
Other languages, drawings, and/or manipulated objects	Proof C	Proof B

Proof A was the type of proof when deductive reasoning was involved and a functional language was used in the course of making a proof. Proof was the type of proof where deductive reasoning was involved and other languages, drawings, and movable objects were used in the course of making a proof. Proof C was the type of proof where inductive reasoning was involved and other languages, drawings, and movable objects were used. Proof D was the type of proof where inductive reasoning was involved and a functional language was used.

Table 3. A combined classification of Moore-Miyazaki category of proof

Miyazaki *(Type of reasoning)*		**Moore** *(Type of difficulties)*						
Type	**Correct Proof**	**Conceptual understanding**					**Mathematical languages and notations**	**Getting started on a proof**
		D1	**D2**	**D3**	**D4**	**D5**	**D6**	**D7**
Proof A	%	AD1	AD2	AD3	AD4	AD5	AD6	AD7
Proof B	%	BD1	BD2	BD3	BD4	BD5	BD6	BD7
Proof C	%	CD1	CD2	CD3	CD4	CD5	CD6	CD7
Proof D	%	DD1	DD2	DD3	DD4	DD5	DD6	DD7

In combining the entries presented in Table 1 and Table 2, the present study applied the coding category presented in Table 3 for conducting an analysis. For example, since Proof A referred to a proof involving deductive reasoning with functional language used for demonstration, and D1 referred to a proof difficulty indicated by prospective teachers' misconception on the definition along with their inability to state the definition, then AD1 referred to a proof performance involving deductive reasoning that indicated difficulties shown by the inability to state definitions. Furthermore, the column correct proof portrays that percentage of the students' correct proofs.

In addition, the coding was carried out to all the prospective teachers' answers. Since there was more than one possibility of coding given to each answer with different categories, the present study selected the most significant feature of the response category that emerged from the answer. Henceforth, each answer only had one code of category. The coding was carried out by the first author and the reliability of the coding was checked through additional coding by an external coder, who was a teacher educator in our university. It was done based on 20 % of 240 prospective teachers' responses in problem proof. 20% of the population chosen randomly became the minimum sample size used in this study that were determined by using Slovin formula with a 10% error margin. In agreement with the multiple coding procedures, this study calculated the inter-rater reliability for each type, which resulted in Cohen's Kappa of 0.69, indicating that the coding was a substantial agreement (Landis & Koch, 1977).

RESULTS AND DISCUSSION

In this section, thc data obtained from the participants were analyzed, discussed, and then presented in Table 4. In accordance with Table 4, there were 38 prospective teachers who were correct in mathematical proof, whereas, the others were wrong in mathematical proof. Table 4 also depicts that 61.66% of the prospective teachers performed Proof A in which this proof required deductive reasoning and functional language used to construct proofs. Meanwhile, 0.84% of the prospective teachers conveyed Proof B with deductive reasoning and manipulated objects or using a sentence without functional language in proof. 31.25% of the prospective teachers showed Proof C in which they used inductive reasoning and other languages, images, and manipulated objects to construct proofs. Moreover, 6.25% of the prospective teachers showed Proof D in which they used inductive reasoning and functional language for constructing proofs. Regarding the correctness of the prospective teacher's responses, the present study found that 15.8% of the prospective teachers' responses were correct and 84.2% of the prospective teachers still experienced difficulties in constructing the proof task. Figure 1 to Figure 15 explain the examples of the results of prospective teachers' proof based on Proof A, Proof B, Proof C, and Proof D.

Proof A

The results showed that Proof A was performed by as many as 15.83% of the prospective teachers, meaning that they worked on the proof task correctly according to the deductive reasoning and functional language in constructing proofs. Meanwhile, 45.83% of the prospective teachers had difficulties in proving caused by several things encompassing less understanding of the concept involved (42.08%), lack of knowledge related to mathematical notations (0.42%), and being stuck in starting the proving process (3.33%). In connection with the less understanding of the concept, the prospective teachers' responses consisted of AD1, AD2, AD3, AD4, and AD5 types, while the difficulties to get started on proving included AD7 type, and the lack of knowledge about mathematical notation and logic included AD6 type. Figure 1 shows the correct examples of a prospective teacher's answer to Proof A in constructing the proof.

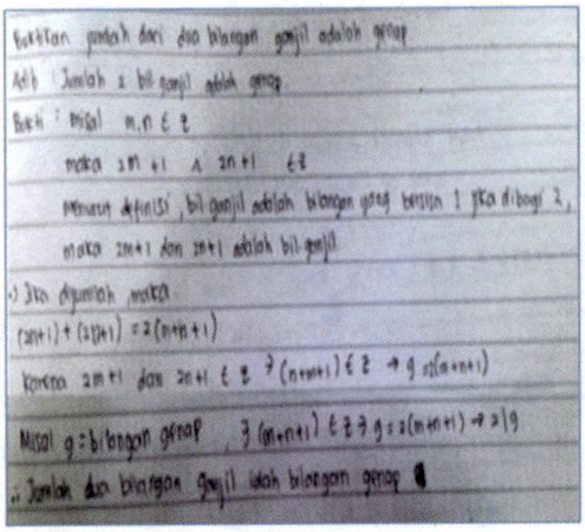

Translation

Prove that the sum of two odd numbers is an even number.

It will be proven: the sum of two odd numbers is an even number.

Proof: If m, n ∈ Z then 2m+1 and 2n+1 ∈ Z. According to definition, an odd number is number that has remainder 1 when divided by 2, and then 2m+1 and 2n+1 are odd numbers. If being summed, then (2m+1) + (2n+1) = 2(m+n+1). Because 2m+1 and 2n+1 ∈ Z ∋ (n+m+1) ∈ Z → g = 2(m+n+1). By the definition, an even number is a number that is dividable by 2, then (m+n+1) is an even number.

Figure 1. Example of Proof A.

Figure 1 shows that prospective teachers worked on proving with the correct response toward the question given. The correct response in this proof, including Proof A, used deductive reasoning and functional language that could be seen in the student work sample (see Figure 1). The deductive reasoning was indicated by the prospective teacher's idea in firstly letting an even number and an odd number with different symbols, which indicated his understanding of the rigorous symbol that had a significant step for being manipulated in the subsequent proving process. Moreover, it also showed some functional languages precisely, such as the symbols of ∈, Z, ∋ and | that indicated their proficiency in dealing with mathematical symbols.

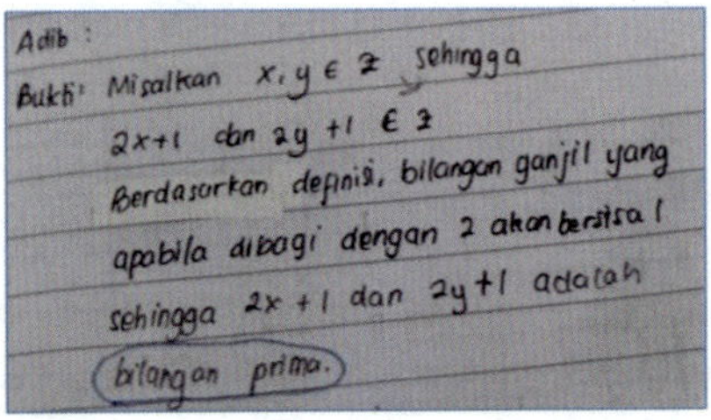

Translation:

It will be proven:

Proof: suppose x,y ∈ Z so that 2x+1 and 2y+1 ∈ Z. By the definition, if an odd number is divided by 2, it will have remainder 1. So, 2x+1 and 2y+1 are prime.

Figure 2. Example of Proof AD1.

Figure 2 explains that the prospective teacher's concept in constructing a proof was not well understood, it could be seen that he could not state the definition correctly. It also indicated in the definition of 2x + 1 and 2y + 1 when he wrote as "prime number". Whereas, based on the definition, the form of '2x + 1'and '2y + 1' were an odd number. By definition, an odd number was a number that had a remainder of 1 when divided by 2 so 2x + 1 and 2y + 1 were an odd number. Based on the data, 6.25% of the total prospective teachers using such typical proving process.

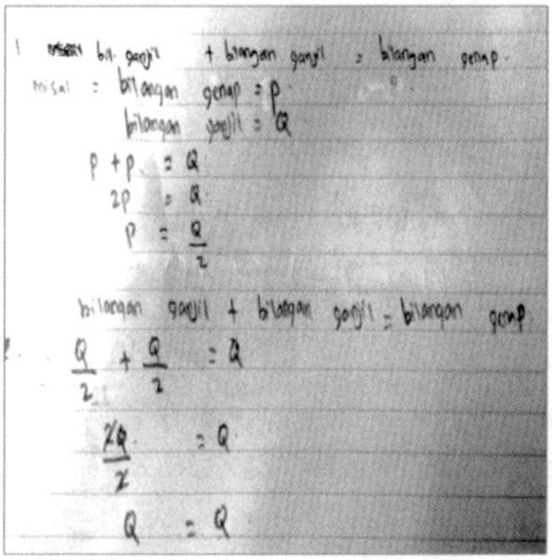

Translation:

Odd number + odd number = even number

Suppose: even number = P

odd number = Q

$P + P = Q$

$2P = Q$

$P = \frac{Q}{2}$

Odd number + odd number = even number

$\frac{Q}{2} + \frac{Q}{2} = Q$

$\frac{2Q}{2} = Q$

Figure 3. Example of Proof AD2.

Figure 3 shows that the prospective teacher was not able to construct proofs because of less understanding of the theorem or the concepts involved. The concept of proof that should be proved was used in the proof. The result of $\frac{Q}{2} + \frac{Q}{2} = Q$, which $\frac{Q}{2}$ should have been proved because it was an even number and it was used in the proof. This happened due to prospective teacher's weak intuitive understanding before starting the proving process, so that, she could not solve proof task formally, logically, and relevantly to the definition. Based on the data, 3.75% of the 240 prospective teachers experienced such proving errors.

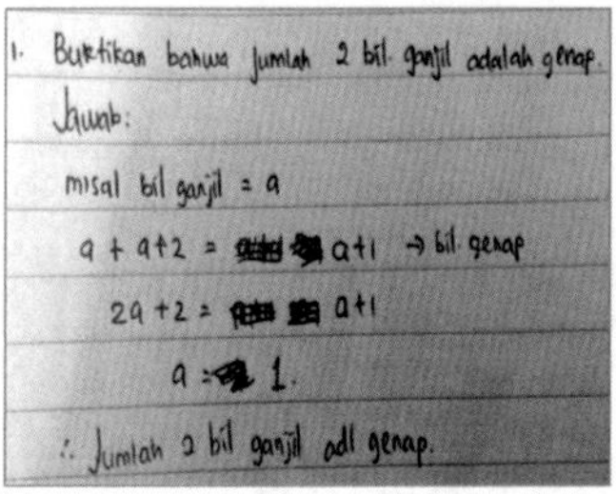

Translation

Prove that the sum of two odd numbers is an even number.

Answer: suppose odd number = a

a+a+2 = a+1 → an even number

2a+2 = a + 1

a = 1

Thus, the sum of two odd numbers is an even number.

Figure 4. Example of Proof AD3.

Table 4. The classification of prospective teachers' answers toward the proof task

Type	Classification: Miyazaki (Types of reasoning) — Correct Proof		Moore (Types of difficulties) — Conceptual understanding: D1		D2		D3		D4		D5		Mathematical language and notation: D6		Getting started on a proof: D7		Total	
	n	%	n	%	n	%	n	%	n	%	n	%	n	%	n	%	n	%
Proof A	38	15.8	15	6.25	9	3.75	6	2.5	5	2.08	66	27.5	1	0.42	8	3.33	148	61.66
Proof B	0	0	1	0.42	0	0	0	0	0	0	0	0	1	0.42	0	0	2	0.84
Proof C	0	0	0	0	6	2.5	6	2.5	63	26.25	0	0	0	0	0	0	86	31.25
Proof D	0	0	12	5	0	0	0	0	0	0	3	1.25	0	0	0	0	4	6.25
Number (n)	38	15.8	28	11.6	15	6.25	12	5	68	28.33	69	28.75	2	0.84	8	3.33	240	100

Figure 4 explained that the prospective teacher's understanding of the concepts in constructing proof was unrevealed, especially on his concept images. The language required to express mathematical ideas in the proof was still insufficient and unclear. He began to construct a proof by letting an odd number as a and then being manipulated in an equation resulting a=1. However, it was not clear to bring the proof into corresponding directions of a valid proof that was the sum of two odd numbers is an even number. Based on the data, 2.5% of the 240 prospective teachers experienced such proving errors.

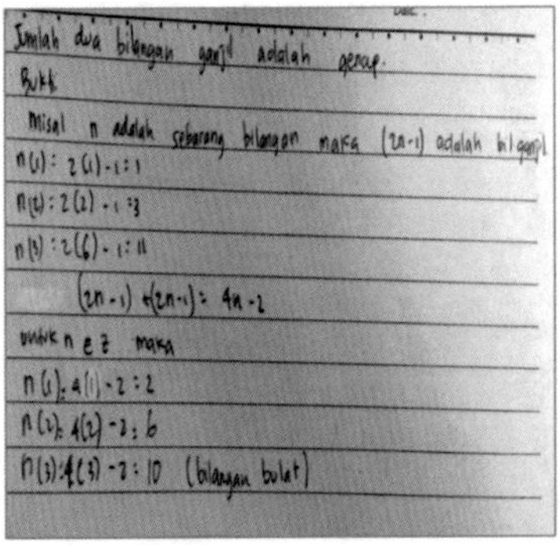

Translation

Prove that the sum of two odd numbers is an even number.

Proof:

Suppose n is an arbitrary number, then (2n-1) is an odd number.

n(1) = 2(1) – 1 = 1

n(2) = 2(2) – 1 = 3

n(3) = 2(6) – 1 = 11

(2n – 1) + (2n – 1) = 4n – 2

For n ∈ Z then n(1) = 4(1) – 2 = 2; n(2) = 4(2) – 2 = 6; n(3) = 4(3) – 2 = 10 (integer).

Figure 5. Example of Proof AD4.

Figure 5 shows that the prospective teacher started to use functional language that was supposed as n, but it was not completed. The results show that the prospective teacher used several examples to understand the concept of constructing a proof. Moreover, he was still unable to build their examples, which were used to construct proofs. Based on the data, 2.08% of the 240 prospective teachers experienced such proving errors.

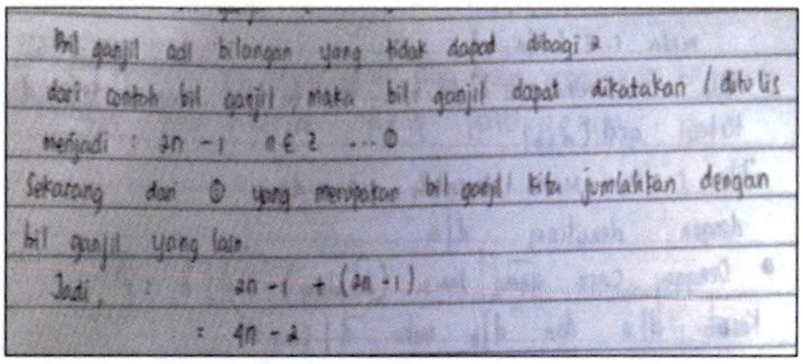
Bil ganjil adl bilangan yang tidak dapat dibagi 2
dari contoh bil ganjil maka bil ganjil dapat dikatakan / ditulis
menjadi : 2n − 1 n ∈ Z ...①
Sekarang dari ① yang merupakan bil ganjil kita jumlahkan dengan
bil ganjil yang lain
Jadi, 2n − 1 + (2n − 1)
= 4n − 2

Translation

Odd number is a number that cannot be divided by 2. Based on this definition, then an odd number can be said or written as 2n-1, n ∈ Z ...(1)

From (1), we will add with the order odd number. Thus, 2n-1 + (2n-1) = 4n-2.

Figure 6. Example of Proof AD5.

Figure 6 shows that the prospective teacher could involve deductive reasoning with a definition, but she still did not know how to use the definition of an odd number correctly. This was indicated by the use of a definition of an odd number namely 2n – 1. Afterwards, she added the other odd number that resulted (2n - 1) + (2n - 1). Based on the sum of these numbers, she used the same variables namely n, however, letting two arbitrary odd numbers with the same variables was not accepted due to the possibility that those two numbers could be different. Based on the data, 27.5% of the total prospective teachers experienced such proving errors.

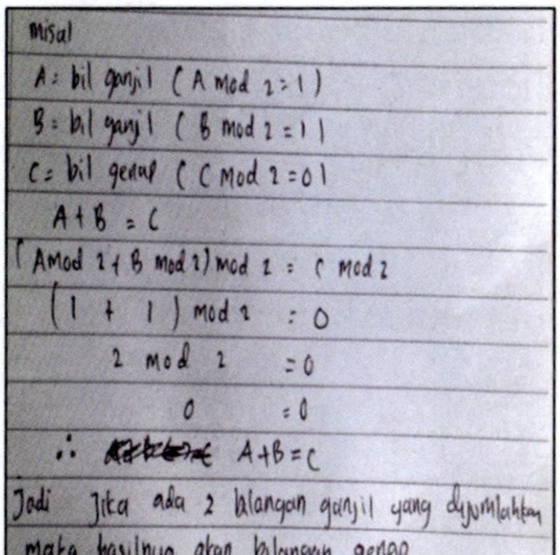
Misal
A: bil ganjil (A mod 2 = 1)
B: bil ganjil (B mod 2 = 1)
C: bil genap (C mod 2 = 0)
A + B = C
(A mod 2 + B mod 2) mod 2 = C mod 2
(1 + 1) mod 2 = 0
2 mod 2 = 0
0 = 0
∴ A + B = C
Jadi jika ada 2 bilangan ganjil yang dijumlahkan maka hasilnya akan bilangan genap.

Translation

Suppose

A = odd number (A mod 2 = 1)

B = odd number (B mod 2 = 1)

C = even number (C mod 2 = 0)

A + B = C

(A mod 2 + B mod 2) mod 2 = C mod 2

(1 + 1) mod 2 = 0

2 mod 2 = 0

0 = 0

Thus, A + B = C

We can conclude that the sum of two odd numbers is an even number.

Figure 7. Example of Proof AD6.

Figure 7 explains that deductive reasoning was correctly used by the prospective teacher's proof. However, the prospective teacher used particular language and mathematical notation incorrectly. This problem could be seen from the notation "=", which meant "is equal" instead of "is equivalent". The symbol was normally used in congruence involving modulo. Therefore, it could be concluded that the prospective teacher did not fully understand the meaning of a notation "=". Based on the data, 0.42% of the total prospective teachers experienced such proving errors.

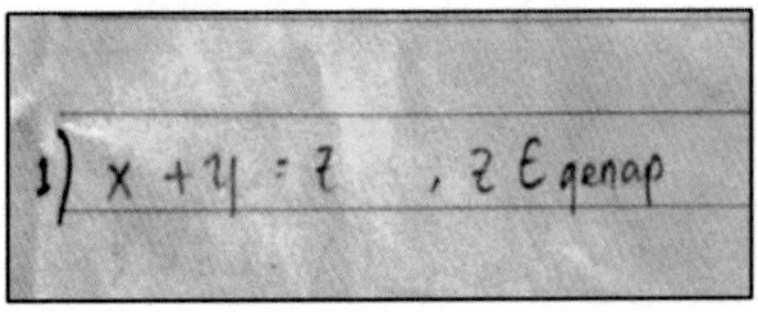

Translation

$x + y = Z$, $Z \in$ even number.

Figure 8. Example of Proof AD7.

Figure 8 shows that the prospective teacher involved deductive reasoning in constructing the proof. The finding $x + y = z$ showed that the prospective teacher used functional language at the beginning of the proof. However, this equation was meaningless due to the poor mathematical argument and interpretation of the symbols given. The prospective teacher in this proof likely had less knowledge of proving so it was difficult to start constructing a proof. Based on the data, 3.33% of the total prospective teachers experienced such typical proving errors.

Proof B

In Proof B, the prospective teachers involved deductive reasoning and other languages, drawings, and movable objects during constructing a proof. In this category, 0.84% of the prospective teachers had difficulties in constructing a proof caused by several things, comprising a less understanding of the concept (0.42%) and less understanding of mathematical notations and language in constructing a proof (0.42%). The following examples show the results of prospective teachers' answers that contained errors in constructing Proof B.

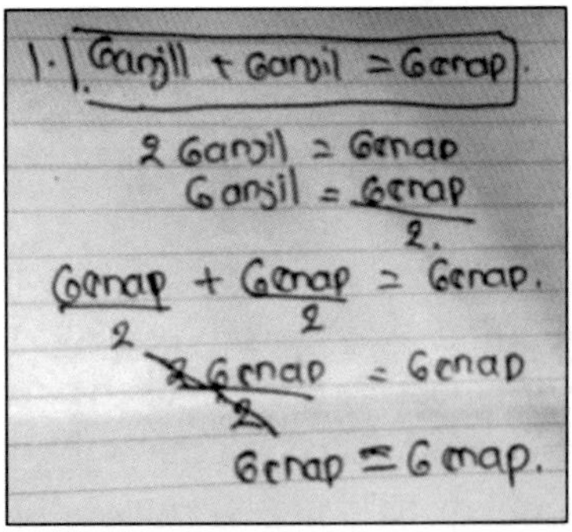

Translation

Odd + odd = even

2 Odd = even

Odd $= \frac{\text{even}}{2}$

$\frac{\text{even}}{2} + \frac{\text{even}}{2} =$ even

$\frac{2\ \text{even}}{2} =$ even

even = even

Figure 9. Example of Proof BD1.

Figure 9 explains that the prospective teacher involved deductive reasoning but did not use functional language in constructing a proof. It could be seen from his work showing that "odd + odd = even". He also could not state the definition correctly as indicated in his definition that "odd $= \frac{even}{2}$". Whereas, when an even number was divided by 2, the result was also an even number. Hence, he performed the proof incorrectly. Based on the data, 0.42% of the total prospective teachers experienced such proving errors.

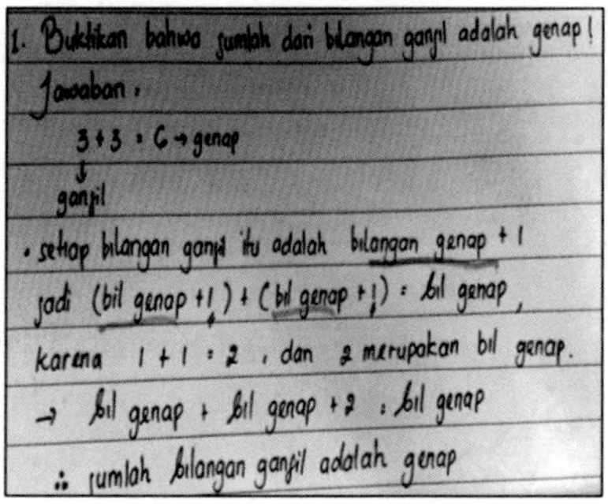

1. Buktikan bahwa jumlah dari bilangan ganjil adalah genap!
Jawaban:
3+3 = 6 → genap
↓
ganjil
• setiap bilangan ganjil itu adalah bilangan genap + 1
jadi (bil genap +1) + (bil genap +1) = bil genap,
karena 1 + 1 = 2, dan 2 merupakan bil genap.
→ bil genap + bil genap + 2 = bil genap
∴ jumlah bilangan ganjil adalah genap

Translation
Prove that the sum of two odd numbers is an even number.
Answer:
3 + 3 = 6, where 3 is an odd number, and 6 is an even number.
Every odd number is even number + 1. Thus, (even number + 1)
+ (even number + 1) = even number, cause 1 + 1 = 2, and 2 is even number.
→ Even number + even number + 2 = even number.
Thus, the sum of odd numbers is an even number.

Figure 10. Example of Proof BD6.

In accordance with Figure 10, it was indicated that the prospective teacher already involved deductive reasoning but she demonstrated the proof by her language without the use of appropriate functional language in constructing a proof. This could be seen from her sentence every odd number is an even number plus one. This sentence should employ some symbols using a functional language, for example, 2n + 1 for an odd number with n integer. Thus, the prospective teacher still did not understand how to use the symbolic language in proof. It could be a result of the limitations of her conceptual understanding about the nature of proof. Based on the data, 0.42% of the total prospective teachers experienced such typical proving errors.

Proof C

In this type of proof, the prospective teachers were not able to prove using inductive reasoning and other languages, drawings, and movable objects. However, 31.25% of the prospective teachers got difficulties in constructing this type of proof caused by less understanding of mathematical concepts. The following example shows student's answer that had difficulties in constructing Proof C.

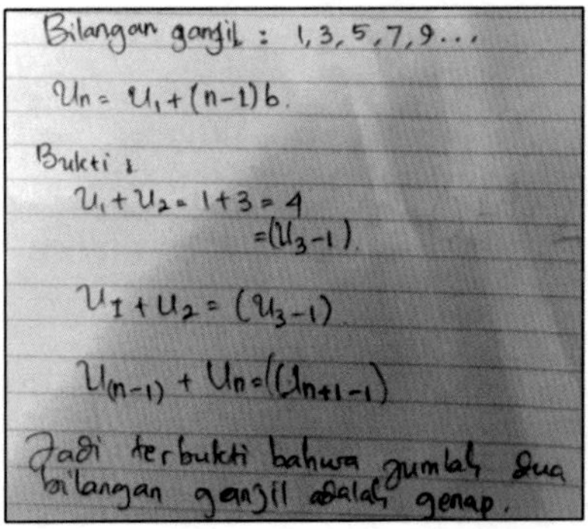

Translation

Odd number = 1, 3, 5, 7, 9,...

$U_n = U_1 + (n-1)b$

Proof:

$U_1 + U_2 = 1 + 3 = 4$

$= (U_3 - 1)$

$U_1 + U_2 = (U_3 - 1)$

$U_{(n-1)} + U_n = (U_{n+1-1})$

Thus, it is proven that the sum of two odd numbers is an even number.

Figure 11. Example of Proof CD2.

Figure 11 shows that the prospective teacher tried to perform inductive reasoning by giving some examples of the number involved in an arithmetic equation at the beginning of stating a proof. Nevertheless, it was unclear that the concept of proof used in constructing a proof was well presented. For example, from the equation U1 + U2 = 1 + 3 = 4 = (U3- 1), the prospective teacher concluded that U(n-1) + U2 = (Un + 1-1). In this case, the prospective teacher still did not understand the whole direction of the proof due to the lack of an intuitive understanding about how a mathematical proof should work. Therefore, he could not finish constructing the proof correctly. Based on the data, 2.5% of the total prospective teachers experienced such typical proving errors.

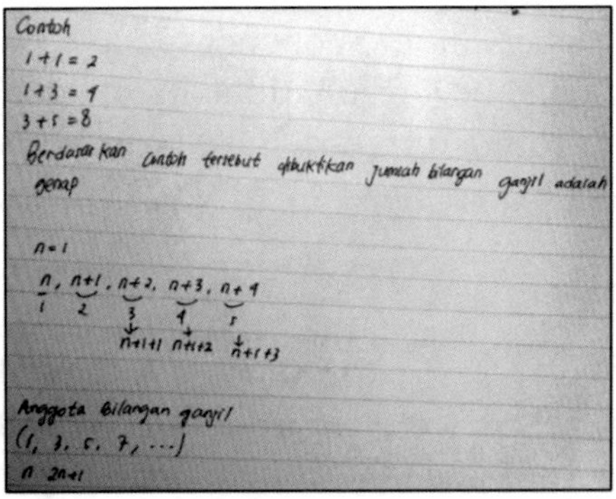

Translation

1 + 1 = 2 1 + 3 = 4 3 + 5 = 8

Based on this example, it is proven that the sum of two odd numbers is an even number.

n = 1

n→1, n+1 → 2, n+2→3→n+1+1, n+3→4→n+1+2, n+4→5→n+1=3

The members of odd number: (1, 3, 5, 7,...)

n 2n+1

Figure 12. Example of Proof CD3.

Figure 12 depicts the prospective teacher performed inductive reasoning by giving an example of numbers. However, it was noted that her understanding of the concept of proof was still unclear. The prospective teacher started from stating the proof by giving some examples, then generalized them into the formal form. Afterwards, she continued to resume her work by providing other examples. It was incompatible with the concept of inductive proof where the valid examples satisfying the condition of a statement should be generalized into a formal conclusion. Based on the data, 2.5% of the total prospective teachers experienced such typical proving errors.

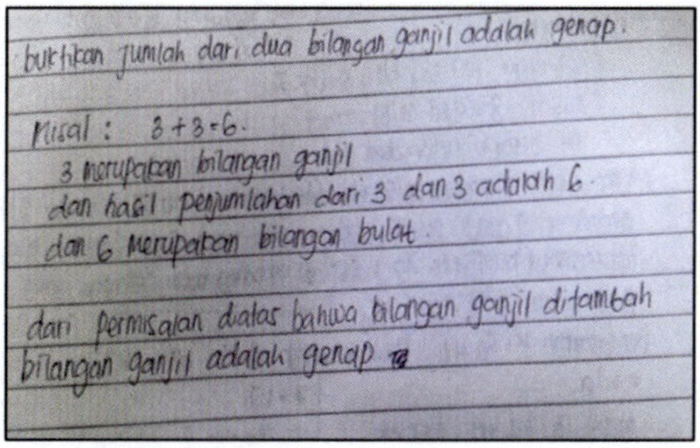

Translation

Prove that the sum of two odd numbers is an even number.

Suppose: 3 + 3 = 6

Where, 3 is an odd number and the result of 3 + 3 = 6, 6 is an integer.

Therefore, the sum of two odd numbers is an even number.

Figure 13. Example of Proof CD4.

Figure 13 portrays that the prospective teacher used examples to construct an inductive proof but she did not complete the proof. She was more inclined to mention some numbers on an arithmetical operation, namely 3 + 3 = 6, but did not conclude her examples of proving to the general form, which was a functional language used in constructing a proof. Based on the data, 26. 25 % of the total prospective teachers experienced such proving errors.

Proof D

In relation with Proof D, the prospective teachers could not perform the proving process correctly based on inductive reasoning and functional language in constructing a proof. The number of prospective teachers who got proving difficulties in this category was 6.25%, in which the problem was caused by a weak understanding of the concept in the proof task. The following examples show the prospective teacher's answers that had difficulties in constructing Proof D.

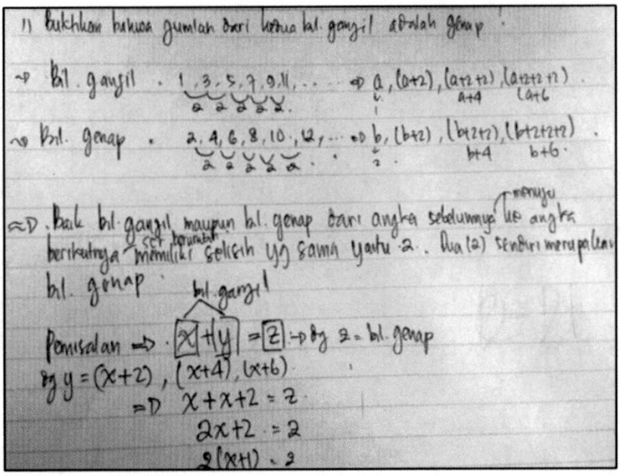

Translation

Prove that the sum of two odd numbers is an even number.

Odd numbers = 1, 3, 5, 7, 9, 11, ... → = a, (a + 2), (a+2+2), (a+2+2+2)
= a, a + 2, a +4, a + 6

Even numbers = 2, 4, 6, 8, 10, 12, ...

→ = b, (b + 2), (b+2+2), (b+2+2+2)

= b, b + 2, b + 4, b + 6

For either odd or even numbers, the difference between every two consecutive number is 2. And, 2 is an even number. Where x, y are odd numbers and z is an even number.

With y = (x+2), (x+4), (x+6)

x + x + 2 = z; 2x + 2 = z; 2(x+1) = z

Figure 14. Example of Proof DD1.

Figure 14 shows that the prospective teacher did not understand how to define an odd number in the form of functional language. It could be seen from the student's answer on the definition of even and odd numbers. The prospective teacher gave a series of example on how odd numbers and even numbers are illustrated as evidence that he started proving inductively. Despite he tried to arrange the general form of the examples, which was the series a, (a + 2), (a+2+2), (a+2+2+2) for even number, and b, (b + 2), (b+2+2), (b+2+2+2) for odd number. However, it did not show how an odd number and an even number should be mathematically symbolized. Thus, his final step, which was 2(x+1) = z yielded a variety of interpretations, did not certainly describe the condition expected in the proof task. Based on the data, 5 % of the total prospective teachers experienced such proving errors.

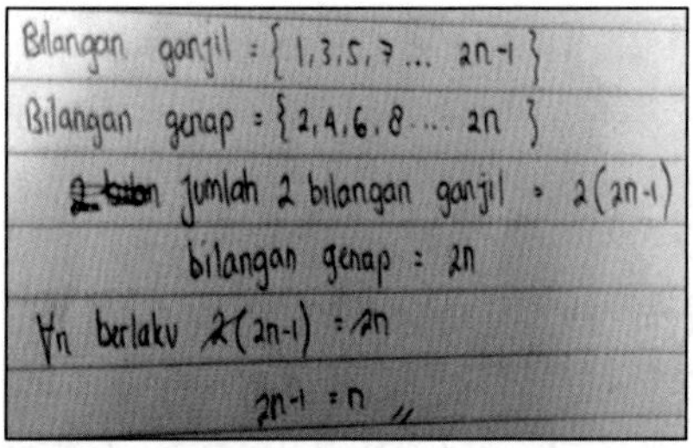

Translation

Odd numbers = {1, 3, 5, 7, … 2n – 1}

Even numbers = {2, 4, 6, 8, … 2n}

The sum of two odd numbers = 2(2n – 1)

Even numbers = 2n

∀ n apply 2(2n – 1) = 2n

2n – 1 = n

Figure 15. Example of Proof DD5.

Figure 15 portrays the prospective teacher used inductive reasoning but he still did not know how to use the definition correctly. The figure also shows (2n-1) = n that indicated that he did not use the definition of odd number correctly. He canceled number 2 on the right and left instead of using the definition of an odd and even number that he had written. Based on the data, 1.25% of the total prospective teachers experienced such proving errors.

Our finding indicates that the prospective teachers apply deductive and inductive methods in constructing a proof. The deductive method consists of two types namely Proof A and Proof B and the inductive method consists of two types covering Proof C and Proof D. Table 4 points out that 62.5% of the prospective teachers use deductive method while 37.5% of them use inductive method. Meaning that, more than half of the prospective teachers' answers use deductive method. Despite some of the prospective teachers having errors in constructing a proof, they already try to construct a proof with deductive and inductive methods. This finding is consistent with the research conducted by Miyazaki (2000) that most students in his study use a deductive method instead of the inductive one in constructing a proof.

Furthermore, the prospective teachers perform Proof A, Proof B, Proof C, and Proof D types with the percentage of 61.66%, 0.84%, 31.25%, and 6.25%, respectively. Therefore, it shows that Proof A is the most commonly found in the prospective teachers' answers than those of other types. When compared to the findings undertaken by Kögce et al. (2010), it does not align with the results of Kögce et al. (2010), in which the study result reports that Proof C is performed by most students than the other types of proof (51.2%). The fact that our study has found many deductive methods in our participants' answers might occur because the proof task given in the present study demands a solver to use a deductive method instead of an inductive one. However, in connection with the results of the prospective teacher's answers, there are some answers indicating an inductive method, all of which are still incorrect. This finding very likely corresponds to Demiray and Bostan (2017) who report that most of the students' incomplete proof yielded incorrect proofs are caused by the unsuitable inductive method the students use in constructing their proofs. Therefore, it indicates that the type of proof task affects the selection of proving methods.

The results of the present study are also consistent with Miyazaki's (2000) study, where Proof A is performed by the highest number of prospective teachers. The results show that prospective teachers can use deductive reasoning and sufficient techniques of proof, however, there are still some errors in constructing a proof. In addition, Miyazaki (2000) points out that Proof C is performed by the least number of prospective teachers. This does not align with the present study that reveals the fact that Proof C is performed by the second-highest number of prospective teachers. This result shows that the prospective teachers still involve inductive reasoning with other languages, drawings, and objects used in the process of constructing a proof. In this regard, promoting deductive argumentation among students in mathematics education is important since many prospective teachers, including those who enrolled postgraduate study program in mathematics education, often perform logically disconnected premises and conclusions drawn within a mathematical proof (Ndemo, 2019).

Regarding the difficulties in proving, the results show that AD5 type is performed by 27.5% of the prospective teachers, which becomes the highest result of proof. Similarly, Edwards and Ward (2004) convey that many prospective teachers cannot use the definition to make mathematical proofs. Aligned with Moore's category, the use of definition is indeed being one of the difficulties in constructing a mathematical proof. In addition, the present study shows that the number of prospective teachers performing incorrect proofs is bigger than those who make the correct ones. That is, most prospective teachers still experience difficulties in constructing a mathematical proof. 79.98% of the prospective teachers are still weak in understanding the concept of proof, 3.33% of them lack of knowledge, and 0.84% of them get limited ability to construct a proof. This study is consistent with Chin and Lin's (2009) study revealing that most prospective teachers have problems in constructing valid algebraic proofs.

The present study not only indicate the performance of prospective teachers regarding types of reasoning and difficulties in proving processes, but also show the potential use of analytical framework of assessing individuals' proving performance. This framework is developed in order to get a broader insight on how to evaluate types of proof and proving difficulties from a written response representing individual proof performance. It is expected that this analytical framework will complement other frameworks in assessing individual cognitive performance related to proof, such as mathematical knowledge for teaching proof (Buchbinder & McCrone, 2020) as the knowledge of different types of proofs becomes one of vital components in this framework. However, the combined Miyazaki-Moore method used in the framework is carried out to only assess students' written responses. In future, it is more beneficial if the data include interview results to confirm the detailed information about the students' difficulties in constructing proofs.

CONCLUSION

The most prospective teachers construct a proof by employing deductive reasoning rather than inductive reasoning. It can be seen in Proof A that has been performed by the highest number of prospective teachers. However, some prospective teachers still experience difficulties in constructing a mathematical proof. The types of difficulties mostly found in the prospective teachers' answers include the fact that they cannot appropriately use the definition in making mathematical proofs.

This study only presents the prospective teachers' responses in constructing a proof because the researchers want to know the trend of mapping models in assessing prospective teachers regarding their knowledge about proof constructions. They also have empirically proven that the framework proposed in this study can work. The advantages of using the framework cover the ability to assess students' types of reasoning and difficulties in constructing proofs simultaneously.

As the constructive feedbacks, the framework can be used as an evaluation tool for the needs of mathematics teacher education program in a university curriculum. For further studies, the present study offers a potentially broader insight on assessing learners' cognitive processes to study learners' reasoning process in mathematical proof regarding proving difficulties and types of reasoning since the framework developed in this study has not covered such issue yet. In addition, this framework is intended to only code the responses based on the participants, meaning that every single response can only get chance to be coded in one category of proof based on types of reasoning and difficulties. Thus, it is suggested that the framework can be developed into covering more than one category of proof since, for example, a response from another mathematical proof task may be categorized in more than one type of difficulties.

REFERENCES

1. Almeida, D. (2000). A survey of mathematics undergraduates' interaction with proof: some implications for mathematics education. International Journal of Mathematical Education in Science and Technology, 31(6), 869-890. https://doi.org/10.1080/00207390050203360
2. Ayalon, M., & Even, R. (2008). Deductive reasoning: in the eye of the beholder. Educational Studies in Mathematics, 69(3), 235-247. https://doi.org/10.1007/s10649-008-9136-2
3. Baki, A. (2008). Mathematics education from theory to practice. Ankara: Harf Educational Publications
4. Balacheff, N. (2010). Bridging knowing and proving in mathematics: A didactical perspective. In G. Hanna, H. N. Jahnke & H. Pulte (Eds.), Explanation and Proof in Mathematics—Philosophical and Educational Perspectives (vol. 45, pp. 115-135). New York: Springer. https://doi.org/10.1007/978-1-4419-0576-5_9
5. Blanton, M. L., Stylianou, D. A., & David, M. M. (2003). The nature of scaffolding in undergraduate students' transition to mathematical proof. In N. Pateman, B. J. Dougherty, & J. Zilliox (Eds.), Proceedings of the 27th Annual Meeting of the International Group for Psychology in Mathematics Education (vol. II, pp. 113-120). Honolulu: University of Hawaii.
6. Buchbinder, O., & McCrone, S. (2020). Preservice teachers learning to teach proof through classroom implementation: Successes and challenges. The Journal of Mathematical Behavior, 58, 100779. https://doi.org/10.1016/j.jmathb.2020.100779
7. Carrillo-yañez, J., Climent, N., Montes, M., Contreras, L. C., Flores-medrano, E., Escudero-ávila, D., ... & ribeiro, M. (2018). The mathematics teacher's specialised knowledge (MTSK) model. Research in Mathematics Education, 20(3), 236-253. https://doi.org/10.1080/14794802.2018.1479981
8. Chin, E., & Lin, F. (2009). A comparative study on junior high school students' proof conceptions in algebra between Taiwan and the UK. Journal of Mathematics Education, 2(2), 52-67.
9. Clark, M., & Lovric, M. (2008). Suggestion for a theoretical model for secondary–tertiary transition in mathematics. Mathematics Education Research Journal, 20(2), 25–37. https://doi.org/10.1007/BF03217475

10. Christou, C., & Papageorgiou, E. (2007). A framework of mathematics inductive reasoning. Learning and Instruction, 17(1), 55-66. https://doi.org/10.1016/j.learninstruc.2006.11.009
11. De Villiers, M. D. (1990). The role and function of proof in mathematics. Pythagoras, 24, 17-24.
12. Demiray, E., & Bostan, M. I. (2017). An investigation of pre-service middle school mathematics teachers' ability to conduct valid proofs, methods used, and reasons for invalid arguments. International Journal of Science and Mathematics Education, 15(1), 109-130. https://doi.org/10.1007/s10763-015-9664-z
13. Dickerson, D. S., & Pitman, D. J. (2016). An examination of college mathematics majors' understandings of their own written definitions. The Journal of Mathematical Behavior, 41, 1-9. https://doi.org/10.1016/j.jmathb.2015.11.001
14. Edwards, B. S., & Ward, M. B. (2004). Surprises from mathematics education research: Student (mis) use of mathematical definitions. The American Mathematical Monthly, 111(5), 411- 424. https://doi.org/10.1080/00029890.2004.11920092
15. Epp, S. S. (2003). The role of logic in teaching proof. The American Mathematical Monthly, 110(10), 886-899. https://doi.org/10.2307/3647960
16. Hanna, G., De villiers, M., Arzarello, F., Dreyfus, T., Durand- Guerrier, V., Jahnke, N.H., (...) Yevdokimov, O. (2009). ICMI Study 19: Proof and Proving in Mathematics Education (Discussion Document). In Lin, Fou-Lai (Eds.). Proceeding of The ICMI Study 19 Conference: Proof and Proving in Mathematics Education (vol.1). Taipei: The Department of Mathematics, National Taiwan Normal University
17. Jones, K. (1997). Student-teachers' conceptions of mathematical proof. Mathematics Education Review, 9, 21–32.
18. Kögce, D., Aydin, M. & Yildiz, C. (2010). The views of high school student about proof and their levels of proof (the case of Trabzon). Procedia Social and Behavioral Sciences, 2, 2544-2549. https://doi.org/10.1016/j.sbspro.2010.03.370
19. Knuth, E. J. (2002a). Teachers' conceptions of proof in the context of secondary school mathematics. Journal of Mathematics Teacher Education, 5(1), 61-88. https://doi.org/10.1023/A:1013838713648

20. Knuth, E. J. (2002b). Secondary school mathematics teachers' conceptions of proof. Journal for Research in Mathematics Education, 33(5), 379–405. https://doi.org/10.2307/4149959

21. Landis, J. R., & Koch, G. G. (1977). The measurement of observer agreement for categorical data. Biometrics, 33(1), 159-174. https://doi.org/10.2307/2529310

22. Lee, H. S. (2002). Optimal consensus of fuzzy opinions under group decision making environment. Fuzzy sets and systems, 132(3), 303-315. https://doi.org/10.1016/S0165-0114(02)00056-8

23. Martin, W. G., & Harel, G. (1989). Proof frames of preservice elementary teachers. Journal for Research in Mathematics Education, 20(1), 41–51. https://doi.org/10.2307/749097

24. Miles, M. B., & Huberman, A. M. (1994). Qualitative data analysis: An expanded sourcebook. Thousand Oaks, CA: Sage

25. Miyazaki, M. (2000). Levels of proof in lower secondary school mathematics. Educational Studies in Mathematics, 41(1), 47-68. https://doi.org/10.1023/A:1003956532587

26. Moore, R.C. (1994). Making the transition to formal proof. Educational Studies in Mathematics, 27, 249-266

27. Moral?, S., U?urel, I., Türnüklü, E. B. ve Ye?ildere, S. (2006). Matematik ö?retmen adaylar?n?n ispat yapmaya yönelik görü?leri. Kastamonu E?itim Dergisi, 14(1), 147-160.

28. Morris, A. K. (2002). Mathematical reasoning: Adults' ability to make the inductive-deductive distinction. Cognition and Instruction, 20(1), 79-118. https://doi.org/10.1207/S1532690XCI2001_4

29. Morselli, F. (2006). Use of examples in conjecturing and proving: An exploratory study. International Group for the Psychology of Mathematics Education, 4, 185.

30. National Council of Teachers of Mathematics [NCTM]. (2000). Principles and standards for school mathematics. Reston, VA: NCTM.

31. Ndemo, Z. (2019). Flaws in Proof Constructions of Postgraduate Mathematics Education Student Teachers. Journal on Mathematics Education, 10(3), 379-396. https://doi.org/10.22342/jme.10.3.7864.379-396

32. Ozdemir, E., & Ovez, F. T. D. (2012). A research on proof perceptions and attitudes towards proof and proving: Some implications for elementary mathematics prospective teachers. Procedia-Social and Behavioral Sciences, 46, 2121-2125. https://doi.org/10.1016/j.sbspro.2012.05.439
33. Özer, Ö., & Arõkan, A. (2002). Students' levels of doing proof in high school mathematics classes. Paper presented at the meeting of 5th National Science and Mathematics Education Congress (pp. 1083-1089). Ankara: Middle East Technical University.
34. Selden, A., & Selden, J. (2007). Overcoming students' difficulties in learning to understand and construct proofs (Report No. 2007-1). Cookeville: Mathematics Department, Tennesse Technological University.
35. Selden, J., Benkhalti, A., & Selden, A. (2014). An analysis of transition-to-proof course students' proof constructions with a view towards course redesign. In Proceedings of the 17th Annual Conference on Research in Undergraduate Mathematics Education (pp. 246-259).
36. Shaker, H., & Berger, M. (2016). Students' difficulties with definitions in the context of proofs in elementary set theory. African Journal of Research in Mathematics, Science and Technology Education, 20(1), 80-90. https://doi.org/10.1080/10288457.2016.1145449
37. Shulman, L. S. (1986). Those who understand: Knowledge growth in teaching. Educational Researcher, 15(2), 4-14. https://doi.org/10.3102%2F0013189X015002004
38. Simon, M. A., & Blume, G. W. (1996). Justification in the mathematics classroom: A study of prospective elementary teachers. The Journal of Mathematical Behavior, 15(1), 3-31. https://doi.org/10.1016/S0732-3123(96)90036-X
39. Smith, J. C. (2006). A sense-making approach to proof: Strategies of students in traditional and problem-based number theory courses. The Journal of Mathematical Behavior, 25(1), 73-90. https://doi.org/10.1016/j.jmathb.2005.11.005
40. Stavrou, S. G. (2014). Common Errors and Misconceptions in Mathematical Proving by Education Undergraduates. In the Undergraduate Mathematics Preparation of School Teachers: The Journal, 1, 1-8

41. Steele, M. D., & Rogers, K. C. (2012). Relationships between mathematical knowledge for teaching and teaching practice: The case of proof. Journal of Mathematics Teacher Education, 15(2), 159-180. https://doi.org/10.1007/s10857-012-9204-5

42. Stylianides, A. J. (2007). The notion of proof in the context of elementary school mathematics. Educational Studies in Mathematics, 65, 1-20. https://doi.org/10.1007/s10649-006-9038-0

Chapter

8

FLAWS IN PROOF CONSTRUCTIONS OF POSTGRADUATE MATHEMATICS EDUCATION STUDENT TEACHERS

Zakaria Ndemo

Bindura University of Science Education, 741 Chimurenga road, Bindura, Zimbabwe

ABSTRACT

Intending to improve the teaching and learning of the notion of mathematical proof this study seeks to uncover the kinds of flaws in postgraduate mathematics education student teachers. Twenty-three student teachers responded to a proof task involving the concepts of transposition and multiplication of matrices. Analytic induction strategy that drew ideas from the literature on evaluating students' proof understanding and Yang and Lin's model of proof comprehension applied to informants' written responses to detect the kinds of flaws in postgraduates' proof attempts. The study revealed that the use of empirical verifications was dominant and in situations. Whereby participants attempted to argue using arbitrary

Citation: (APA): Ndemo, Z. (2019). Flaws in Proof Constructions of Postgraduate Mathematics Education Student Teachers. Journal on Mathematics Education, 10(3), 379-396.(18 pages).

Copyright: © Creative Commons Attribution 4.0 International License (http://creativecommons.org/licenses/by/4.0/).

mathematical objects, the cases considered did not represent the most general case. Flawed conceptualizations uncovered by this study can contribute to efforts directed towards fostering strong subject content command among school mathematics teachers.

Keywords: Mathematical proof; transpose and multiplication of matrices; flawed conceptualisations; levels of proof comprehension

Over the past decades research on mathematical proof has gained increased attention and most research studies have revealed that student teachers have a fragile understanding of mathematical proof (Bieda, 2010; Lesseig, 2016; Mejia-Ramos, & Inglis, 2009; Maya & Sumarmo, 2011; Noto, et al. 2019). However, a huge number of those studies were based on convincement issues. In other words, studies sought to determine how convincing a given argument would be to the participant (e.g., Bleiler, Thompson, & Krajčevski, 2014; Martin & Harel, 1989). Hence, there is scarcity of research that examines how students construct proofs of mathematical statements, particularly on students' competencies in resolving proof tasks.

Precisely, the argument is that little is known about students' reasoning as they immerse themselves with proof tasks as instructional and assessment strategies have tended to promote memorization and regurgitation of lecture notes (Ndemo, Zindi, & Mtetwa, 2017; Stylianou, Blanton, & Rotou, 2015). Hence, if we conceive proving as a problem solving process then arguments generated by students when they engage with proof tasks should illuminate the kinds of students' thoughts about mathematical proof (Lee & Smith, 2009). I stark contrast, many school and university students and even teachers of mathematics have only a superficial grasp of the idea of a mathematical argument (Jahnke, 2007; Brown & Stillman, 2009; Saleh, et al. 2018). Yet prospective secondary mathematics teachers need to exit teacher preparation with a firm grasp of the concepts of the concepts of transposition and multiplication of matrices (Brown & Stillman, 2009).

The notion of mathematical proof has persistently caused severe discomforts among teachers and learners at many scholastic levels. Hence, it is necessary that teachers engage in substantial learning of the concept of mathematical proof. Furthermore, mathematics education that encourages student teachers to engage in autonomous proof constructions is crucial for their learning in order to build their capacity to explain the concept in a persuasive manner to their future students (Jones, 1997).

To bring the research problem in proper perspective the researcher refers to Harel, Selden and Selden's (2006) comment about students' struggles with the notion of mathematical proof. Harel, et al. (2006) wrote:

We know where the students are, we know where the mathematicians are, but we just don't know how to get mathematics students from where they are to where we want them to be (p. 148).

Harel et al.'s quote points to an undesirable gap between students' and experts' understandings of mathematical proof. Hence, one of the primary goals of mathematics instruction at tertiary level is to promote among student teachers conceptions of mathematical proof held by expert mathematicians. Therefore, mathematics education instruction should aim to enhance expert conceptions of the concept of mathematical proof among student teachers. In this regard, the researcher was impelled to explore student teachers' thinking as reflected in in-class mathematics problem solving tasks as student teachers engage with concepts deemed to be within their conceptual reach. The research problem is explained in the next section.

Statement of the Problem

In-service student teachers do not have a firm grasp and appreciation of the idea that proofs that explain can be more elucidating and help to foster justification. Doing proofs at school level and even at undergraduate level has been characterised by students and instructors resorting to rote memorisation and regurgitation of instructor notes. The researcher argues that if comprehension tests only ask students to regurgitate memorised facts then such students are likely to develop a superficial understanding of those mathematical facts (Mejia-Ramos, et al. 2012). Further, such students are likely to emphasise form over substance, that is, the ritual proof scheme becomes dominant (Harel & Sowder, 2007). Yet, teachers need a flexible and firm understanding of mathematics content they are supposed to teach (Shahrill, et al. 2018; Prahmana & Suwasti, 2014). The researcher reiterates that the notion of mathematical proof has been reported to develop reasoning, that is, the ability to think rationally and logically among learners. However, the ability to reason can be impeded by flaws in students' mental representations of the concept of mathematical proof (Garret, 2013).

Promoting argumentation skills can illumine the kinds of flaws in students' thoughts about the concepts of transposition and multiplication of matrices. It is in an argument that we likely to find the most significant way in which higher order thinking can manifest during mathematics learning

(Jonassen & Kim, 2010; Putri & Zulkardi, 2018; Ahmad, et al. 2018). The goal of the current study is to gain insights into the kinds of limitations in postgraduate students' argumentation schemes__chunks of reasoning with respect to the notions of transpose and multiplication of matrices. Pertinent questions that therefore, come to mind in this respect are: how can we develop an understanding of the flaws in student teachers' mental representations of the idea of proof? What is the nature of these limitations among in-service mathematics teachers?

Generating answers to these questions can contribute to useful ideas for teacher education in Zimbabwe. While many studies on students' discomforts with the concept of a mathematical proof have been carried out most of such studies have been based on arguments participants find convincing from those availed by researchers (e.g., Bleiler, et al. 2014; Matin & Harel, 1989). Therefore there is scarcity of empirical studies based on students' own proof constructions, that is, their own actual voices which in turn could illuminate their thinking processes as they engage with proof and proving (Duval, 2006; MejiaRamos & Inglis, 2009; Ndemo, Zindi, & Mtetwa, 2017; Mumu, et al. 2018). This study intends to respond to this dearth in studies grounded in students' own proof constructions by addressing the research question, such as what kinds of flaws characterise postgraduate students' conceptions of the concepts of transposition and multiplication of matrices?

By addressing this question the study may provide mathematics teachers with a clearer picture of what is needed to help students to develop a good command of the concept of mathematical proof in order to teach it effectively to their future students. Furthermore, it was anticipated that evaluating postgraduate students' understandings of the notions of transposition and multiplication of matrices could in turn inform teachers and mathematics educators of what specific aspects they understand and what aspects they do not understand.

Furthermore, proof serves as a vehicle for discovering new mathematical ideas and if learners properly grasp the notion of proof then they learn from it (Mejia-Ramos, et al. 2012). Mathematical proof also serves the purpose of promoting reasoning skills (Weber & Mejia-Ramos, 2015), which in turn would contribute to the student's cognitive development about the concept of mathematical proof. In this regard, the study aims to inject new ideas into the growing body of theoretical frameworks and methodologies for understanding mathematical concepts.

Matrix Theory

If A is a mxn matrix of a field K, of scalars then the transpose of matrix A is the nxm matrix whose rows are the columns of A in same order (Goodaire, 2014). The transpose of matrix A is denoted A . Hence, if $A = (a_{ij})$ then $A^t = (a_{ji})$ (Lipschutz, 1991). In other words, if

$$A = \begin{pmatrix} a_{11} & a_{12} & \cdots & a_{1n} \\ a_{21} & a_{22} & \cdots & a_{2n} \\ \cdots & \cdots & \cdots & \cdots \\ a_{m1} & a_{m2} & \cdots & a_{mn} \end{pmatrix} \text{ then } A^t = \begin{pmatrix} a_{11} & a_{21} & \cdots & a_{m1} \\ a_{12} & a_{22} & \cdots & a_{m2} \\ \cdots & \cdots & \cdots & \cdots \\ a_{1n} & a_{2n} & \cdots & a_{mn} \end{pmatrix}.$$

First, we observe that the product AB of two matrices is somewhat complicated and hence there is need to describe prerequisite ideas for the definition of the product of matrices. The product $A \cdot B$ of a row matrix $A = (a\ )$ and a column matrix $B = (b_i)$ is defined as:

$$A \cdot B = \begin{pmatrix} a_1 \\ \vdots \\ a_n \end{pmatrix} \begin{pmatrix} b_1 \\ \vdots \\ b_n \end{pmatrix} = a_1 b_1 + a_2 b_2 + \cdots + a_n b_n = \sum_{k=1}^{n} a_k b_k$$

Second, we observe that the row matrix A and the column B should have the same number of elements for the product $A \cdot B$ to be defined. Finally, the product $A \cdot B$ is a scalar or a $1x1$ matrix (Lipschutz, 1991).

The researcher now describes the definition of a product of two matrices. Suppose $A = (a_{ij})$ and $B = (b_{ij})$ are matrices over the field of scalars K, the product AB is defined if the number of columns of matrix A is equal to the number of rows of matrix B. If the number of columns of matrix A is equal to the number of rows of B we say the two matrices A and B are of compatible sizes (Goodaire, 2014). Thus, if matrices A and B are compatible say A is an mxp and B is a pxn matrix then the product AB is an mxn matrix whose ij-th entry is obtained by multiplying the i −th row A_i of the matrix A by the j −th column, B , of the matrix B. That is: $AB = \begin{pmatrix} A_1 \cdot B^1 & A_1 \cdot B^2 & \cdots & A_1 \cdot B^n \\ A_2 \cdot B^1 & A_2 \cdot B^2 & \cdots & A_2 \cdot B^n \\ \cdots & \cdots & \cdots & \cdots \\ A_m \cdot B^1 & A_2 \cdot B^2 & \cdots & A_m \cdot B^n \end{pmatrix}$. Alternatively, we can write the product as:

$$AB = \begin{pmatrix} a_{11} & a_{12} & \cdots & a_{1p} \\ a_{21} & a_{22} & \cdots & a_{2p} \\ \cdots & \cdots & \cdots & \cdots \\ a_{m1} & a_{m2} & \cdots & a_{mp} \end{pmatrix} \begin{pmatrix} b_{11} & b_{12} & \cdots & b_{1n} \\ b_{21} & b_{22} & \cdots & b_{2n} \\ \cdots & \cdots & \cdots & \cdots \\ a_{p1} & a_{p2} & \cdots & a_{pn} \end{pmatrix} = \begin{pmatrix} c_{11} & c_{12} & \cdots & c_{1n} \\ c_{21} & c_{22} & \cdots & c_{2n} \\ \cdots & \cdots & \cdots & \cdots \\ c_{m1} & c_{m2} & \cdots & c_{mn} \end{pmatrix}, \text{ where } c_{ij} =$$

$a_{i1}b_{1j} + a_{i2}b_{2j} + \cdots + a_{ip}b_{pj}$ for $i = 1, 2, \ldots, m$ and $j = 1, 2, \ldots, n$

The researcher emphasizes that if A and B are not compatible then the product AB is not defined. So if A is an mxp matrix and B is a qxn matrix then the product AB exists only if $p = q$. This study sought to explore how postgraduate students could apply the notions of transpose and multiplication of matrices to determine whether the product $A\,{}^{t}A$ exists.

Yang and Lin's (2008) Model

The goal of the study was to uncover the sorts of flaws in postgraduate students' proof attempts. To assess in-service teachers' understandings of mathematical proof the researcher drew ideas from a model for assessing comprehension of mathematical proof by Yang and Lin (2008). Yang and Lin introduced what has come to be known as a model of reading comprehension of geometric proof (RCGP). The model consists of four levels which represent increasing levels of cognition. Next, the researcher briefly describes these levels.

First, there is the surface level of the RCGP model whereby a prover acquires basic knowledge regarding the meaning of the statements and symbols in the proof. For example, for the theorem: A real sequence converges to a real number L if given $\varepsilon > 0$, the interval $(L - \varepsilon, L + \varepsilon)$ contains all but infinitely terms of the sequence (a_n), surface level understanding of this theorem can include the conception of ε as a small radius, an understanding of a finite set and the basic idea that a sequence is a mapping with domain the set of natural numbers, ($\aleph$) and range in real numbers, ($\mathbb{R}$). The second level has been called recognising elements. At this level a prover should be capable of identifying the logical status of statements that are explicitly or implicitly involved in the proof construction exercise. The researcher now uses the example given at the surface level that concerns a theorem on the characterisation of converging sequences in $\mathbb{R}$ to describe proof understanding anticipated at this level. At this second level of Yang and Lin's YCGP model, a prover should recognise that for every $\varepsilon > 0$, there is a natural number, $N(\varepsilon)$, dependent on ε such that if $n > N(\varepsilon)$ then $|a_n - L| < \varepsilon$. A prover's chunk of reasoning at the second level should show also awareness that the interval $(L - \varepsilon, L + \varepsilon)$ contains infinitely many terms of the sequence $(a\ )$.

At the third level there is what Yang and Lin refer to as chaining of elements. Central at this level is the fact a prover demonstrates his/her understanding of the way in which different statements are connected in the proof by identifying the logical relations between them (Mejia-Ramos et al., 2012). Following up on the example given to illustrate the YCGP model, the researcher describes the logical relations between statements a prover should depict in his/her argumentation. A prover should make the connection that since $\varepsilon > 0$ then there is a natural number, $N(\varepsilon)$, such that if $n > N(\varepsilon)$ then

$|a_n - L| < \varepsilon$

$\Rightarrow -\varepsilon < a_n - L < \varepsilon$

$\Rightarrow L - \varepsilon < a_n < L + \varepsilon$

$\Rightarrow a_n \epsilon\ (L - \varepsilon, L + \varepsilon)$ for $n > N(\varepsilon)$.

Hence, a prover who would have attained the third level of Yang and Lin's model of reading of comprehension of geometric proof (RCGP) should make a series of logical inferences illustrated. The third level of Yang and Lin's model is similar to Azarello's (2007) conceptualisation of proof as a sequence of inferences. Azarello (2007) in Mejia-Ramos and Weber (2014) view proof as a series of claims of the form $P_1 \rightarrow P_2 \rightarrow P_3 \rightarrow \cdots \rightarrow P_n$, where Pn is the consequent statement or conclusion while, $P_1, P_2, \ldots, P_{n-1}$ constitute the premises of the proof construction. In this conceptualisation of composing proofs the focus is on how each new inference is derived from the previous inference, for instance, how does claim P_3 lead to claim P_4 (Mejia-Ramos & Weber, 2014).

Finally, Yang and Lin's RCGP model has a fourth level called encapsulation whereby a prover is anticipated to interiorize a proof holistically and develop an appreciation and understanding of its application in other contexts (Mejia-Ramos & Weber, 2014). The researcher illustrates the encapsulation phase using the example of real sequence already given. A prover at the fourth level of proof understanding can draw ideas from this theorem and use them together with ideas from the completeness property of $\mathbb{R}$ to prove our first criterion for convergence that: A bounded monotone sequence converges. In their study, Yang and Lin focused on the first three levels of their model. The current study that involves postgraduate students seeks to explore those students' understanding of ideas drawn from Elementary Linear Algebra of Matrices. The concepts had been covered during their studies at undergraduate level. Hence, the study seeks to explore elements of the encapsulation level of Yang and Lin's model in postgraduates' solution attempts to the task.

Other Useful Ideas on Evaluating Students' Understanding of Mathematical Proof

Mejia-Ramos and Weber (2014) and Mamon-Downs and Downs (2013) concur that one way to evaluate whether a student understands a proof is by determining the student's proof behaviour or problem solving behaviour. In addition to scrutinising one's proof behaviour, Hanna and Barbeau (2008) suggest that a student's proof construction competence can be ascertained by determining the extent to which the student applies pertinent ideas to the proof task or theorem in other situations. It can be, therefore, noted that Hanna and Barbeau's suggestion on how to evaluate proof understanding is similar to the encapsulation level of proof conceptualisation proposed by Yang and Lin (2008).

Conradie and Frith (2000) stated that students often fail to grasp the meaning of terms when trying to comprehend a proof thereby hindering their ability to understand other aspects of the proof. To get a sense of students' grasp of key terms to a theorem, Conradie and Frith (2000) suggest that a researcher can ask students to define the key terms or sentences. This technique of determining understanding features at the surface level of Yang and Lin's (2008) model. Alternatively, students can be asked to identify examples that illustrate a given theorem or a term in a theorem. Generating examples that illustrate a theorem or ideas embedded in that theorem is similar to Yang and Lin's second level of their RCGP model. Finally, a prover should develop a firm grasp of the logical relationships of the statement being proven and the major assumptions and conclusions of the proof. In other words, proof understanding involves grasping the proof framework (Selden & Selden, 2009).

Morselli (2006) writes that generating examples has several benefits in proving. The benefits of generating examples to illustrate key concepts involved in constructing proof of a given theorem include illuminating the defining property of pertinent mathematical ideas to the proof being constructed, revealing logical connections in ideas embedded in the mathematical statement that can form the crux of the proof, and providing a pictorial representation of the mathematical ideas especially in proof situations where graphical instantiations can be found. However, despite these benefits, empirical verifications have a severe limitation that the statement can be true for particular examples a prover could have considered but can be false for just one instance not considered by the prover. This is so called notion of a counter example (Stylianides, 2011). Hence, empirical explorations do not provide complete and conclusive evidence about the truth-value of a statement a prover may seek to establish.

Finally, another useful idea in evaluating an individual's ability to compose proof relates to the structure of a mathematical argument. A mathematical argument consists of a connected sequence of assertions in which the consequent statement is called the conclusion and the rest of which is called the premises (Curd, 1992). The premises provide valid reasons for inferring that the conclusion is true. Furthermore, a mathematical proof is said to be valid if it is deductive, contains no errors and provides complete and conclusive evidence about the truth-value of a mathematical statement (Weber & Mejia-Ramos, 2015). The ideas presented in this section were important in evaluating students' responses to the task. For instance, arguments generated by mathematics education post graduate students were checked for logical consistency between premises and conclusions drawn.

METHOD

Research Design

A cross-sectional survey research design with a qualitative bent was used in this study (Flick, 2011). The intent of the study was to capture the in-service teachers' state of knowledge structures about the notions of multiplication of matrices and transpose of a matrix. Further, the study sought to investigate mathematical connections built by the postgraduate students which would in turn allow them to determine whether the product $A\,{}^{t}A$ was defined. Hence, the study was designed to get what Flick (2011) calls "a picture of the moment" (p. 67). Such picturing was based on the assumption that concepts examined were considered to be within the conceptual reach of the participants since they teach these at secondary school level and at teacher training colleges.

Study Informants

The study involved in-service mathematics education students who were studying towards a master degree in mathematics education. The cohort consisted of 23 members: 15 males and 8 females. Of the 15 male participants, 4 were lecturers from primary teacher training colleges while the rest were high school mathematics teachers with more than 10 years teaching experience. One out of 8 female in-service teachers was a lecturer at a primary teacher training college while the other 7 female student teachers were high school mathematics teachers with more than 8 years of teaching experience. Both college and high school mathematics curricula cover subject content on Elementary Linear Algebra of Matrices. Hence, the researcher had assumed on reasonable grounds that the concepts were within the conceptual reach of the participants.

The postgraduate programme in mathematics education offered by the university that served as the study site is now described. The master degree programme has duration of 2 years during which the in-service teachers major in mathematics content and professional courses. Professional courses deal with 3 mathematics education modules and a module covering foundations in science education

In addition professional studies also include Research Methods and Statistics module that is designed to prepare postgraduate students for research projects during the second and final year of their studies. There are 5 modules under the Professional component of the postgraduate programme just described. Mathematics content courses include: Metric Spaces and Topology, Functional Analysis and Nonlinear Differential Equations. Subject content courses drawn from the learning area of Statistics are: Multivariate Statistics, Survey Sampling Methods, and Operations Research. Mathematics content and Professional modules are covered during the first two semesters of year one.

During the third and fourth semesters, that is, during year two, student teachers engage in research projects. The content of the postgraduate programme articulated here supports this researcher's assumption that theory of matrix multiplication and the notion of transpose of a matrix were within students' conceptual reach. The study involved all the 23 students who had enrolled for the master degree studies. Data collection took place during week 10 of semester 2 of year one and this period was deemed strategic for data collection because lectures had been completed and students were doing individual studies and hence pressure had eased.

Research Instrument and Data Collection Procedure

A proof task with an open instruction: If A is a mxn matrix, determine whether the product $A\,{}^{t}A$ is defined. Justify your answer, was posed. Although the task involved elementary concepts in Matrices, it was anticipated that the answer generated would not be a result of applying standard procedure involving a known mathematical result or a known theorem. In other words, an attempt was made to avoid assessing proven results in literature on Linear Algebra such as: (i) $(kA)^t = kA^t$ where k is a scalar (ii) $(AB)^t = B^tA^t$.

The argument here is that assessing students' competences at composing such proofs could possibly lead to regurgitation of proofs from textbooks and so would serve very little purpose with respect to the research goal of establishing the kinds of flaws in postgraduate mathematics education students' conceptualizations of mathematical proof. Instead of reproducing memorised theorems such as those stated here, the postgraduate students were expected to tap from their knowledge of matrix multiplication and the notion of a transpose of a matrix to decide whether the product A^tA is defined. It was also considered that these are well known mathematical ideas for the postgraduates. Hence, students could face no difficult in establishing the connections between these ideas. Thus, the task was intended to measure students' competence at autonomous proof writing.

Following Dahlberg and Housman (1997) task-based interviews were used to gather data. To collect data a task sheet with space for writing the answer was provided. Students responded to the task individually in the mathematics lecture room. No time restrictions were imposed. The participants took about 15 minutes working on the task. All task sheets distributed were returned with some text scribbled on by the students. During data collection process, participants were encouraged to document their thinking as much as possible and request for extra answer sheets when necessary.

Data Analysis

The data analysis procedure employed the analytic induction strategy (Punch, 2005). Analytic induction comprises a series of alternating inductive and deductive steps whereby data driven inductive codes are followed by deductive examination as described in the following steps. First, a marking guide was devised to evaluate postgraduate students' proof efforts. Second, the researcher then performed content analysis (Berg, 2009) of the students' written efforts. Content analysis was facilitated by a refinement of a classification originally developed by Stylianides and Stylianides (2009). The refinement was driven by the desire not only to identify correct proofs but to also explicate the different thinking styles displayed by the postgraduate student teachers. Following Stylianides and Stylianides' (2009) classification as well as a scrutiny of the students' written efforts, a data matrix with the following format was then constructed. Column 1 entries consist of categories identified from students' written responses, in column 2 each category is described, column 3 entries are frequencies of the categories. Steps 1 and 2 so far described led to a data matrix. The creation of the data matrix constituted the induction analysis part of the analytic induction strategy employed in this study.

Finally, the emerging categories from content analysis of postgraduates' proof attempts were mapped to levels of proof understanding in Yang and Lin's (2008) model and other ideas about proof understanding explained in Section 2.3 of this paper in order to ascertain postgraduate students' level of grasp of the concept of mathematical proof. The mapping of results of inductive content analysis constituted the deductive analysis part of the analytic induction strategy. Furthermore, in-vivo codes (Corbin & Strauss, 2008; Varghese, 2009) were used to support inferences made about postgraduate students' proof construction competences.

Ethical Considerations

Flick (2011) suggests that research should involve participants who have been informed about the aim of the study and that participation should be voluntary. Hence, with regards to informed consent the researcher explained to the postgraduate mathematics education students why research into the nature of students' flawed conceptualisations of mathematical proof was necessary as it could promote conceptual teaching of the mathematical ideas. The students were asked to complete informed consent forms. Further, the researcher emphasised that consent was to be given voluntarily (Flick, 2011).

Another ethical concern was about anonymization of data. Anonymizing data involved removing any identifiers from the students' responses (Punch, 2005). Furthermore, during research reporting pseudonyms were used to describe postgraduates' flaws in their conceptualisations of mathematical proof.

RESULT AND DISCUSSION

Inductive content analysis of postgraduate students' written responses revealed the categories summarised in Table 1.

Table 1. Emerging categories from content analysis of postgraduate students' proving attempts ($n = 23$)

Category	Description	Frequency
C1	Correct proof	4
C2	Empirical argument used	8
C3	Argument adduced does not represent the most general case	10
C4	Consequent statement missing or wrongly formulated	7
C5	Argument has algebraic slips	1
	Total	**30**

Table 1 shows that the dominant scheme of argumentation was one in which student teachers produced justifications that failed to offer complete and conclusive evidence about the fact that the product $A\,{}^tA$ is defined for any matrix $A = (a_{ij})$ over a field of scalars K. From the same Table 1 it can be seen that postgraduate students' efforts were also dominated by use of specific examples, denoted by C_2 with 8 out of 30 responses and a very significant number of responses (7 out of 30) were in the category in which the conclusion did not follow logically from the premises__ represented by code C4. Finally, the same Table 1 illustrates that very few (4 out of 30) student teachers managed to justify the existence of the product $A\,{}^tA$ __ a worrisome result at postgraduate level, more so in light of the fact that elementary concepts of transpose and multiplication of matrices were explored. Presented next is a discussion of thinking styles displayed in each category summarised in Table 1 for the purpose of illuminating flaws detected in postgraduate mathematics education student teacher informants.

C1: Correct proof constructed

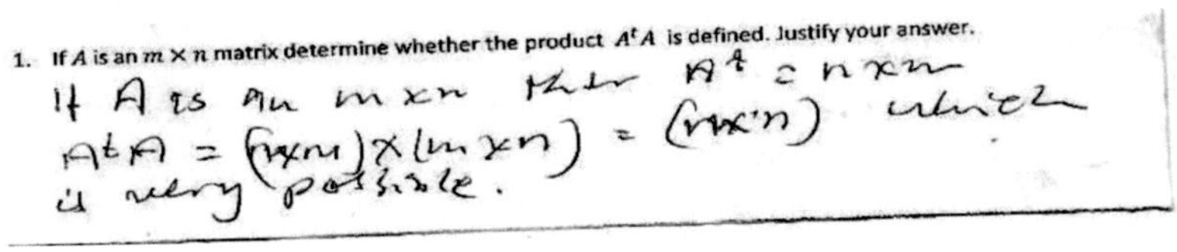

Figure 1: Trevor's written response to the task on multiplication of matrices.

Figure 1 illustrates that Trevor could state the order of the transpose matrix A^t correctly as nxm. Trevor then tested for compatibility of the two matrices by determining whether the number of columns of A^t was equal to the number of rows of matrix A and then reached the conclusion that the product $A\,{}^tA$ is defined. Hence, Trevor's written response shows that he had a firm of grasp of the concepts of transposition and multiplication of matrices. Further, there was proper chaining of these knowledge elements which then led to the valid conclusion that the product $A\,^tA$ exists (Yang & Lin, 2008).

C2: Empirical arguments used

Typical examples in this category were produced by Munya and Tauya. Munya and Tauya's efforts are shown in Figures 2 and 3 respectively

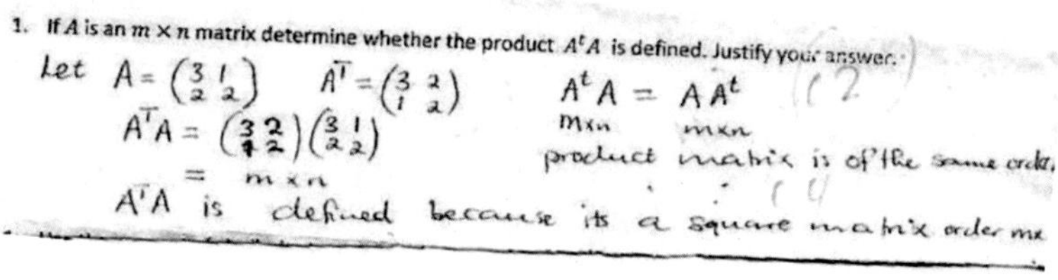

Figure 2. Munya' written response to the task.

Figure 2 illustrates that Munya chose a specific example of a 2 −square matrix A and proceeded to write the transpose of matrix A correctly. The product $A\,^tA$ is then equated to mxn, presumptively referring to the order. Munya went on to write $A\,^tA = AA$, which is a false assertion because a matrix A and its transpose are not commutative over the binary operation multiplication over the field K. It is a serious flaw in reasoning displayed by Munya. Further, Munya represented the order of the product $A\,^tA$ to be mxn and the order of $AA^{\,t}$ was written also as mxn. It was yet another flawed reasoning as Munya could not identify that the operation of transposing entails interchanging the rows and columns of a matrix.

Finally, Munya concluded that the "product matrix is of the same order." This is a flawed argument because for m by n matrix A the product $A\,^tA$ is a n −square matrix while the product $AA^{\,t}$ is a m −square matrix. The discussion of Munya's proof effort reveals that he had not even attained the second level of Yang and Lin's (2008) model of comprehension of geometric proof. Munya was operating at surface level of Yang and Lin's model as he could state the transpose of the 2 −square matrix he had written. Furthermore, Munya concluded on the basis of the specific example of the 2 −square matrix that "$AA^{\,t}$ is defined because it a square matrix order m." The claim shows that Munya had not grasped the fundamental limitation that empirical verifications cannot be elevated to the status of a proof (Ndemo, Zindi, & Mtetwa, 2017; Stylianides, 2011). In other words, Munya exhibited a weak command of the notion of counter-argumentation in mathematics.

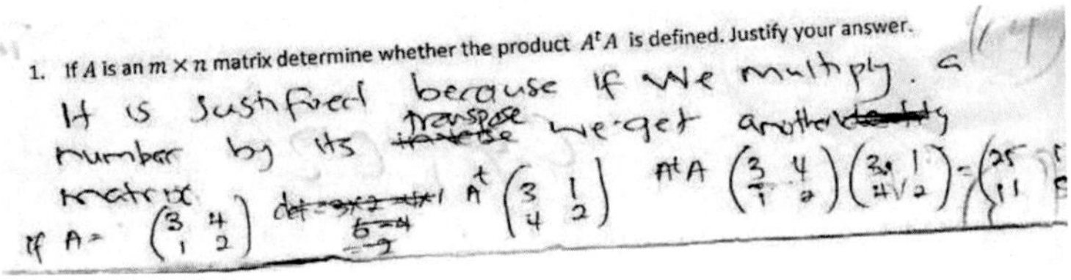

Figure 3. Tauya's written response to the task.

From Tauya's solution attempt the word number was used to refer to a matrix when he wrote "when we multiply a number by its transpose we get another matrix." This claim by Tauya reveals lack of precision in the manner he used the word number which he used interchangeably with the word matrix. Further, Tauya's effort also reveals lack of deep grasp of the basic limitation that specific examples cannot be used to represent general matrix multiplication. In other words, although matrix multiplication held in the single instantiation considered, the specific example picked by Tauya should not be regarded as a proof. In terms of Yang and Lin's (2008) model of geometric proof comprehension, Tauya's effort reveals that he had not interiorised matrix multiplication. According to Stylianides and Stylianides (2009) an argument is deemed to be valid if it is deductive and the premises logically imply the consequent statement. A true deductive argument once constructed offers complete and conclusive evidence about the truth of a mathematical statement. Hence, it becomes superfluous to look for further evidence about the truth-value of the mathematical statement (Weber & Mejia-Ramos, 2014). It can, therefore be, inferred from Tauya's use of a single example to resolve the proof task that he lacked a good grasp of these fundamental ideas about proving and proof in the area of Elementary Algebra.

C3: Argument produced does not represent the most general case

Table 1 shows that this category that emerged from inductive content analysis of students' was the most dominant among mathematics education postgraduates with 10 out of 30 responses. Next, the researcher now presents typical students' responses in this category and discusses these results within the perspective of Yang and Lin's model and other ideas about students' conceptions of mathematical proof. First Ticha's response is considered.

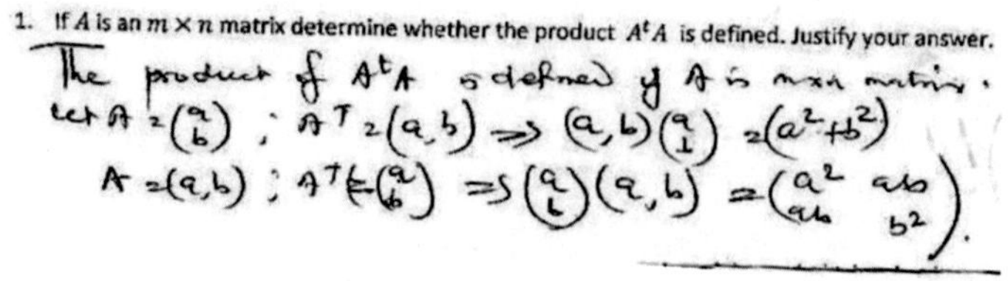

Figure 4. Ticha's written attempt to the task.

Figure 3 shows that whilst Ticha produced his argument in terms of arbitrary mathematical objects by using the column matrix $A = \begin{pmatrix} a \\ b \end{pmatrix}$ whose transpose was correctly written as $A^t = (a\ b)$, the argument cannot be considered to represent the general mxn matrix A. Similarly, a row matrix $A = (a\ b)$ whose transpose $A^t = \begin{pmatrix} a \\ b \end{pmatrix}$ was also considered and the product $A^tA = \begin{pmatrix} a^2 & ab \\ ab & b^2 \end{pmatrix}$ was then determined.

Hence, although in both cases Ticha's efforts involved manipulating arbitrary mathematical objects, the two cases cannot be deemed to represent the most general case of the product AB of two matrices A and B since Ticha's arguments involved column and row matrices. However, there was proper chaining of the elements (Azarello, 2007; Mejia-Ramos et al., 2012) as Ticha could carry out matrix multiplication correctly that led to the conclusion that "A^tA is defined if A is mxn matrix." It can be argued that traces of the encapsulation phase of proof comprehension were missing because Ticha could not conceive matrix multiplication in terms of the broader and more general case. Another example in this category by Mutaka is now presented and discussed.

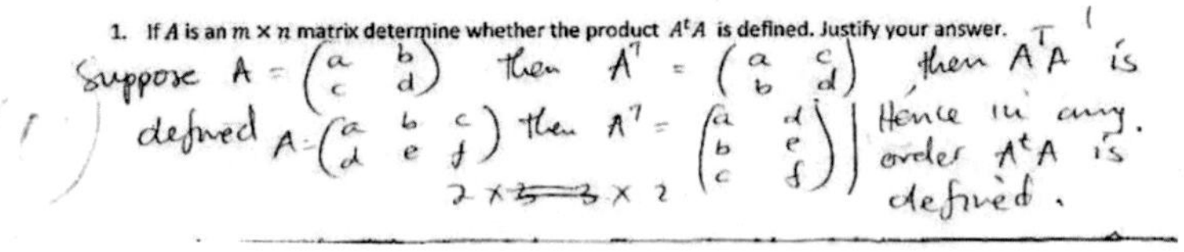

Figure 5. Mutaka's written effort to resolve the task.

First, Mutaka considered the special case of a 2 −square matrix $A = \begin{pmatrix} a & b \\ c & d \end{pmatrix}$ and wrote the transpose of the matrix A correctly as $A^t = \begin{pmatrix} a & c \\ b & d \end{pmatrix}$. Mutaka's argument up to this stage indicated that the surface level of Yang and Lin's (2008) model of proof comprehension had been attained as shown by correct use of the definition of the idea of a transpose. However, chaining of the elements of $A\ t$ and A was not evident because after transposing the matrix A, Mutaka just wrote the conclusion that "then $A\ {}^{t}A$ is defined." Second, Mutaka considered another special case of a $2x3$ matrix A with arbitrary entries and as before he managed to interchange rows and columns of A to get the transpose A . Similar to his proof behavior in the first example discussed, there was no chaining of elements observed here that led to the conclusion stated as "Hence in any order $A\ {}^{t}A$ is defined." In a similar fashion to Ticha's example the 2 −square and $2x3$ matrices used by Mutaka to support his conclusion that the product $A\ {}^{t}A$ is defined for any mxn matrix A over a field of scalars K cannot be considered to be representative of the most general case of matrix multiplication. Hence, the encapsulation level (Yang & Lin, 2008) was not reached by Mutaka and Ticha. Next, the researcher focuses on informants' written responses in which the conclusion was either missing or wrongly formulated.

C4: Consequent statement missing or wrongly formulated

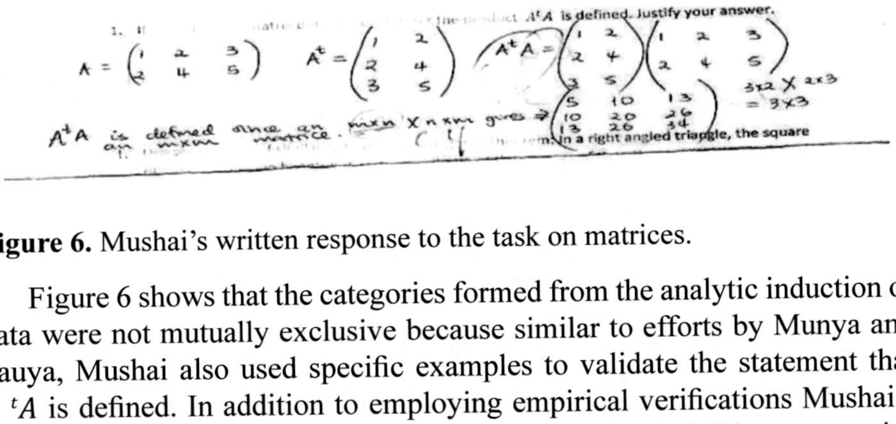

Figure 6. Mushai's written response to the task on matrices.

Figure 6 shows that the categories formed from the analytic induction of data were not mutually exclusive because similar to efforts by Munya and Tauya, Mushai also used specific examples to validate the statement that $A\ {}^{t}A$ is defined. In addition to employing empirical verifications Mushai's arguments also revealed the following limitation. Mushai did not recognise that matrix multiplication is not commutative as he referred to $A\ {}^{t}$ as a mxn matrix yet matrix A has been given as a mxn matrix. Furthermore, he identified A as nxm in stark contrast to the assertion that A was given as a mxn matrix. This chaotic proof behaviour led to the conclusion that $A\ {}^{t}A$ is "a mxm matrix," instead of an n −square matrix. Hence, whilst Mushai later on chained the elements correctly his failure earlier on to correctly identify the order of the matrix A led to the wrong conclusion. Another example in this category is now examined.

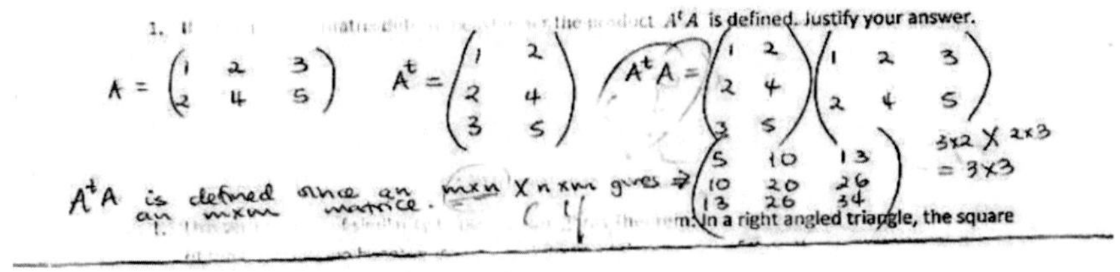

Figure 7. Maunja' written response to task on matrix multiplication.

The product $A^{t}A$ denotes that the matrix A is post-multiplying the transpose matrix A . Hence, for compatibility of the multiplication operation the number of columns of A^{t} should be equal to the number of rows of A. Figure 7 illustrates that Maunja's argument contradicts to the assertion just stated concerning matrix multiplication. He wrote that "$A^{t}A$ will be defined ... the no of columns in matrix A will always be equal to the no of rows in the 2 *nd* matrix." The second matrix mentioned by Maunja presumptively referred to the post-multiplying matrix which in this case should be the matrix A. Maunja's conclusion is a typical example of many such conclusions (7 out 30) drawn by postgraduate students that did not logically follow from the premises. Finally, the researcher focuses on a typical example of a flawed argument caused by an algebraic slip made by Mujuru.

C5: Argument has algebraic slips

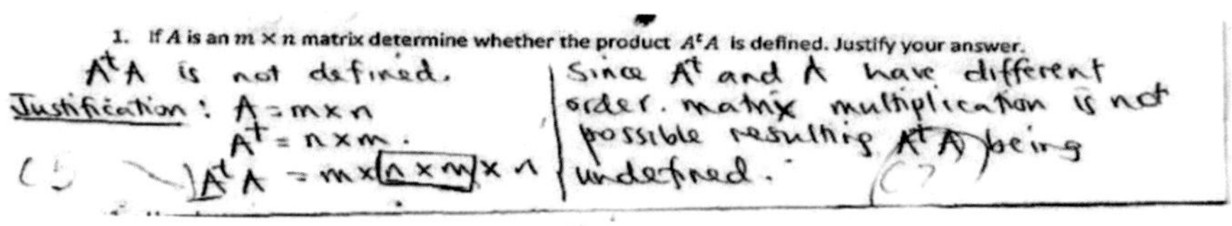

Figure 8. Mujuru's proof attempt to the task on matrices.

Figure 8 shows that Mujuru started very well as she was able to write the transpose matrix A^{t} correctly as nxm matrix. However, her woes with the proof task manifested at the chaining stage (Yang & Lin, 2008) whereby her representation of matrix A^{t} as mxn led to the assertion that "$A^{t}A = mxn\ X\ mxn$." This statement made it impossible for her to determine the product $A^{t}A$ and led to the conclusion that "$A^{t}A$ is not defined." It can be observed also from the assertion "$A^{t}A = mxn\ X\ mxn$," that the two matrices A^{t} and A have the same order, which is a false assertion. Hence, the wrong conclusion drawn can be attributed to the algebraic slip made by Mujuru.

CONCLUSION

Postgraduate had not grasped the fundamental limitation that empirical verifications do not count as proofs. A case in point was the use of a single instantiation by Tanya. Furthermore, students produced arguments which were not typical of the most gencral. Although student teachers' efforts to justify that the product is defined were terms of arbitrary mathematical objects, those objects did not represent the most general case. For instance, a column matrix $A = \binom{a}{b}$ was used to prove that the product $A\,{}^tA$ exists. The cases used by postgraduate were not representative of the general matrix multiplication discussed in Section on multiplication of matrices. However, the use arbitrary objects was a huge step forward in current efforts to promote deductive argumentation among students in mathematics education. Lastly, postgraduate students' written responses revealed that the premises and the conclusion drawn were not logically connected. Further, in other cases the definition of the transpose of a matrix was not properly grasped as shown by proof behaviour such as referring to $A^{\,t}$ as a mxn matrix when the matrix was stated as a mxn matrix. In addition, in some cases for the product, $A\,{}^tA$, the students focused on the number of columns of the matrix A instead of the number of rows since A was the post-multiplying matrix.

As concluding remarks, the researcher emphasizes that postgraduate students" flawed conceptions uncovered by this study have important implications for teacher preparation in Zimbabwe. Subject content mastery by students was fragile and hence the need for continuing professional development of in-service mathematics on subject content knowledge. The need to promote good grasp of concepts of Linear Algebra implies that mathematics educators and researchers need to find ways of ameliorating flawed conceptions of the concept of a mathematical proof. Furthermore, teacher preparation needs to include content and instructional strategies that foster and enhance prospective secondary mathematics teachers' explanatory role. Such content and strategies should also develop an appreciation of the function of a mathematical proof in justifying why a given mathematical assertion is true or false.

REFERENCES

1. Ahmad, S., Prahmana, R.C.I., Kenedi, A.K., Helsa, Y., Arianil, Y., & Zainil, M. (2018). The instruments of higher order thinking skills. Journal of Physics: Conference Series, 943(1), 012053. https://doi.org/10.1088/1742-6596/943/1/012053.
2. Azarello, F. (2007). The proof in the 20th century: From Hilbert to automatic theorem proving. In P. Boero (Ed.), Theorems in schools: From history, epistemology, and cognition to classroom practice. Rotterdam: Sense Publishers.
3. Berg, B.L. (2009). Qualitative Research Methods for the Social Sciences Topics. Boston: Allyn Brown.
4. Bieda, K.N. (2010). Enacting proof-related in middle school mathematics: Challenges and opportunities. Journal for Research in Mathematics Education, 41(4), 351-382.
5. Bleiler, S.K., Thompson, D.R., & Kraj?evski, M. (2014). Providing written feedback on students' mathematical arguments: proof validations of prospective secondary mathematics teachers. Journal of Mathematics Teacher Education, 17(2), 105-127. https://doi.org/10.1007/s10857-013-9248-1.
6. Brown, J., & Stillman, G. (2009). Preservice teachers' competencies in proof. In F.-L Lin, F-J., Hsieh, G. Hanna, M. de Villiers (Eds), Proceedings of the ICMI Study 19 Conference: Proof and Proving in Mathematics Education: Vol. 2. (pp.196-201). Taipei, Taiwan: National Taiwan University.
7. Conradie, J., & Frith, J. (2000). Comprehension tests in mathematics. Educational Studies in Mathematics, 42(3), 225-235. https://doi.org/10.1023/A:1017502919000.
8. Corbin, J., & Strauss, A. (2008). Basics of Qualitative Research. Thousand Oaks: Sage.
9. Curd, M. (1992). Arguments and Analysis: An Introduction to philosophy. London: West Publishing.
10. Dahlberg, R.P., & Housman, D.L. (1997). Facilitating learning events through example generation. Educational Studies in Mathematics, 33(3), 283-299. https://doi.org/10.1023/A:1002999415887.
11. Duval, R. (2006). A cognitive analysis of problems of comprehension in a learning of mathematics. Educational Studies in Mathematics, 61(1), 103-131. https://doi.org/10.1007/s10649-006-0400-z.

12. Flick, W. (2011). Introducing Research Methodology: A Beginner's Guide to Doing a Research Project. London: Sage Publications.
13. Garret, L. (2013). Flawed mathematical conceptualizations: Marlon's dilemma. Journal of Developmental Education, 37(2), 2-4, 6-8.
14. Goodaire, E.D. (2014). Linear Algebra: Pure and Applied. Toh Tuck: World Scientific Publishing.
15. Hanna, G., & Barbeau, E. (2008). Proofs as bearers of mathematical knowledge. ZDM, 40(3), 345-353. https://doi.org/10.1007/s11858-008-0080-5.
16. Harel, G., and Sowder, L. (2007). Toward a comprehensive perspective on proof. In F. Lester (Ed.), Second Handbook of Research on Mathematics Teaching and Learning (pp 805-842). Charlotte, NC: Information Publishing.
17. Harel, G., Selden, A., & Selden, J. (2006). Advanced mathematical thinking: Some PME perspectives. In A. Gutierrez and P. Boero (Eds.), Handbook of Research on the Psychology of Mathematics Education: Post, Present and Future. Rotterdam: Sense, pp. 147-172.
18. Jahnke, H.N. (2007). Proofs and hypotheses. ZDM, 39(1-2), 79-86. https://doi.org/10.1007/s11858-006-0006-z.
19. Jonassen, D.H., & Kim, B. (2010). Arguing to learn and learning to argue: Design justifications and guidelines. Educational Technology Research and Development, 58(4), 439-457. https://doi.org/10.1007/s11423-009-9143-8.
20. Jones, K. (1997). Student teachers' conceptions of mathematical proof. Mathematics Education Review, 9, 21-32.
21. Lee, K., & Smith, P. (2009). Cognitive and linguistic challenges in understanding proving. In F-L. Lin, F-J. Hsieh, G. Hanna, & M. de Villiers (Eds.), Proceedings of the ICMI Study 19 Conference on Proof and Proving in Mathematics Education: Vol. 2. (pp. 15-21). Taipei, Taiwan: National Taiwan Normal University.
22. Lesseig, K. (2016). Investigating mathematical knowledge for teaching proof in professional development. International Journal of Research in Education and Science, 2(2), 253-270.
23. Lipschutz, S. (1991). Schaum's Outline of Theory and Problems of Linear Algebra. New York: McGraw-Hall.

24. Mamona-Downs, J., & Downs, M. (2013). Problem solving and its elements in forming proof. The Mathematics Enthusiast, 10(1), 137-162.
25. Martin, G., & Harel, G. (1989). Proof frames of pre-service elementary teachers. Journal for Research in Mathematics Education, 20(1), 41-51.
26. Maya, R., & Sumarmo, U. (2011). Mathematical understanding and proving abilities: experiment with undergraduate student by using modified Moore learning approach. Journal on Mathematics Education, 2(2), 231-250. http://doi.org/10.22342/jme.2.2.751.231-250.
27. Mejia-Ramos, J.P., & Inglis, M. (2009). Argumentative and proving activities in mathematics education research. In F-L. Lin, F-J. Hsieh, G. Hanna, & M. de Villiers (Eds), Proceedings of the ICMI Study 19 Conference on Proof and Proving in Mathematics Education: Vol. 1. (pp. 88-93). Taipei, Taiwan: National Taiwan Normal University.
28. Mejia-Ramos, J.P., & Weber, K. (2014). Why and how mathematicians read proofs: Further evidence from a survey study. Educational Studies in Mathematics, 85(2), 161-173. https://doi.org/10.1007/s10649-013-9514-2.
29. Mejia-Ramos, J.P., Fuller, E., Weber, K., Rhoads, K., & Samkoff, A. (2012). An assessment model for proof comprehension in undergraduate mathematics. Educational Studies in Mathematics, 79(1), 3-18. https://doi.org/10.1007/s10649-011-9349-7.
30. Morselli, F. (2006). Use of examples in conjecturing and proving: An exploratory study. In J. Novotna, H. Moraova, M. Kratka, & N. Stehlikova (Ed), Proceedings of the 30th conference of the International Group for the Psychology of Mathematics Education: Vol. 4. (pp. 185-192). Prague: PME.
31. Mumu, J., Prahmana, R.C.I., & Tanujaya, B. (2018). Construction and reconstruction concept in mathematics instruction. Journal of Physics: Conference Series, 943(1), 012011. https://doi.org/10.1088/1742-6596/943/1/012011.
32. Ndemo, Z., Zindi, F., & Mtetwa, D. (2017). Mathematics Undergraduate Student Teachers' Conceptions of Guided Inductive and Deductive Teaching Approaches. Journal of Curriculum and Teaching, 6(2), 75-83.
33. Noto, M.S., Priatna, N., & Dahlan, J.A. (2019). Mathematical proof: The learning obstacles of preservice mathematics teachers on

transformation geometry. Journal on Mathematics Education, 10(1), 117-126. https://doi.org/10.22342/jme.10.1.5379.117-126.

34. Prahmana, R.C.I., & Suwasti, P. (2014). Local instruction theory on division in mathematics GASING. Journal on Mathematics Education, 5(1), 17-26. https://doi.org/10.22342/jme.5.1.1445.17-26.
35. Punch, K.F. (2005). Introduction to Social Science Research: Quantitative and Qualitative Approaches. London: Sage.
36. Putri, R.I.I., & Zulkardi. (2018). Higher-order thinking skill problem on data representation in primary school: A case study. Journal of Physics: Conference Series, 948(1), 012056. https://doi.org/10.1088/1742-6596/948/1/012056.
37. Saleh, M., Prahmana, R.C.I., Isa, M., & Murni. (2018). Improving the reasoning ability of elementary school student through the Indonesian Realistic Mathematics Education. Journal on Mathematics Education, 9(1), 41-54. https://doi.org/10.22342/jme.9.1.5049.41-54.
38. Selden, J., & Selden, A. (2009). Understanding the proof construction process. In F.-L Lin, F-J . Hsieh, G. Hanna, M. de Villiers (Eds), Proceedings of the ICMI Study 19 Conference: Proof and Proving in Mathematics Education: Vol. 2. (pp.196-201). Taipei, Taiwan: National Taiwan University.
39. Shahrill, M., Putri, R.I.I., Zulkardi, & Prahmana, R.C.I. (2018). Processes involved in solving mathematical problems. AIP Conference Proceedings, 1952(1), 020019. https://doi.org/10.1063/1.5031981.
40. Stylianides, A.J. (2011). Towards a comprehensive knowledge package for teaching proof: A focus on the misconception that empirical arguments are proofs. Pythagoras, 32(1), 1-10.
41. Stylianides, A.J., & Stylianides, G.J. (2009). Proof constructions and evaluations. Educational Studies in Mathematics, 72(3), 237-253. https://doi.org/10.1007/s10649-009-9191-3.
42. Stylianou, D., Blanton, M.L., & Rotou, O. (2015). Undergraduate students' understanding of proof: Relationships between proof conceptions, beliefs, and classroom experiences. International Journal of Research in Mathematics Education, 1(1), 91-134. https://doi.org/10.1007/s40753-015-0003-0.
43. Varghese, T. (2009). Secondary-level teachers' conceptions of mathematical proof. Issues in the Undergraduate Mathematics Preparation of School Teachers: The Journal, 1, 1-13.

44. Weber, K., & Mejia-Ramos, J.P. (2014). Mathematics majors' beliefs about proof reading. International Journal of Mathematics Education in Science and Technology, 45(1), 89-103. https://doi.org/10.1080/0020739X.2013.790514.
45. Weber, K., & Mejia-Ramos, J.P. (2015). On relative and absolute conviction in mathematics. For the Learning of Mathematics, 35(2), 3-18.
46. Yang, K., & Lin, F.L. (2008). A model of reading comprehension of geometry proof. Educational Studies in Mathematics, 67(1), 59-76. https://doi.org/10.1007/s10649-007-9080-6.

Chapter

9

MATHEMATICAL UNDERSTANDING AND PROVING ABILITIES: EXPERIMENT WITH UNDERGRADUATE STUDENT BY USING MODIFIED MOORE LEARNING APPROACH

Rippi Maya[1], Utari Sumarmo[2]

[1]State Islamic University - Sunan Gunung Jati, Bandung
[2]Indonesia University of Education, Bandung

ABSTRACT

This paper reports findings of a posttest experimental control group design conducted to investigate the role of modified Moore learning approach on improving students' mathematical understanding and proving abilities. Subject of study were 56 undergradute students of one state university in Bandung, who took advanced abstract algebra course. Instrument of study were a set test of mathematical understanding ability, a set test

Citation: (APA): Maya, R., & Sumarmo, U. (2011). Mathematical understanding and proving abilities: experiment with undergraduate student by using modified Moore learning approach. Journal on Mathematics Education, 2(2), 231-250.(20 pages).

Copyright: © Creative Commons Attribution 4.0 International License (http://creativecommons.org/licenses/by/4.0/).

of mathematical proving ability, and a set of students' opinion scale on modified Moore learning approach. Data were analyzed by using two path ANOVA. The study found that proof construction process was more difficult than mathematical understanding task for all students, and students still posed some difficulties on constructing mathematical proof task. The study also found there were not differences between students' abilities on mathematical understanding and on proving abilities of the both classes, and both abilities were classified as mediocre. However, in modified Moore learning approach class there were more students who got above average grades on mathematical understanding than those of conventional class. Moreover, students performed positive opinion toward modified Moore learning approach. They were active in questioning and solving problems, and in explaining their works in front of class as well, while students of conventional teaching prefered to listen to lecturer's explanation. The study also found that there was no interaction between learning approach and students' prior mathematics ability on mathematical understanding and proving abilities, but there were quite strong association between students' mathematical understanding and proving abilities.

Keywords: modified Moore learning approach, mathematical understanding ability, mathematical proving ability.

INTRODUCTION

Some studies reported that mathematical proving was a difficult task for many high school and undergraduate students. Whereas, possessing mathematical proving ability was a certainty ability, because it is an essential ability that should be possess by all students who learn mathematics. Moreover, that ability was needed for pursuing further mathematics contents. That statement was in line with Solow's opinion (1990) that the truth of each proposed mathematical statement must be tested before the statement was used as a basic or reference for testing the truth of other mathematical statement. Afterward, the tested mathematical truth was represented in mathematical language through a proof.

Some researchers (see Arnawa (2006); Barnard (2000); Downs and Downs in Arnawa (2006); Kusnandi (2008); Moore (1994); Moore in Weber (2003); Senk in Hanna and Jahnke (1996); Tall (1999)) conducted studies according to proving abilities. Moore (1994) proposed in detailed seven difficulties on mathematical proving namely:

1) Students did not understand definitions, or they could not state definitions. They consider that definition was an abstract thing.
2) Students had only few intuitive understanding on mathematical concepts.
3) Students' concept images were not enough for carrying out a proof;
4) Students were unable to generate and use their own examples.
5) Students did not know how to use definitions for acquiring structured proof entirely.
6) Students were unable to understand and to use mathematical language and its symbol.
7) Students did not know how to begin proof.

Selden & Selden (1995) cited Moore statements that knowing a definition and could conver example and non example, it didn't mean that the students could master language and logical structure for writing proofs directly. Teaching how to compose a proof was a difficult task. Finding of Senk (in Hanna & Jahnke (1996)) strengthened that statement. Senk reported that from 1520 high school students only 30% students who mastered 75% writing proof in Euclid geometry and only 3% of them who obtained ideal score. Further Tall (1999) strengthened supposition of difficluty to teach proof as well. He stated that proving was an essential mathematical ability, however it was often difficult to teach. Wahyudin (1999) gave similar statement that mathematics teachers only mastered 62,88% mathematics contents and only 50% teachers could prove a rule by using mathematical induction, and only 20 % students mastered mathematics concept correctly. At higher education level, students often say that they can follow a proof that explained by their lecturer in class, but they are unable to compose proofs by themselves when required to do so for homework (Barnard, (2000)). That difficulty was caused by students' unability in investigating mathematical statements deeply.

Furthermore, Moore (Weber, 2003) proposed that sometimes students could state definition of a concept, but they did not understand meaningfully. That disadvantages caused students posed difficulty when they were asked to explain the concept by their own words or to generate an example from a previous concept. Weber (2003), stated that in general students knew what had to do in composing a proof, they could reason deductively, restate and manipulate definition and draw a valid conclussion. However, to know logical rules and definition did not guarantee that students could reason the concept meaningfully. Students should understand a concept intuitively before they could compose a proof, accompanied with habits of active learning in understanding a proof. If students accustomed to cope lecturer's proof but they never composed proof by themselves, then they would pose difficulty to solve mathematical proof. Selden and Selden (Weber, 2003) stated that students' difficulty in proving was caused by their lack of determination for validating proofs, and they did not know whether a proof was true or false. Moore (Weber, 2003) was sure that students would get a few knowledge of advanced mathematics if they only coped lecturer's proof passively. On the other hand students would learn more mathematics concepts and its proof if they tried to compose a mathematical statement by themselves.

According to Moore's findings about students' difficulties on proving, most of them were caused by their lack of understanding on mathematical concepts and definitions so that they were unable to construct a mathematical proof and to write mathematical notation or to use mathematical language correctly. Therefore, before students able to construct a proof, they should have to master all relevant definitions and theorems. In fact, it is obvious that students' mathematical proving ability associated with their mathematical understanding ability.

The words structure or notation in a proof was standard. Sometimes, it was not easy for some students to understand the proof well, so they might need help to comprehend proof meaningfully. This was happened because some of them considered that proof was only manipulation of unmeaningfull mathematical symbols (Downs & Downs in Arnawa (2006)). Students did not aware that actually proof was really composed by mathematical words and symbols included in a theorem or previous theorems. Researchers observed that students had a little attention on important aspect of mathematical proof, so as long as a theorem could be used for solving mathematical problem, then proof was not a focus of their attention anymore. Besides that, the lack of mathematical understanding used by students did not have ilustration how to begin a mathematical proof. Likewise, mathematics concepts were mutual related to each other. Understanding on a new concept was related by understanding on previous concepts, so it was understandable that without understanding previous concepts students would experience difficulty to explain or even to begin a proof. Arnawa (2006) reported that students, who learned Structure Algebra based on APOS theory, obtained better proving ability than students were taught by conventional teaching.

According to mathematical understanding ability, Polya (Sumarmo, 1987 and 2002) proposed four level of understanding namely, mechanical, inductive, rational, and intuitive understanding. Mechanical understanding happened when a person only memorized rules and implemented it correctly; inductive understanding happened when he had tried a rule in simple cases and he knew that the rule operated correctly. While rational understanding was obtained when a person knew a rule meaningfully or accompanied with its reason. Furthermore, intuitive understanding was obtained when he was sure on the truth of a rule without doubtful.

Alfeld (2004) stated that a person understood mathematics when he was able to explain mathematics concept in other simpler form of concept, then he was able to connect logically among facts and different concepts; and he could recognize relation between a new concept with previous concepts. When a person mastered all those things, it was called he had good mathematical understanding. According to Polya and Alfeld and some findings that presented before, it was interpreted that in proving a theorem a person should related a new concept or theorem with previous ones. To overcome difficulties on mathematical proving problem, modified Moore method offered learning approach which motivated students to learn mathematical proving actively. Students were motivated to think independently, start with a simple problem for improving a solution accompanied with its supporting reason and to communicate their ideas writtenly or orally so that it could be understood by other students. In written communication students could write it down on a board or in an artickel form, while in oral communication students presented and defended it in front of class.

To consider the characteristics of modified Moore learning approach, it was predicted that the learning approach would train students to have self regulated thinking on solving problem, had abilities of composing relevant reason, and convincing other students through a written or oral presentation. With little lecturer's guidance, this learning approach would able to help students overcame their difficulties on mathematical understanding and proving tasks. In order to obtain optimal result it was suggested that this learning approach was implemented to no more than 24 students. When there were more than 24 students so we could form into groups which the members had various prior mathematics abilities.

Abstract algebra was classified as an advanced and difficult course which contains more proving tasks. Many students failed this course. To overcome this problem, Arnawa (2006) and Nurlaelah (2009) conducted studies which implemented modified APOS theory to improve students' proving ability and mathematical power. Both studies reported that modified APOS learning approach was more successful than conventional teaching in improving students' proving ability and mathematical power.

Those arguments and some findings of studies which implemented modified Moore and modified APOS learning appproaches motivated researcher to conduct an experiment by implementing modified Moore learning approach to improve students' mathematical understanding and proving abilities in advanced abstract algebra course. Considering that mathematics as a systematic science, it was predicted that besides learning approach, students' ability in Structure Algebra would have important role in improving those abilities in advanced structure algebra as well.

METHODOLGY

The main goals of this study namely: 1) Were abilities on mathematical understanding and proving abilities of students taught by using modified Moore learning approach better than those abilities of students of conventional class? 2) Were there interaction between learning approach and students' prior mathematics ability on students' mathematical understanding ability and on mathematical proving ability? 3) Was there association between mathematical understanding and proving abilities? 4) What was students' disposition on modified learning approach? 5) What kinds of difficulties did students experience in solving mathematical proving problems?

This study was a post-test experimental control group design as follow.

X	O
--------------------	--------------------
	O

Note:

O : mathematical understanding test and mathematical proving test.

X : modified Moore learning approach

Subject of this study were 56 students from two classes of mathematics department of a state university in Bandung, who took Advanced Structure Algebra course. Instruments of this study were: two tests namely a mathematical understanding essay tests consisted of 6 items and a mathematical proving essay test consisted of 5 items, and a disposition scale Likert model consisted of 16 statements. The both tests were composed by using Arikunto (2002) as a guide, while the disposition scale was modified from Sumarmo (2002). The reliability test were calculated with Cornbah alpha and it were 0,66 and 0,62 for understanding test and proving test respectively, while item validity were between 0,32 and 0,55 and between 0,69 an 0,80 for understanding test and proving test respectively. Further, learning materials for modified Moore learning approach were modified from Mahavier, May & Parker (2006), Cohen (1982), Chalice (1995), Mahavier (1999) and for the Structure Algebra material was modified from Gallian (2006).

Before experiment was conducted and data were analized, students were classified according to the rule as in Table 1.

Table 1. Classification Rules of Students according to MUA and MPA

Classification	PMA	MUA	MPA
High	$70 \leq PMA \leq 100$	$70 \leq MUA \leq 100$	$70 \leq MPA \leq 100$
Medium	$55 \leq PMA < 70$	$55 \leq MUA < 70$	$55 \leq MPA < 70$
Low	$0 \leq PMA < 55$	$0 \leq MUA < 55$	$0 \leq MPA < 55$

Note:

PMA: prior mathematics ability

MUA: mathematical understanding ability

MPA : mathematical proving ability

Based on the rule in Table 1, there was only a person with high PMA and the rest were classified as medium and low levels of PMA. Afterwards data were analyzed by using two path ANOVA which preceded by certain statistics testing relevant to characteristics of the data

In the following we presented sample of instruments of this study.

1) Sample item of mathematical understanding test

Observe commutative ring Z_{10} and Z_{12} . Suppose M and N were maximum ideal of Z_{10} and suppose P and Q were maximum ideal of Z_{12}.

Determine those are maximum ideal.

2) Sample item of mathematical proving test

Suppose R was commutative ring with unit element, and suppose I was ideal of R. Prove R/I was integral domain if and only if I was prime ideal.

FINDINGS AND DISCUSSION

1. Students' Mathematical Understanding and Proving Abilities

Students' Mathematical Understanding and Proving Abilities according to learning approach and level of students' PMA are presented in Table 2.

a) According to students' PMA, in conventional class there were only two groups namely medium (3) and low level (27) of PMA; while in MLA (Modified Moore Learning Approach) class there were three groups namely high level (1) PMA, medium level (6) PMA, and low level (19) PMA. This findings pointed out that students' achievement in structure algebra (namely PMA) were clasified as low or many students failled in this course. Because there was only 1 student with high PMA in MLA class, so we could not compare their abilities on MUA and MPA of for both classes.

Table 2. Students' MUA and MPA according to Learning Approach and Level of PMA

PMA	Mathematical Understanding Ability (MUA)						Mathematical Proving Ability (MPA)					
	MLA			CNV			MLA			CNV		
	n	Mean	sd	n	Mean	sd	N	Mean	sd	n	Mean	sd
Low	19	60,58	11,99	27	67,15	16,14	19	50,53	11,29	27	53,70	14,72
Med	6	76,00	13,18	3	83,33	4,51	6	70,83	12,07	3	68,33	16,07
High	1	100		0			1	80,00		0		
Total	26	65,65	15,19	30	68,77	16,11	26	56,35	14,80	30	55,17	15,23

Note:

Ideal score was 100; n : number of subject; PAM: Prior mathematics ability;

MLA: Modified Moore learning approach; CNV: conventional teaching

b) For entirely students, MUA of students in MLA class (65,65) was lower than MUA of students in CNV class (68,77). However according to the result of two path ANOVA in Table 3, there was no difference between students' MUA of MLA class and of CNV class. Similar findings of students' MPA namely there was no difference of students' MPA as well (MPA of MLA class was 56,35 and MPA of CNV class was 55,17).

Table 3. Two Path ANOVA of Students' MUA and Students' MPA according to Learning Approach

Source	Sum of Squares	Df	Mean Square	F	Sig.	Ho
MUA:						
Between groups	134,963	1	134,963	0,548	0,462	Accepted*
Inter groups	13289,251	54	246,097			
Total	13424,214	55				
MPA:						
Between groups	19,377	1	19,377	0,086	0,771	Accepted **
Inter groups	12202,051	54	225,964			
Total	12221,429	55				

Note:

(*) Ho: there was no difference of MUA of MLA class and of CNV class;

(**) Ho: there was no difference of MPA of MLA class and of CNV class;

Table 4. Two Path ANOVA of MUA and MPA of Students with Low PMA

Source	Sum of Squares	Df	Mean Square	F	Sig.	Ho
MUA:						
Between groups	481,265	1	481,265	2,26	0,14	Accepted
Inter groups	9362,039	44	212,774	2	0	*
Total	9843,304	45				
MPA:						
Between groups	112,59	1	112,59	0,62	0,43	Accepted
Inter groups	7924,366	44	180,099	5	3	**
Total	8036,957	45				

Note:

(*) Ho: there was no difference of MUA of MLA class and of CNV class with low PMA;

(**) Ho: there was no difference of MPA of MLA class and of CNV class with low PMA ;

c) Based on level of PMA, on medium and low level of PMA there were no difference of students' MUA of MLA class and of CNV class. Similar findings for students' MPA, there were no difference of students' MPA of MLA class and of CNV class as well (see Table 4 and Table 5.).

Table 5. Two Path ANOVA of MUA and MPA of Students with Medium PMA

Source	Sum of Squares	Df	Mean Square	F	Sig.	Ho
MUA:						
Between groups	107,556	1	107,556	0,82	0,39	Accepted
Inter groups	908,667	7	129,810	9	3	*
Total	1016,222	8				
MPA:						
Between groups	12,5	1	12,5	0,07	0,79	Accepted
Inter groups	1237,5	7	176,786	1	8	**
Total	1250,0	8				

Note:

(*) Ho: there was no difference of MUA of MLA class and of CNV class with medium PMA;

(**) Ho: there was no difference of MPA of MLA class and of CNV class with medium PMA ;

d) Based on classification of students' score on MUA and on MPA in Table 6., it was found that number of students who obtain medium and hig scores on MLA class (73 %) was higher than those number students in CNV class (66,6%). However on students' MPA, number of students who obtained medium and high scores on CNV (43,3%) was higher than those number of students in MLA class. Those findings indicated that MLA was little more effective on achieving students' MUA, while CNV was little more effective on obtaining students' MPA.

Table 6. Classification of Students' MUA and MPA on MLA and CNV Classess

	Mathematical Proving Ability (MPA)							
MUA	High		Medium		Low		Total	
	MLA	CNV	MLA	CNV	MLA	CNV	MLA	CNV
High	5 (19,2%)	4 (13,3%)	2 (7,6%)	5 (16,6%)	3 (11,5%)	9 (30%)	10 (38,4%)	18 (60%)
Medium	0	0	2 (7,6%)	0	7 (26,9%)	2 (6,6%)	9 (34,6%)	2 (6,6%)
Low	0	0	1 (3,8%)	4 (13,3%)	6 (23,1)	6 (20%)	7 (26,9%)	10 (33,3%)
Total	5 (19,2%)	4 (13,3%)	5 (19,2%)	9 (30%)	16 (61,5%)	17 (56,6%)	26 (100%)	30 (100%)

e) By using two path ANOVA on Table 7 and Table 8, it was interpreted there were no interaction between learning approach and level of PMA on students' MUA, and on students' MPA. The graph of those interaction were ilustrated on Diagram 1 and Diagram 2.

Table 7. Two Path ANOVA between Learnng Approaches and Level of PMA on Students' MUA

Source	Sum of Squares	Df	Mean Square	F	Sig.	H_0
Learning approach (A)	327,78	1	327,78	1,628	0,208	Accepted *
Level of PMA (B)	1694,09	1	1694,09	8,412	0,005	Rejected **
A x B	0,99	1	0,99	0,005	0,94	Accepted***

Note:

(*) Ho: there was no difference of MUA between MLA class and of CNV class

(**) Ho: there was no difference of MUA between medium PMA and low PMA

(***)Ho: there was no interaction between learning approaches (MLA and CNV) and level of PMA on students' MUA

Table 8. Two Path ANOVA between learnng Approaches and Level of PMA on Students' MPA

Source	Sum of Squares	Df	Mean Square	F	Sig.	H_0
Learning approach (A)	0,778	1	0,778	0,004	0,948	Accepted *
Level of PMA (B)	2069,92	1	2069,92	11,52	0,001	Rejected **
A x B	54,66	1	54,66	0,304	0,584	Accepted***

Note:

(*) Ho: there was no difference of MUA between MLA class and of CNV class

(**) Ho: there was no difference of MUA between medium PMA and low PMA

(***)Ho: there was no interaction between learning approaches (MLA and CNV) and level of PMA on students' MPA

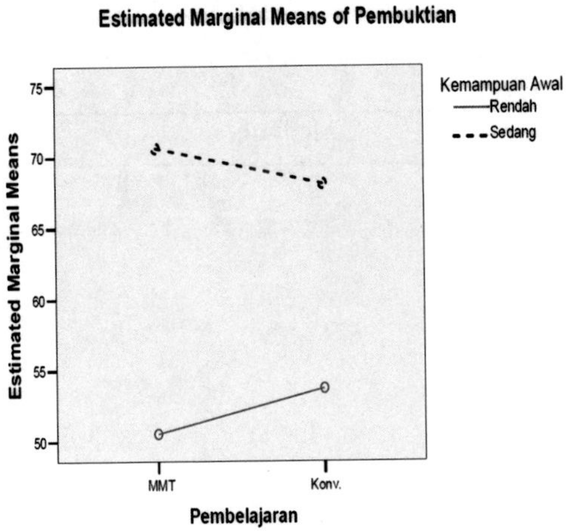

Diagram 1. Interaction between learning approach and level PMA on students' MPA.

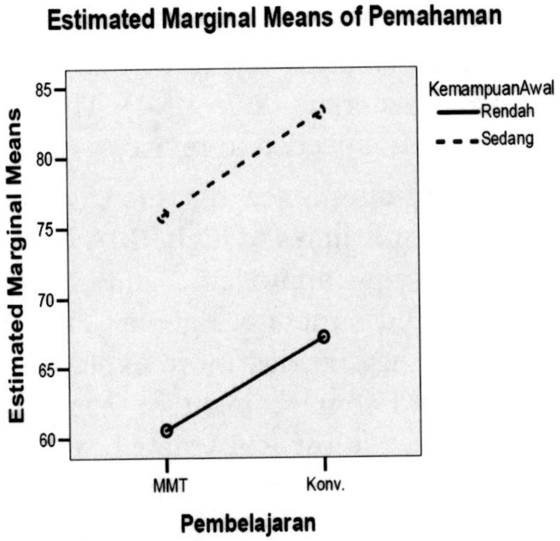

Diagram 2. Interaction between learning approach and level PMA on students' MUA.

f) The result from association analysis by using contigency table and statistics χ^2 cal , on (Tabel 9.), it were found χ^2 cal = 10,8 and χ^2 tab, = 9,48 with dk = 4 and $\alpha = 0{,}05$. Because of χ^2 cal > χ^2 tab. so H0 was rejected. That was meant there was fairly high association between MUA and MPA with coefficient of contigency C = 0,54, from C_{max} = 0,816, or C = 0,66 C_{max}.

Table 8. Association between MUA and MPA on MLA Class

MUA	MPA			
	High	Medium	Low	Total
High	5 (19,2%)	2 (7,6%)	3 (11,5%)	10 (38,4%)
Medium	0	2 (7,6%)	7 (26,9%)	9 (34,6%)
Low	0	1 (3,8%)	6 (23,1)	7 (26,9%)
Total	5 (19,2%)	5 (19,2%)	16 (61,5%)	26 (100%)

g) Students disposition on MLA was classified as positive (2,99 of 4). According to its components, students' positive attitude was on disposition on MLA (3,04 of 4); on learning in small group (3,08 of 4), on presentation task (3,09 of 4), while neutral attitude was on learning materials (2,75 of 4). Those findings indicated that learning material needed to revise.

h) Based on result of questioner of some students, it was disclosed that students with medium and high PMA were pleased and could follow MLA learning approach , and stated to obtain higher ability on advanced structure algebra. They liked presentation task as well, but they needed more explanation about exercises they had done and some content in learning materials. In the contrary students with low PMA stated to dislike, were less able to follow the lesson, and obtained less gain on MUA and MPA abilities, exercise tasks were too difficlut, and they needed some explanation about task had to be done. However they felt to get assisstance in small group learning, and liked presentation task.

i) From analysis of students' work, it were found some students' difficulties namely: 1) Students were not able to generate an example. 2) Students were not able to explain a concept into simpler form of concepts 3) Students did not understand standard mathematical notation and mathematical language 4) Students did not know how to start a proof. 5) Students' concept of understanding were not enough for starting a proof. 6) Students' lack of understanding toward mathematical notation caused them used unexact or confusing mathematical language. 7) Students were not able to seek relation among concepts, definition, theorems, and among theorem and relevant definition.

j) Analysis toward findings of this study compared to previous findings among other things were as follow. 1) Findings of this study on students'understanding and proving on advanced stucture algebra were lower than Kusnandi's findings (2008) on number theory course, and Nurlaelah's findings on mathematical power on structure algebra, and were lower than Dasari's findings (2009) on basic statistics. Those were understandable according to some reason such as course of this study was more complex than all previous courses, and this study was conducted during a half semester while the previous study was conducted during one semester. Those argument indicated that for improving advanced abilities such proving task needed longer time and needed to mastered prerequisite contents. Those pointed out that modified Moore learning approach could be implemented better for advanced courses when it was conducted in a longer time and needed to strengthen students understanding on prerequicite courses. 2) The previous studies with modified Moore method were conducted to medium-high students, while this study was conducted to medium-low students. So it was rational that findings of this study was lower than findings of previous studies. However there were similarity findings on students' diffculties between previous studies and this study. Those argument pointed out that modified Moore learning approach could be implemented for various level of students' PMA by revising and adding some explanation and excersices task for students with low level of PMA.

CONCLUSSION AND RECOMMENDATION

1. Conclusion

Based on findings of this study and discussion, it was obtained some conclusion as follow.

There were no differences on mathematical understanding and proving abilities of students taught by modified Moore learning approach and taught by conventional teaching either entirely or in each level of students' prior mathematics ability. Those abilities were classified as fairly good, however according to number of students who obtained medium and high scores on mathematical understanding ability, in modified Moore learning approach there more students than number of students with medium and high mathematical understanding ability of conventional teaching. Those condition pointed out that modified Moore learning approach was a little more effective compared to conventional teaching on improving mathematical understanding and proving abilities.

Moreover, during modified learning approach students performed active learning independently, were unafraid to pose question and to present and to explain their ideas in front of class, while students on conventional class were more passive in solving problems and tended to wait lecturer's explanation. Although there were no difference of students' mathematical and proving abilities, but according to learning process, the modified Moore learning approach gave more chances for students to learn actively. Those ilustration supported that modified learning approach was better than conventional teaching in improving mathematical understanding and proving abilities and habits of good learning. The last phrase was very importance for learning further advanced mathematics courses.

Other conclussion of this study was there were no interaction between learning approach and prior mathematics ability on sudents' mathematical understanding and proving abilities. Moreover there was quite strong association between mathematical understanding ability and mathematical proving ability. According to students' opinion on mathematics learning, students of modified Moore learning approach performed positive disposition, namely: they were pleased on modified Moore learning approach, they liked to learn in small group and to present and to explain their work in front of class, and they felt to obtain gain on mathematical understanding and proving abilities on an Advanced Structure Algebra course. However they proposed more explanation about exercises they had done and some contents in learning materials.

Besides those conclussion, there were still some students' difficulties on solving mathematical proof problem namely:

1) students were unable to generate an example.
2) Students were unable to explain a concept into simpler form of concepts
3) Students did not understand standard mathematical notation and mathematical language.
4) Students did not know to start a proof.
5) Students' concept understanding did not satisfy for starting a proof.
6) Students' lack of understanding toward mathematical notation caused they used unexact or confusing mathematical language.
7) Students were unable to seek relation among concepts, definition, theorems, and among theorem and relevant definition.

2. Recommendation

According to discussion of study findings and those conclussion, it were proposed some recommendation as follow.

Considering that mathematical understanding and proving abilities were essensial and difficult tasks and they needed more longer time to learn, so it was recommended that in implementing modified Moore learning approach lecturer should be more patient in giving guidance and presenting excersices task so that students were motivated to compose mathematical proof by themselves, and conduct the lesson in adequate time, namely in one semester. Besides that, lecturer should have cultivated students habits of positive learning disposition continuosly that was needed for learning further advanced mathematics courses.

Previous learning materials of Advanced Structure Algebra should be completed with more examples, ilustration, and excercises with various level of difficulty, and it should be accompanied with relevant guidance and questions that motivated students to learn actively. In implementing the lesson besides students' presentation task, it was also recommended to carried out discussion on students' presentation and some selected excercises tasks in learning material. Part of difficult proof problems that needed more time to solve could be given as home work or group task.

Besides to complete learning material for conducting a similar study on proving ability, it was also recommended to conduct study by implementing modified Moore learning approach for improving other high mathematical thinking such as mathematical critical and creative thinking, communication, reasoning, and problem solving and improving self regulated learning, habits of positive learning such as critical and creative disposition either in abstract algebra or other advance mathematics courses.

REFERENCES

1. Alfeld, P. (2004). Understanding Mathematics, a Study Guide. [Online]. Available at: http://www.math.utah.edu/ [May 20, 2010]
2. Arikunto, S. (2002). Dasar-dasar Evaluasi Pendidikan. Jakarta: PT. Bumi Aksara.
3. Arnawa, I M. (2006). Meningkatkan Kemampuan Pembuktian Mahasiswa dalam Aljabar Abstrak melalui Pembelajaran berdasarkan Teori APOS. Disertation at Post Graduate Studies at Indonesia University of Education. Bandung, Indonesia: not published.
4. Artemov, S. N. (2001). Explicit Provability and Constructive Semantic. In The Bulletin of Symbolic Logic [Online], Vol. 7, No. 1, March, (2001), 1-3. Available at: http://www.jstor.org/ [March 6, 2007]
5. Baker, D. & Campbell, C. (2004). Fostering The Development of Mathematical Thinking: Observations from A Proofs Course. In Primus: Problem, Resources, and Issues in athematics Undergraduates Studies. [Online]. Available at: http://findarticles.com/ [February 13, 2007]
6. Barnard, T. (2000). Why Are Proofs Difficult? In The Mathematical Gazette [Online], Vol. 84, No. 501, November, (2000), 415-422. Available at: http://www.jstor.org/ [February 13, 2007]
7. Chalice, D.R. (1995). How to Teach a Class by The Modified Moore Method. In The American Mathematical Monthly [Online], Vol. 102, No. 4, April, (1995), 317-321. Available at: http://www.jstor.org/ [July 11, 2007]
8. Cohen, D.W. (1982). A Modified Moore Method for Teaching Undergraduate Mathematics. In The American Mathematical Monthly [Online], Vol. 89, No. 7, August- September, (1982), 473-474 & 487-490. Available at: http://www.jstor.org/ [July 11, 2007]
9. Dasari, D. (2009). Meningkatkan Kemampuan Penalaran Statistik melalui Pendekatan Model PACE. Disertation at Post Graduate Studies at Indonesia University of Education. Bandung, Indonesia: not published.
10. Dancis, J. & Davidson, N. (1970) Texas Method and the Small Group Discovery Method. [Online]. Available at: http://www.discovery.utexas.edu/ [June 20, 2007]
11. Devlin, K. (1996). Mathematical Proofs in The Computer Age. In The Mathematical Gazette [Online], Vol. 80, No. 487, Centenary Issue,

March., (1996), 149-162. Available at: http://www.jstor.org/ [February 13, 2007]

12. Gallian, J. A. (2006). Contemporary Abstract Algebra. Boston, MA: Houghton Mifflin Company.
13. Hanna, G. & Jahnke, H.N. (1993). Proof and Application. In Educational Studies in Mathematics [Online], Vol. 24, 421-438. Available at: http://www.jstor.org/ [March 6, 2007]
14. ______________________. (1996). Proof and Proving. In A.J. Bishop et.al (Eds), International Handbook of Mathematics Education, Kluwer Academic Publishers, Netherland.
15. Hanna, G. & Barbeau, E. (No Year). Proof in Mathematics. [Online]. Available at: http://www.math.utoronto.ca/barbeau/hannajoint.pdf/ [April 25, 2007]
16. Hersh, R. (1993). Proving is Convincing and Explaining. In Educational Studies in Mathematics [Online], Vol. 24, No. 4, Aspects of Proof, (1993), 389-399. Available at: http://www.jstor.org/ [February 20, 2007]
17. Kusnandi. (2008). Pembelajaran Matematika dengan Strategi Abduktif-Deduktif untuk Menumbuhkembangkan Kemampuan Membuktikan pada Mahasiswa. Disertation at Post Graduate Studies at Indonesia University of Education. Bandung, Indonesia: not published.
18. Lamport, L. (1995). How to Write a Proof. In The American Mathematical Monthly [Online], Vol. 102, No. 7, August - September, (1995), 600-608. Available at: http://www.jstor.org/ [March 6, 2007]
19. Leron, U. (1983). Structuring Mathematical Proofs. In The American Mathematical Monthly [Online], Vol. 90, No. 3, March, (1983), 174-185. Available at: http://www.jstor.org/ [February 13, 2007]
20. Mahavier, W.S. (1999). What is The Moore Method? In Primus [Online], Vol. 9, December, (1999), 339-254. Available at: http://www.discovery. utexas. edu/ [June 20, 2007]
21. Mahavier, W.T., May, E.L., & Parker, G.E. (2006). A Quick-Start Guide to the Moore Method. [Online]. Available at: http://www.discovery.utexas.edu/ [June 20, 2007]
22. Mahavier, W.T. & Mahavier, W.S. (No Year). Analysis. In MathNerds Course Guide Collection [Online]. Available at: http://www.jiblm.org/ [May 6, 2008]

23. Maya, R. (2006). Pembelajaran dengan Pendekatan Kombinasi Langsung-Tidak Langsung untuk Mengembangkan Kemampuan Berpikir Matematik Siswa SMA. Thesis at Post Graduate Studies at Indonesia University of Education. Bandung, Indonesia: not published.
24. Maya, R. (2011). Pengaruh Pembelajaran dengan Metode Moore Termodifikasi terhadap pencapaian kemampuan Pemahaman dan Pembuktian Matematik Mahasiswa. Disertation at Post Graduate Studies at Indonesia University of Education. Bandung, Indonesia: not published.
25. Moore, R. C. (1994). Making The Transition to Formal Proof. In Educational Studies in Mathematics [Online], Vol. 27, No. 3, October, (1994), 249-266. Available at: http://www.jstor.org/ [February 20, 2007]
26. Neuberger, J.W. (2003). Analysis. In Class Notes: The Legacy of R.L. Moore Project [on CD]. The Educational Advancement Foundation: Austin – Texas.
27. Nurlaelah, E. (2009). Pencapaian Daya dan Kreativitas Matematik Mahasiswa Calon Guru melalui Pembelajaran Berdasarkan Teori Apos. Disertation at Post Graduate Studies at Indonesia University of Education. Bandung, Indonesia, not published.
28. Selden, J. & Selden, A. (1995). Unpacking The Logic of Statement. In Educational Studies in Mathematics [Online], Vol. 29, No. 2, Advanced Mathematical Thinking, September, (1995), 123-151. Available at: http://www.jstor.org/ [March 6, 2007]
29. Selden, J., Selden, A. & McKee, K. (2009). Improving Advanced Students' Proving Abilities. [Online]. Available at: http://tsg.icme11.org/document /get/729/ [May 21, 2009]
30. Solow, D. (1990). How to Read and Do Proofs: An Introduction to Mathematical Thought Processes. Kanada: John Wiley & Sons, Inc.
31. Sumarmo, U. (1987). Kemampuan Pemahaman dan Penalaran Matematika Siswa SMA Dikaitkan dengan Kemampuan Penalaran Logik Siswa Dan Beberapa Unsur Proses Belajar Mengajar. Disertation at Post Graduate Studies at Indonesia University of Education. Bandung, Indonesia: not published.
32. __________. (2002). Daya dan Disposisi Matematik: Apa, Mengapa dan Bagaimana Dikembangkan pada Siswa Sekolah Dasar dan Menengah. Paper presented to One day Seminar at ITB Mathematics Departement at Oktober 2002.

33. Tall, D. (1999). The Cognitive Development of Proof: Is Mathematical Proof For All or For Some? [Online]. Available at: http://www.warwick.ac.uk/ [June 20, 1999]
34. Wahyudin. (1999). Kemampuan Guru Matematika, Calon Guru Matematika, dan Siswa dalam Mata Pelajaran Matematika. Disertation at Post Graduate Studies at Indonesia University of Education. Bandung, Indonesia: not published.
35. Weber, K. (2001). Student Difficulty in Constructing Proofs: The Need for Strategic Knowledge. In Educational Studies in Mathematics [Online], Volume 48: 101- 119. Available at: http://www.jstor.org/ [March 6, 2007]
36. _________. (2003). Students' Difficulties with Proof. [Online]. Available at: http://www.maa.org/ [February 7, 2007]

Chapter

UNDERSTANDING ON STRATEGIES OF TEACHING MATHEMATICAL PROOF FOR UNDERGRADUATE STUDENTS

10

Syamsuri, Indiana Marethi, and Anwar Mutaqin

FKIPUniversitas Sultan Ageng Tirtayasa, Indonesia

ABSTRACT

Many researches revealed that many students have difficulties in constructing proofs. Based on our empirical data, we develop a quadrant model to describe students' classification of proof result. The quadrant model classifies a students' proof construction based on the result of mathematical thinking. The aim of this article is to describe a students' comprehension of proof based on the quadrant model in order to give appropriate suggested learning. The research is an explorative research and was conducted on 26 students majored in mathematics education in public university in Banten province,

Citation: (APA): Syamsuri, S., Marethi, I., & Mutaqin, A. (2018). Understanding on strategies of teaching mathematical proof for undergraduate students. Jurnal Cakrawala Pendidikan, 37(2). (12 pages).

Copyright: © Creative Commons Attribution-ShareAlike 4.0 International License (http://creativecommons.org/licenses/by-sa/4.0/).

Indonesia. The main instrument in explorative research was researcher itself. The support instruments are proving task and interview guides. These instruments were validated from two lecturers in order to guarantee the quality of instruments. Based on the results, some appropriate learning activities should be designed to support the students' characteristics from each quadrant, i.e: a hermeneutics approach, using the two column form method, learning using worked-example, or using structural method.

Keywords: proof, proving learning, undergraduate, quadrant model

INTRODUCTION

Proof and proving in mathematics education are an important part of mathematics, as a pillar of the mathematics building. Therefore, mathematics education especially in mathematics learning in university emphasizes the constructing of mathematical proof. Students who enter college level should develop a formal mathematical knowledge.

Many researchers have investigated about mathematical proving and its learning. Some researchers wrote about students' proof schemes for mathematical proving (Iannone, Inglis, MejíaRamos, Simpson, & Weber, 2011; Lee, 2016; Syamsuri & Santosa, 2017). And another ones wrote about thinking process in proof production (Weber & Alcock, 2004; Sean Larsen & Zandieh, 2008; Stylianides & Stylianides, 2009; Syamsuri, Purwanto, Subanji, & Irawati, 2017) and its material learning (Fadillah & Jamilah, 2016). It is indicated that teaching mathematical proof is important in mathematics education.

Students are introduced to formal proof in the study of mathematics at university. Formal proof as a process begins from explicit quantified definitions and deduces that other properties hold as a consequence (Tall et al., 2012). Learning how to construct formal proof is given to make sense a formal definition that can be used in building the basis of deduction theorem. It is indicated that undergraduate students must develop their mathematics knowledge formally. A formal proof is based on rigorous logic and can be communicated like a discussion of mathematicians on scientific forum. Therefore, undergraduate students need proving exercises so that they will able to understand a formal mathematics structure.

The process of proving a mathematical proposition is a sequence of mental and physical actions, such as writing, thinking tobegin the proof, draw diagrams, to reflect on previous actions or trying to remember the example. The process of proof formation of a theorem of or statement is more complex than the proof itself (Selden & Selden, 2003). Therefore, teacher is needed for facilitating in learning mathematics thus facilitate students in mathematical proofs. In mathematics, one of the teachers'aid to help the students in order to make it easy to perform mathematical proofs is to make it into a tangible proof (Sowder & Harel, 2003). So, teaching assistance on mathematical proof to the students can be done gradually and trying to create proof into something tangible.

Many researches revealed that many students have difficulties in constructing proofs (Moore, 1994; Baker & Campbell, 2004; Weber, 2001). Moore stated that there are 7 difficulties in mathematical proving, i.e.: the students (1) did not know the definitions, (2) had little intuitive understanding of the concepts, (3) had inadequate concept images for doing the proofs, (4) were unable, or unwilling, to generate and use their own examples, (5) were unable to understand and use mathematical language and notation, (6) did not know how to use definitions to obtain the overall structure of proofs, and (7) did not know how to begin proofs. Meanwhile, Baker and Campbell (2004) explained that students' difficulties in proving were caused by some factors, i.e.: (1) understanding of the rules and nature of proof, (2) conceptual understanding, (3) proof techniques and strategies, and (4) cognitive load. In addition, Weber (2001) asserted that: (1) students' difficulties included their misunderstanding of a theorem or a concept and misapplying it, (2) students' inadequate conception knowledge about mathematical proof, and (3) students' inadequacy in developing strategies for proof. Therefore, the difficulties that students face consist of uncompleted conceptual understanding and incorrect proving strategy.

Another research was revealed a proof assessment instruments. Andrew created The Proof Error Evaluation Tools (PEET) to assess a students' proof. A PEET consists of proof structure and conceptual understanding (Andrew, 2009). And also, Mejia-Ramos et al. (2012) developed an assessment model for proof comprehension in undergraduate mathematics. They stated that the model described ways to assess students' understanding of seven different aspects of a proof. These types of assessment are: (1) Meaning of terms and statements, (2) Logical status of statements and proof framework, (3) Justification of claims, (4) Summarizing via high-level ideas, (5) Identifying the modular structure, (6) Transferring the general ideas or methods to another context, and (7) Illustrating with examples. Therefore, the assessment consists of conceptual understanding and proving strategy.

Based on our empirical data, we develop a quadrant model to describe students' classification of proof production. Figure 1 describes the quadrant model.

The quadrant model classifies a students' proof construction based on result of mathematical thinking. Therefore, investigation of thinking process or outcome of thinking process is necessary for selecting an appropriate learning strategies. Many prior researches revealed outcome thinking process about proof construction. For instance, a students have difficulties in proof construction (Moore, 1994; Gibson, 1998; Baker & Campbell, 2004; Weber, 2006) and so students made errors in proof construction (Selden & Selden, 2003; Sowder & Harel, 2003). Nevertheless, researches which reveal mental structures and mental mechanisms are rarely found. Whereas knowing students' thinking process that consisting of both mental structures and mental mechanisms can help teachers or lecturers in order to give an appropriate learning assistance. Some suitable learning activities should be designed to support the construction of this thinking process. In addition, if students' thinking process is incorrect, then refinement thinking process can be easy in order to it does not occur in next learning. In this article, we will describe a students' comprehension of proof based on quadrant model in Figure 1.

RESEARCH METHOD

Participants

The research was conducted on 26 students majored in mathematics education in public university at Banten province, Indonesia. The consideration of that was because the students were able to think a formal proof in mathematics. And also, the students are more easily managed by the researcher to follow the procedures empirically planned, so that the data obtained is a students' actual of reflection thinking. Giving a proving-question in the beginning, aim to select students who will be the subjects, which further deepened with the interview. The diagram of selecting research subject is Figure 2.

Instruments

The main instrument in explorative research was researcher itself. The support instruments are proving-task and interview guides. These instruments were evaluated and validated from two lecturers in order to guarantee the quality of instruments. The interview is open and it's needed to reveal students' response about proof comprehension. Procedures to obtain data are 1) subject is given the task proving and asked him/her to accomplish the task by think-aloud. And then 2) subject is interviewed based onthe-task. Therefore, the scratch of proving-task and transcript of the interview is obtained. The proving-task is in the following.

Prove: For any positive integer m&n, if m^2 and n^2 are divisible by 3, then m+n is divisible by 3.

We used this task because some methods can be used for solving, i.e.: direct proof, contradiction, and contrapositive. Besides, we would like to test students' comprehension about mathematical induction method, because some students have an opinion that using mathematical induction to prove a number which is "divisible by 3".

The interview guides is created for confirmation and clarification about students' proof comprehension. According to MejiaRamos et al. (2012) we compile the interview questions, i.e.: (1) What mathematical concepts that are used in the proving task? Explain. (2) How do you accomplish the task? (3) Why do you argue with this step?, (4) What is the big idea in your proof construction?, (5) What mathematical proposition that supports your proof construction?, (6) Would you like to give an example of the task?, and (7) In the task, if number "3" is replaced by "4", what is your opinion?

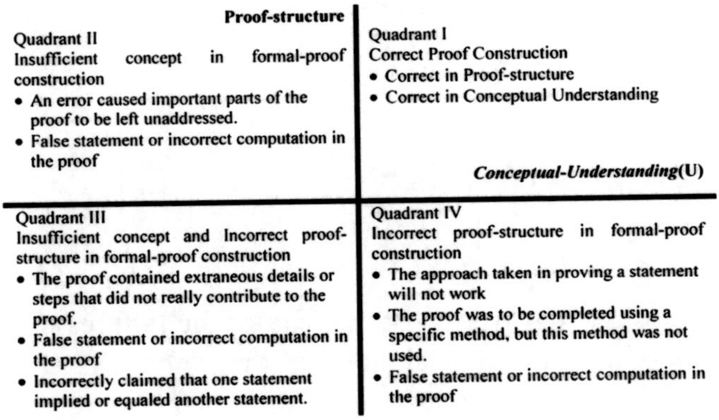

Figure 1. Quadrant model of students formal-proof construction (Syamsuri, Purwanto, Subanji, & Irawati, 2016).

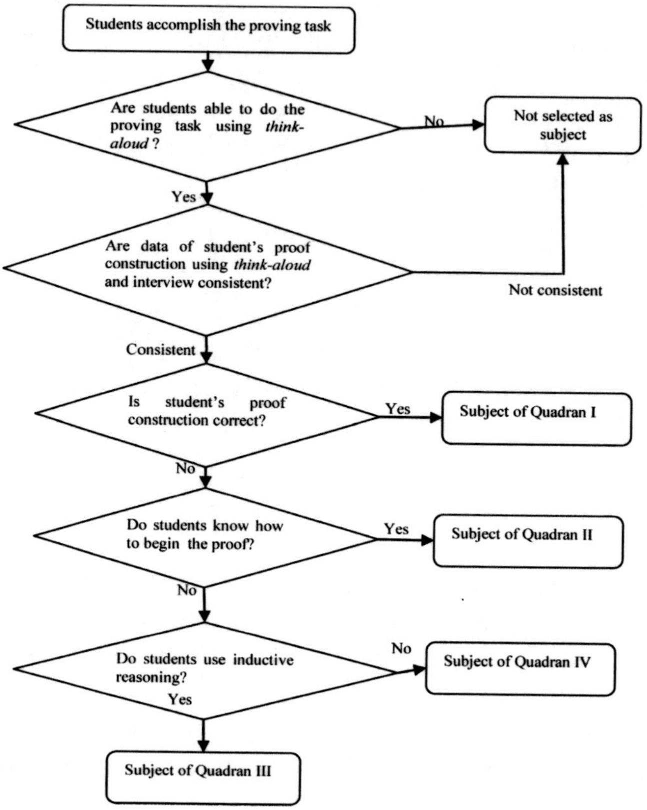

Figure 2. Process of selecting subject.

RESULTS AND ANALYSIS

Results

The following are the description of the data of the failures experienced by students. The tool used in the evaluation of the proofs was the Error Proof Evaluation Tools (PEET) developed by Andrew (2009).

The most frequent errors were an error in claiming a statement, the error in making a statement, and one in choosing a proof type. Errors in claims occurred at the beginning of proof by stating $n^2 = (3a)^2$, with a is an integer. The error in choosing a proof type was by choosing an incorrect method.

Based on PEET results, students' response can be further classified into three categories. First, there are 6 students who are wrong in extracting a concept so become wrong in expressing formal proof (corresponding to U4-2 and U5-1 on Table 1 and 2). Secondly, as many as 10 students less likely to complete in deepening the concept, so there is a missing piece of concept (corresponding to S2, S3, U4-1, U5-1 and U5-2 on Table 1 and 2). Finally, there are 5 students who committed errors in selecting methods of proof in the initial stage of logic verification (corresponding to S1-2, S2, and S7-1 on Table 1 and 2). Therefore, of the 26 students who were given the task of mathematical proof, there were 5 students, 6 students, 10 students and 5 students in Quadrant I, Quadrant II, Quadrant III and Quadrant IV, respectively. We will describe characteristics of proof comprehension of S1, S2, S3, and S4 as subjects in Quadrant I, Quadrant II, Quadrant III and Quadrant IV, respectively. The students' proof construction is below.

Table 1. Students' error on proof-structure (S) in constructing formal proof

No	Element of Proof Structure	Component Descriptions	Number of Students who made error
1	Proof setup	S1. Introduced variables without defining them or performed operations that were undefined.	8
		S1. The approach taken in proving a statement will not work.	10
		S1. The proof was to be completed using a specific method, but this method was not used.	10
2	Assumption	S2. Made a false assumption somewhere in the proof.	15
3	Linear/sequential order	S3. Didn't proceed through the proof in a linear fashion, and ideas were not in logical order.	10
4	Stray details/ conciseness	S4. The proof contained extraneous details or steps that did not really contribute to the proof.	6
		S4. The length of the proof was unnecessarily long and thus extremely difficult to follow.	2
5	Neat presentation	S5. The write-up was illegible at times, making it difficult to read and/or understand.	0
6	Technology's place	S6. Relied too much on calculator or computer-generated information in one step of the proof.	0
7	Proof type	S7. Needed to show p → q but did not show directly, or by -q →-p, or by contradiction	15
		S7. Only gave an example to establish the truth of a mathematical statement.	10
8	Correct use of symbols/notation	S8. Used nonstandard or confusing notation.	0

Table 2. Students' error on conceptual-understanding (U) in constructing formal proof

Table 2. Students' error on conceptual-understanding (U) in constructing formal proof

No	Element of Mathematical Concept	Component Descriptions	Number of Students who made error
1	Sufficient details	U1. Wrote a statement that was not justified, explained, or verified.	10
2	Clarity	U2. Wrote a statement or paragraph that was ambiguous, confusing, and/or unnecessarily complex.	3
3	Pictures in the proof	U3. Failed to include an illustrative picture that would make the proof easier to understand.	0
		U3. Relied too much on a picture to prove something was true.	0
4	Crucial step/main idea	U4. Did not sufficiently justify a crucial step in the proof.	18
		U4. An error caused important parts of the proof to be left unaddressed.	10
5	Correct implications and statements	U5. Made a false statement or incorrect computation in the proof.	21
		U5. Incorrectly claimed that one statement implied or equaled another statement.	21
6	All cases present	U6. Included some cases but not others (which were not trivial).	1

Students' proof construction depends on their understanding about the proving task. A different proof construction among students indicated that the understanding of proof is different too. Subject S1 knew how to begin the proof. And also, he could connect to appropriate mathematical concepts well. Subject S2 knew how to begin the proof, but he fails in connecting mathematical concept. Subject S3 didn't know how to begin the proof, and so she solves the proving task using inductive reasoning. Subject S4 didn't know how to begin the proof, and so she fails in beginning a proof. Therefore, we suggest that the proof comprehension of these students is different. Furthermore, the different proof comprehension influences a learning strategy for students in order to afford a mathematics proving.

Table 3. Students' proof comprehension

No.	Component of Proof Comprehension	Subject S1	Subject S2	Subject S3	Subject S4
1	Logical status of statements and proof framework	Direct proof Using congruent modulo number	Direct proof Using congruent modulo number	Inductive reasoning	Direct proof Failure in beginning a proof
2	Justification of claims	Create 5 correct claims	Create a correct claim and an incorrect claim	Create only a clarified claim	Create an incorrect claim in beginning proof
3	Meaning of terms and statements	- Integer - quadratic number - number that divisible by 3 - congruent modulo number	- integer - quadratic number - number that divisible by 3 - congruent modulo number	- integer - number that divisible by 3 - simple description of concepts	- integer - number that divisible by 3 - adequacy of concept image

4	Summarizing via high-level ideas	- Showed that if both m^2 and n^2 are divisible by 3 then m^2-n^2 is divisible by 3 - Showed that if m^2-n^2 is divisible by 3 then $(m+n)(m-n)$ is divisible by 3 - Showed that if $(m+n)(m-n)$ is divisible by 3 then $m+n$ is divisible by 3	- Showed that if both m^2 and n^2 are divisible by 3 then m^2-n^2 is divisible by 3 - Showed that if m^2-n^2 is divisible by 3 then $(m+n)(m-n)$ is divisible by 3	- Not appeared	- Define that m=3a - Define that n=6a - Verified that m^2 is divisible by 3 - Verified that n^2 is divisible by 3 - Verified that $m+n$ is divisible by 3
5	Identifying the modular structure	There are some theorems - If a|b and a|c then a|b-c, a|b+c - If c|ab then c|a or c|b	- Not appeared	Not appeared	- Not appeared
6	Illustrating with examples	- Give an example that m^2=6^2 and n^2=3^2	- Failure in giving an example; m=2 and n=7, because he believe that the proposition is incorrect	- m=6 and n=3, but she determined m and n before m^2 and n^2.	- She stated that the proposition is incorrect because not all of m^2 and n^2 is divisible by 3.
7	Transferring the general ideas or methods to another context	Before reflective thinking, he stated that "If both m^2 and n^2 are divisible by 4 then $m+n$ is divisible by 4" is correct proposition, nevertheless when asked to give an example, he gave m=2 and n=4 that is a counter-example of the proposition. Immediately, S1 aware a failure and stated that the proposition is incorrect	S2 stated that "If both m^2 and n^2 are divisible by 4 then $m+n$ is divisible by 4" is incorrect proposition, because the before proposition is incorrect. In addition, S2 tried m=2 and n=4 that is a counter-example of the proposition.	S3 stated that "If both m^2 and n^2 are divisible by 4 then $m+n$ is divisible by 4" is correct proposition. She argued using several example.	S4 stated that "If both m^2 and n^2 are divisible by 4 then $m+n$ is divisible by 4" is incorrect proposition. She argued using incorrect reason.

Characteristics of Proof Comprehension and Appropriate Learning in Quadrant I

Subject S1 has solved the proving task. He knew how to begin the proof. And also, he could connect it to appropriate mathematical concepts well. According to Table 3, Subject S1 could construct a proof in thought-experiment level (Balacheff, 1988;Varghese, 2011) and thought level 2 (Van Dormolen, 1977). This level encouraged student to construct a proof using definition and rigorous logics (Weber, 2004), and also using deductive reasoning and symbolic. Therefore, according to the three worlds, the thinking process of S1 is in axiomatic-symbolic development (Tall, 2010).

According to Table 1 above, characteristics of proof comprehension is a complete comprehension. It indicated that both a local proof comprehension (component 1-3) and a holistic proof comprehension (component 4-7) are complete. Therefore, Subject S1 was able to construct proof correctly. According to Polya (Meel, 2003), this student's understanding is at intuitive level. It showed that in his mind, there are three worlds of knowledge in mathematical understanding, i.e.: applications, meanings, and logical relationship (Lehman, 1977). Meanwhile, following Skemp's understanding, Subject S1 has a relational understanding. And so, type of understanding of the subject is inventing layer (Pirie & Kieren, 1989). Therefore, proof comprehension in this subject is called completed comprehension.

According to students' characteristics in this quadrant, we suggest that all learning strategy is appropriate for students in the quadrant. Mathematics is not only a subject to be learned and taught, but it is to be produced. However, with hermeneutics, it would be easy to develop our own idea and produce mathematics. Thus, some appropriate learning strategies for students in this quadrant are hermeneutics approach. This was the best opportunity for students to learn from the pioneers how to develop a new idea and create something new. Only with hermeneutics, teaching and learning mathematics and also research in mathematics could be flourishing and fruitful (Djauhari, 2015).

Characteristics of Proof Comprehension and Appropriate Learning in Quadrant II

Subject S2 knew how to begin the proof, but he failed in connecting mathematical concept. According to Table 3, Subject S2 could construct a proof in thought-experiment level (Balacheff, 1988;Varghese, 2011) and thought level 2 (Van Dormolen, 1977). This level encouraged student to construct a proof using definition and rigorous logics and also using deductive reasoning and symbolic.

According to Moore, Subject S2 has difficulty in understanding a concept, so his concept image is not enough for constructing a proof (Moore, 1994). Following Gibson's suggestion, Subject S3 has some difficulties in conceptual understanding, proof techniques and strategies. According to Weber (2004) opinion's, S2's difficulties included their misunderstanding of a theorem or a concept and misapplying it and his inadequate conception knowledge about mathematical proof. Thus, Subject S2's fault is in connecting mathematical concept.

According to Table 1 above, characteristics of proof comprehension is an uncompleted comprehension. It indicated that both a local proof comprehension (component 1-3) and a holistic proof comprehension (component 4-7) are incomplete. According to Polya (Meel, 2003), this student's understanding is at rational level. Meanwhile, following Skemp's understanding, type of understanding of Subject S2 is a relational understanding (Skemp, 1978). Whereas, another suggestion stated that type of understanding of the subject is layer observing (Pirie & Kieren, 1989). Therefore, proof comprehension in this subject is called uncompleted comprehension.

We suggest that one of appropriate leaning for this quadrant is using the two-column form method. This method can play a dual role, so student can connect some mathematical concept easily. While in many cases teachers do use the form in a way that limits students ability to think flexibly when formulating an argument, in other cases teachers can use the form in a way that enables greater flexibility in reasoning and proving (Weiss, Herbst, & Chen, 2009).

Besides, students' characteristic is weak validation skill. It indicated that proof comprehension component of "transferring the general ideas or methods to another context" is a wrong validation. Finally, the students may not be able to distinguish proofs from supplementary or explanatory and so may not be able to distinguish proofs from supplementary or explanatory comments. It might be good to present material in a way that makes these comments. For example, explanations of definitions, illustrations of relevant concepts, cautionary remarks, and even remarks of proofs, except those needed to organize their linear presentation and help readers with validation, might best be treated as annotations. This might simultaneously provide prototypical examples for enhancing students' conceptions of proof and also encourage them to validate proofs carefully (Selden & Selden, 1995).

Characteristics of Proof Comprehension and Appropriate Learning in Quadrant III

Subject S3 didn't know how to begin the proof. According to Table 3, Subject S3 could construct a proof in naïve-empirism level (Balacheff, 1988;Varghese, 2011) and ground level (Van Dormolen, 1977). This level encourage student to construct a proof using inductive reasoning. Therefore, according to Tall, the thinking process of S3 is in embodiment world (Tall, 2010).

According to Moore, Subject S3 has difficulty in understanding a concept, so his concept image is not enough for constructing a proof. Subject S3 has a misunderstanding a theorem or a concept and misapplying it (Moore, 1994).

Subject S3 has a difficulty to think deductive reasoning (Recio & Godino, 2001). And so, these difficulties encourage her to construct a proof using inductive reasoning.

According to Table 1 above, characteristics of proof comprehension is uncompleted comprehension. It indicated that both a local proof comprehension (component 1-3) and a holistic proof comprehension (component 4-7) are incomplete. According to Polya (Meel, 2003), this student's understanding is at inductive level. Meanwhile, following Skemp's term, type of understanding of Subject S1 is instrumental understanding (Skemp, 1978). Whereas, another suggestion stated that type of understanding of the subject is layer image having (Pirie & Kieren, 1989). Therefore, the proof comprehension in this subject is called uncompleted comprehension.

The students' characteristics have no proof-structure, and also have a little conceptual understanding. One of learning strategies for this quadrant is asking a generate example. Many mathematics education researchers have suggested that asking learners to generate examples of mathematical concepts is an effective way of learning about novel concept (Iannone et al., 2011).

Of course, the students need assistance to refine a proof-structure and conceptual understanding about proposition. Another method is learning using worked-example (Retnowati, Ayres, & Sweller, 2010; Margulieux & Catrambone, 2016; McLaren, Van Gog, Ganoe, Karabinos, & Yaron, 2016)and what kind of assistance to provide, is a much-debated problem in research on learning and instruction. This study presents two multi-session classroom experiments in the domain of chemistry, comparing the effectiveness and efficiency of three high-assistance (worked examples, tutored problems, and erroneous examples. Weber stated that proving is a problem solving activity (Weber, 2005). According to Retnowati et al. that students could understand the material more easily using worked examples than when solving problems. And also, Margulieux and Catrambone stated that worked-example as guided instruction is important for novices because it helps them to organize and use new informa tion more effectively. The students who are unable to construct a formal proof are novice students. Worked-example is example how to proving a proposition, and involved arguments in every step. Worked-example can refine students' knowledge, 1) how to begin a proof, 2) how to understand about end of proof, 3) how to give argumentation for each step, and 4) how to select mathematical concept needed. Point 1) and 2) related to refine proof-structure, in addition point 3) and 4) related to refine a conceptual understanding. Therefore, we suggest that the proving process will be generated through this method.

Characteristics of Proof Comprehension and Appropriate Learning in Quadrant IV

Subject S4 didn't know how to begin the proof, so she failed in beginning a proof. According to Table 3, Subject S4 could construct a proof in thought-experiment level (Balacheff, 1988;Varghese, 2011) and thought level 2 (Van Dormolen, 1977). This level encourage student to construct a proof using definition and also using deductive reasoning. Therefore, according to Tall, the thinking process of S4 is in axiomatic symbolic development (Tall, 2010).

According to Moore, Subject S4 did not know how to use definitions to obtain the overall structure of proofs (Moore, 1994). Following Gibson's suggestion, Subject S4 has difficulties in conceptual understanding and proof techniques and strategies. According to (Weber, 2004) Weber's opinion, S4' difficulties included their misunderstanding of a theorem or a concept and misapplying it and his inadequate conception knowledge about mathematical proof. Therefore, Subject S4 failed in beginning a proof.

According to Table 1 above, characteristics of proof comprehension is uncompleted comprehension. It indicated that both a local proof comprehension (component 1-3) and a holistic proof comprehension (component 4-7) are incomplete. According to Polya(Meel, 2003), student's understanding is at rational level. Meanwhile, following Skemp's term, type of understanding of Subject S1 is relational understanding. Whereas, following other theory about understanding stated that type of understanding of the subject is formalizing layer (Pirie & Kieren, 1989). Therefore, proof comprehension in this subject is called uncompleted comprehension.

The students' characteristics have no proof structure, so that she/he failed in beginning a proof. One of learning strategies for this quadrant is structural method. Among attempts to improve students' proof comprehension in constructing a proof, one can distinguish two broad approaches: (a) changing the presentation of the proof and (b) changing the way a student engages with it (Leron, 1983). The aim of changing the presentation is to make it easier in understanding proof-structure. A structured proof is arranged in levels, with the main ideas and approach given at the top level and subsequent levels giving details and justifications of each of the steps in the preceding levels.

When the presentation is changed, explanations must be provided by the instructor. The explanations can help students who construct incorrect-proof to understand how constructing correct-proof. In terms of Mejía-Ramos et al.'s framework, a structured proof is designed to facilitate understanding higher-level ideas and identifying modular structure, though it does so at the expense of separating some claims from their supporting data and warrant. These changes are reflected in empirical research on the efficacy of structured proofs. Fuller et al. found that, compared to those who read a traditional proof, students who read structured proofs were more successful at summarizing the key ideas of the proof (Fuller et al., 2011). These way can help students who are in Quadrant-IV.

CONCLUSION

We develop a quadrant model to describe students' classification of proof production. Characteristics of students' proof comprehension in first quadrant are completed comprehension. According to students' characteristics in this quadrant, we suggest that all learning strategies are appropriate for students in the quadrant. Nevertheless, we suggest that appropriate learning strategy for students in this quadrant is a hermeneutics approach. Characteristics of students' proof comprehension in second quadrant are uncompleted comprehension, especially in connecting mathematical concepts and validation skill. We suggest that one appropriate leaning for second quadrant is using the two-column form method. This method can play a dual role, so a student can enable greater flexibility in reasoning and proving.

Characteristics of students' proof comprehension in third quadrant are uncompleted comprehension, especially the difficulty to generate deductive reasoning. The students' characteristics have no proof-structure, and also have a little conceptual understanding. One of learning strategies for this quadrant is asking to generate example and learning using worked-example. Many mathematics education researchers have suggested that asking learners to generate examples of mathematical concepts is an effective way of learning about novel concept. Characteristics of students' proof comprehension in fourth quadrant are uncompleted comprehension, especially in beginning a proof. The students' characteristics have no proof-structure, so that she/he failed in beginning a proof. One of learning strategies for this quadrant is structural method.

REFERENCES

1. Andrew, L. (2009). Creating a proof error evaluation tool for use in the grading of student-generated "Proofs." *PRIMUS : Problems, Resources, and Issues in Mathematics Undergraduate Studies*, *19*(5), 447–462. https://doi.org/10.1080/10511970701765070
2. Baker, D., & Campbell, C. (2004). Fostering The Development of Mathematical Thinking : Observations from a Proofs Course. *PRIMUS : Problems, Resources, and Issues in Mathematics Undergraduate Studies*, *14*(4), 345–353. https://doi.org/10.1080/10511970408984098
3. Balacheff, N. (1988). Aspects of Proof in Pupils' Practice os School Mathematics. *Mathematics, Teacher and Children*, 216–235.
4. Djauhari, M. A. (2015). Hermeneutics in teaching and learning mathematics: Revitalising mathematics education. *ASM Science Journal*, *9*(2), 1–8.
5. Fadillah, S., & Jamilah. (2016). Pengembangan Bahan Ajar Struktur Aljabar Untuk Meningkatkan Kemampuan Pembuktian Matematis Mahasiswa. *Jurnal Cakrawala Pendidikan*, *1*, 106–113.
6. Fuller, E., Mejía-Ramos, J. P., Weber, K., Samkoff, A., Rhoads, K., Doongaji, D., & Lew, K. (2011). No TitleComprehending Leron's structured proofs. In S. Larsen, K. Marrongelle, & M. Oehrtman (Eds.), *Proceedings of the 15th Annual Conference on Research in Undergraduate Mathematics Education* (pp. 84–102).
7. Iannone, P., Inglis, M., Mejía-Ramos, J. P., Simpson, A., & Weber, K. (2011). Does generating examples aid proof production? *Educational Studies in Mathematics*, *77*(1), 1–14. https://doi.org/10.1007/sl0649-01
8. Larsen, S., & Zandieh, M. (2008). Proofs and refutations in the undergraduate mathematics classroom. *Educational Studies in Mathematics*, *67*(3), 205–216. https://doi.org/10.1007/s10649-007-9106-0
9. Lee, K. S. (2016). Students' proof schemes for mathematical proving and disproving of propositions. *Journal of Mathematical Behavior*, *41*, 26–44. https://doi.org/10.1016/j.jmathb.2015.11.005
10. Lehman, H. (1977). On Understanding Mathematics, *27*(2), 111–119.
11. Leron, U. (1983). Structuring Mathematical Proofs. *The American Mathematical Monthly*, *90*(3), 174–185. https://doi.org/10.2307/2975544

12. Margulieux, L. E., & Catrambone, R. (2016). Improving problem solving with subgoal labels in expository text and worked examples. *Learning and Instruction*, *42*, 58–71. https://doi.org/10.1016/j.learninstruc.2015.12.002
13. McLaren, B. M., Van Gog, T., Ganoe, C., Karabinos, M., & Yaron, D. (2016). The efficiency of worked examples compared to erroneous examples, tutored problem solving, and problem solving in computer-based learning environments. *Computers in Human Behavior*, *55*, 87–99. https://doi.org/10.1016/j.chb.2015.08.038
14. Meel, D. (2003). Models and Theories of Mathematical Understanding : Comparing Pirie and Kieren's Model of Growth of Mathematical Understanding and APOS Theory. *CBMS Issue in Mathematics Education American Mathematical Society*, *12*.
15. Mejia-Ramos, J. P., Fuller, E., Weber, K., Rhoads, K., & Samkoff, A. (2012). An assessment model for proof comprehension in undergraduate mathematics. *Educational Studies in Mathematics*, *79*, 3–18. https://doi.org/10.1007/s10649-011-9349-7
16. Moore, R. C. (1994). Making the transition to formal proof*. *Educational Studies in Mathematics*, *27*, 249–266.
17. Pirie, S., & Kieren, T. (1989). A Recursive Theory of Mathematical Understanding. *For the Learning of Mathematics*, *9*(3), 7–11. https://doi.org/10.2307/40248156
18. Recio, A. M., & Godino, J. D. (2001). Institutional and personal meanings of mathematical proof. *Educational Studies in Mathematics*, *48*, 83–99. https://doi.org/10.1017/CBO9781107415324.004
19. Retnowati, E., Ayres, P., & Sweller, J. (2010). Worked example effects in individual and group work settings. *Educational Psychology*, *30*(3), 349–367. https://doi.org/10.1080/01443411003659960
20. Selden, A., & Selden, J. (2003). Validations of proofs considered as texts: Can undergraduates tell whether an argument proves a theorem? *Journal for Research in Mathematics Education*, *34*(1), 4–36. https://doi.org/10.2307/30034698
21. Selden, J., & Selden, A. (1995). Unpacking the Logic of Mathematical Statements. *Educational Studies in Mathematics*, *29*(2), 123–151.
22. Skemp, R. R. (1978). Relational Understanding and Instrumental Understanding. *The Arithmetic Teacher*, *26*(3), 9–15. https://doi.org/10.1017/CBO9781107415324.004

23. Sowder, L., & Harel, G. (2003). Case studies of mathematics majors' proof understanding, production, and appreciation. *Canadian Journal of Science, Mathematics and Technology Education*, *3*(January 2015), 251–267. https://doi.org/10.1080/14926150309556563
24. Stylianides, G. J., & Stylianides, A. J. (2009). Facilitating the Transition from Empirical Arguments to Proof. *Journal for Research in Mathematics Education*, *40*(3), 314–352. https://doi.org/10.2307/40539339
25. Syamsuri, Purwanto, Subanji, & Irawati, S. (2016). Characterization of students formal-proof construction in mathematics learning. *Communications in Science and Technology*, *1*(2), 42–50.
26. Syamsuri, Purwanto, Subanji, & Irawati, S. (2017). Using APOS Theory Framework : Why Did Students Unable to Construct a Formal Proof ? *International Journal on Emerging Mathematics Education*, *1*(2), 135–146.
27. Syamsuri, & Santosa, C. (2017). Karakteristik pemahaman mahasiswa dalam mengonstruksi bukti matematis. *Jurnal Review Pembelajaran Matematika*, *2*(2), 131–143. https://doi.org/https://doi.org/10.15642/jrpm.2017.2.2.131-143
28. Tall, D. (2010). Perception, Operations and Proof in Undergraduate Mathematics. *Community for Undergraduate Mathematics Sciences Newsletter University of Auckland*, pp. 21–28.
29. Tall, D., Yevdokimov, O., Koichu, B., Whiteley, W., Kondrieteva, M., & Cheng, Y. H. (2012). Cognitive Development of Proof. In G. Hanna & M. De Villiers (Eds.), *Proof and proving in mathematics education, The 19th ICMI Study* (pp. 13–49). New York: Springer. https://doi.org/10.1007/978-94-007-2129-6
30. Van Dormolen, J. . (1977). Learning to Understand What Giving a Proof Really Means. *Educational Studies in Mathematics*, *8*(1), 27–34.
31. Varghese, T. (2011). Balacheff ' s 1988 Taxonomy of Mathematical Proofs. *Eurasia Journal of Mathematics, Science & Technology Education*, *7*(3), 181–192.
32. Weber, K. (2001). Student Difficulty in Constructing Proofs : The Need for Strategic Knowledge. *Educational Studies in Mathematics*, *48*(1), 101–119.
33. Weber, K. (2004). A framework for describing the processes that undergraduates use to construct proofs. In *Proceedings of the Annual*

Meeting of the North American Chapter of the International Group for the Psychology of Mathematics Education (Vol. 4, pp. 425–432).

34. Weber, K. (2005). Problem-solving, proving, and learning: The relationship between problem-solving processes and learning opportunities in the activity of proof construction. *Journal of Mathematical Behavior*, *24*, 351–360. https://doi.org/10.1016/j.jmathb.2005.09.005
35. Weber, K., & Alcock, L. (2004). Semantic and syntactic proof productions. *Educational Studies in Mathematics*, *56*, 209–234. https://doi.org/10.1023/B:EDUC.0000040410.57253.a1
36. Weiss, M., Herbst, P., & Chen, C. (2009). Teachers Perspectives on " Authentic Mathematics " and the Two-Column Proof Form. *Educational Studies in Mathematics*, *70*(3), 275–293. https://doi.org/10.1007/sl0649-008-9144-2

Chapter 11

APPLICATION OF DISCOVERY LEARNING METHOD IN MATHEMATICAL PROOF OF STUDENTS IN TRIGONOMETRY

Windia Hadi, Ayu Faradillah

University of Muhammadiyah Prof. DR. HAMKA, Indonesia

ABSTRACT

Trigonometry is a part of mathematics in learning that is related to angles. The purpose of this study was to determine the effect of the application of discovery learning methods in students' mathematical proof ability on trigonometry. The research method used in this study is quasi-experimental. The population in this study was the second semester of 2016/2017. The sample was 66 people who were determined by purposive sampling. The instrument used in this study was a mathematically proof ability test. Analysis of the data used is the t-test. The results of this study are (1) based on an average score of mathematical proof ability The student's mathematical proof ability in applying the Discovery Learning Method to trigonometry is not higher than the mathematical proof ability of students

Citation: (APA): Hadi, W., & Faradillah, A. (2020). Application of Discovery Learning Method in Mathematical Proof of Students in Trigonometry. Desimal: Jurnal Matematika, 3(1), 73-82.(10 pages).

Copyright: © Creative Commons Attribution-ShareAlike 4.0 International License (http://creativecommons.org/licenses/by-sa/4.0/).

who do not use the discovery method (2) there is no significant influence in the application of learning methods Discovery in students' mathematical proofs on trigonometry.

Keywords: Discovery Learning Method, Mathematical Proof Ability, Trigonometry

INTRODUCTION

Trigonometry comes from the Greek trigon, which is three angles and metro, namely measurement. Trigonometry is a branch of mathematics that deals with the edges of triangles and also deals with geometry. Trigonometry is the basis for the comparison of acute angle trigonometry, which developed into the concept of non-acute angles — solving problems related to trigonometry considered very difficult so it requires a more in-depth understanding of studying trigonometry, especially about geometry.

A Mathematics Education student must have the mathematical proof ability in solving a mathematical problem. Mathematical proof ability is essential to be possessed by a Mathematics Education student more accurately. They will become teacher candidates for their students — the proof ability, which is a fundamental ability in understanding trigonometry. Students are said to have the mathematical proof ability if students can validate or criticize evidence and construct evidence related to the types of evidence that often arise.

One of the subjects that require students' skills in the mathematical proof is trigonometry. Trigonometry is one of the compulsory subjects that must be taken by students of mathematics education study programs as teacher candidates. It is said to be a compulsory subject because the concept of material becomes the basic concept that must teach for the next level of topics such as calculus 1.2 and so on, as well as the foundation in High School. The many trigonometric formulas that cause this material are demanding for high school students to deal with in examinations. Facts in the field show trigonometry is not a natural material for students. Not only students, but students also experienced difficulties when developing evidence on the content at the University (Minggi, Paduppai, & Assagaf, 2016)

One reason for the challenge of developing mathematical proof ability, according to Demir ((Ilmi & Rosyidi, 2016) it is not easy for students to build understanding based on trigonometric relationships. In the trigonometry course, there are still many students who are less able to make evidence between trigonometric formulas (Himmi, 2017). The analysis shows that the majority of the sample teacher candidates do not understand quadrants or radians, similar to those found by (Tuna & Kacar, 2012), namely mathematics teacher candidates in Turkey have a misconception about trigonometry.

The difficulties experienced by prospective teachers in research and middle-level students who complain about the problem of trigonometry solving is a cohesive alignment in trigonometric problems. Prospective teachers are fully qualified individuals who, or will teach mathematics at the secondary level. Therefore, misconceptions, procedural approaches, and underdeveloped visualization skills conveyed to their students. If the findings of this study are not addressed, the problems identified in trigonometry will likely continue (Walsh, Fitzmaurice, & O'Donoghue, 2017). Problem is the focus of research on whether there is the effect of discovery learning methods in the ability of mathematical proof in trigonometry.

Also, one of the materials in trigonometry at the high school level is to prove the identity of trigonometry. Students cannot find/prove the ways/ strategies used in solving trigonometric problems (Ilmi & Rosyidi, 2016).

Thus, the evidence related to trigonometry material becomes one of the elements that are difficult for students. Research conducted by (Jingga, Mardiyana, & Setiawan, 2017) done with high school students there are still many shortcomings in understanding the concept of trigonometry and the mistakes made by all groups are the mistakes made describing each of the trigonometric comparison relationships resulting in calculations to be complicated, the cause is the inability of students to determine the relationship between formulas on trigonometric identities, besides that Medium and low group errors are students not thinking about other ideas, lack understanding of the concept of arithmetic operations, lack of skill in doing algebraic manipulation, students' inability to determine the relationship between formulas on trigonometric identities and misinterpreting writing.

Results based on the above problems, the right learning method to improve mathematical proof ability is the discovery learning method. Discovery learning method is a learning method that provides an opportunity for students to construct their knowledge by discovering for themselves a concept, principle, procedure, algorithm and the like that has been learned with the experience obtained and not through notification but its discovery either partially or entirely with a trial error, experiment, explore and guess

(Fajri, Ikhsan, & Subianto, 2018). Based on research (Hadi, 2016) Discovery methods can improve students' mathematical abilities. Thus this study, the author intends to examine "the application of Discovery learning methods in the Mathematical Proof of Student Ability in Trigonometry." Based on the background that has described, then the problem can be formulated as follows: (1) Is the mathematical proof ability of students in applying Discovery learning methods on trigonometry higher than the mathematical proof ability of students who do not use Discovery teaching methods? (2) whether there is a significant influence in the application of Discovery learning methods in students' mathematical proof ability on trigonometry.

RESEARCH METHODS

The approach used in this study uses a quantitative approach. The mathematical proof ability of students of mathematics education study program will explore how the mathematical proofing ability of students after being given the discovery learning method treatment and those not treated in the form of discovery learning method.

Research conducted in this study is a quasi-experimental study. In this study, subjects not randomly grouped, but researchers accepted the subject's condition as it is, with the Posttest-Only Control Design (Sugiyono, 2010). Then the research design can be written as follows.

Table 1. Posstest-Only Control Design

Class	treatment	Posttest
Experimental	X	0
Control	-	0

The population in this study were all semester 2 Mathematics Education study program students, 2016/2017 school year consisting of six classes. Meanwhile, the samples in this study are two classes of semester two students. So, the two types that have to examples are 33 students as the experimental class and 33 students as the control class. The total sample was 66 students. They are taking the sample by purposive sampling.

Data collection techniques used in this study is to carry out the test. The test conducted to measure the mathematical proof ability of the experimental class and control class students.

The test questions consist of 5 indicators of mathematical proof ability, namely: The first indicator criticizes the proof by adding, subtracting or rearranging a mathematical proof, the second indicator reads mathematical proof, the third indicator directly proving, the fourth indicator arranges the evidence and the fifth indicator is validating the form of evidence. Data analysis can be seen in Figure 1 as follows.

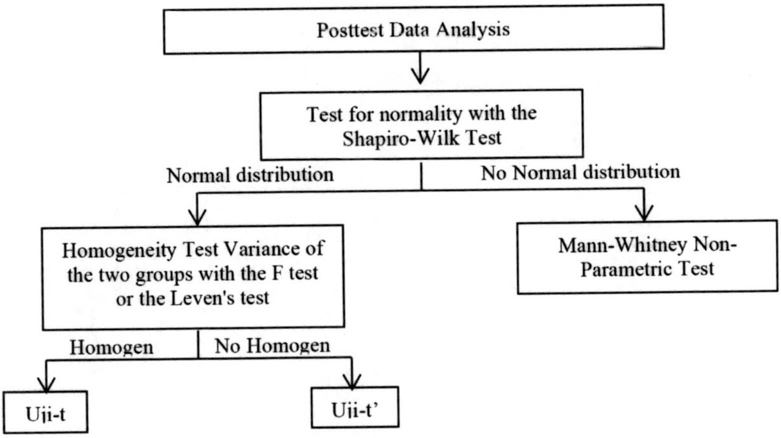

Figure 1. Data Analysis Procedure Chart.

RESULTS AND DISCUSSION

The results of the research data obtained an average score of UHAMKA students 'mathematical proof ability that applied the Discovery learning method on trigonometry was 12.39 with a standard deviation of 4.93, while the average rating of UHAMKA students' mathematical proof abilities that did not use the discovery method on trigonometry was 13.12 with a standard deviation of 2.91.

Based on the calculation of the average experimental class with the control class obtained $thitung = -0.72$ with a significance level of 0.05 and a degree of freedom (dk) of 64 earned $t_{tabel} = 1,66$. When compared, it can be seen the value $t_{hitung} = -0,72 < 1,66 = t_{tabel}$, then H_0 is accepted.

The acceptance of $H0$ concluded that the Students' Mathematical Proof Ability in applying the Discovery Learning Method to trigonometry was not higher than the mathematical proofing ability of students who did not use the discovery learning method. Thus, there is no significant effect of discovery learning methods on the mathematical proof ability of trigonometry.

The following is an analysis of the students' answers from each indicator in the item about the mathematical proof ability of UHAMKA students.

Indicators Criticize Proof by Adding, Subtracting or Rearranging a Mathematical Proof

In this study, what measured in indicator one is the ability of students to criticize the proof by adding, subtracting, or rearranging a mathematical proof. The items that use this indicator are item 1. The following is one of the students' answers from the accurate confirmation ability test number 1 that matches sign one in the experimental class.

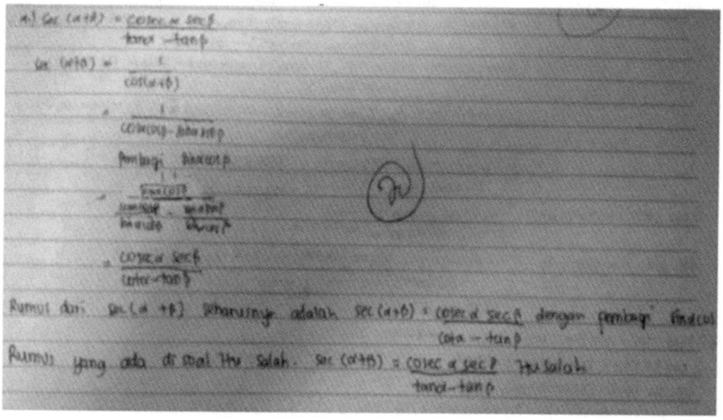

Figure 2. Correct Answer Item No. 1 Experiment Class.

From the results of student answer number 1 in Figure 2, it has been that experimental students can criticize the evidence in problem number 1. Starting from students prove correctly and criticize the proof for problem number 1. In the innovative class, 26 students can describe very well the number problems 1, and some students are wrong in terms of proving, what is meant by students is not careful enough in writing proof. By research (Himmi, 2017) revealed that many students were less able to do tests between trigonometric formulas. There is one student who gets a score between 1, there are three students who get a score of 2, there are two students who get a score of 3. and there is one person who receives a score of 0. The rest of the students managed to answer the indicator questions one well that is getting a score of 4. In class In the control, six students got a score of 3 and 1 student who got a rating of 2. It is seen that similarly, seven students lacked depth to answer the question instrument on the first indicator; the difference is the lowest score in the experimental class and the control class. Seen from the first indicator score in the lowest innovative level 0 and the lowest score control class is 2.

Based on the analysis of student answers, it appears that the experimental class is superior to the control class in which many innovative class students answer correctly than the control class students. So based on the analysis of the answer, the experimental type is better than the control class. That is means that the use of the discovery method affects the mathematical proof ability of students of Mathematics Education Study Program in UHAMKA.

The Second Indicator Reads Mathematical Proof

What is measured in the second indicator is the ability of proof to read the mathematical proof. Below is an example of an experimental class student learning mathematically correct tests.

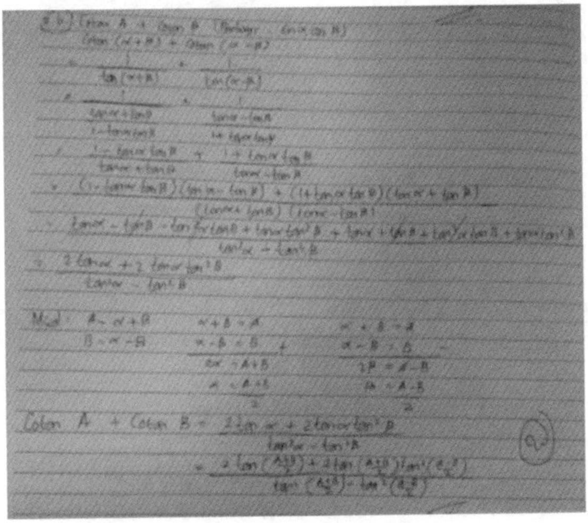

Figure 3. Correct Answer Item No. 2 Experiment Class.

From the results of student answer number 2, in Figure 3., it can be seen that experimental students can read mathematical proofs well. That can see from the way students can describe evidence well. That is consistent with the findings from (Dereu, 2019) that moving towards a measurement-oriented and constructivist approach can provide students with a deeper understanding of trigonometry.

The Third Indicator is Direct Proof

The third indicator is direct proof. The following are the answers of students in the third indicator. This third indicator describes how students in proving trigonometric formulas with different versions whose results will be the same.

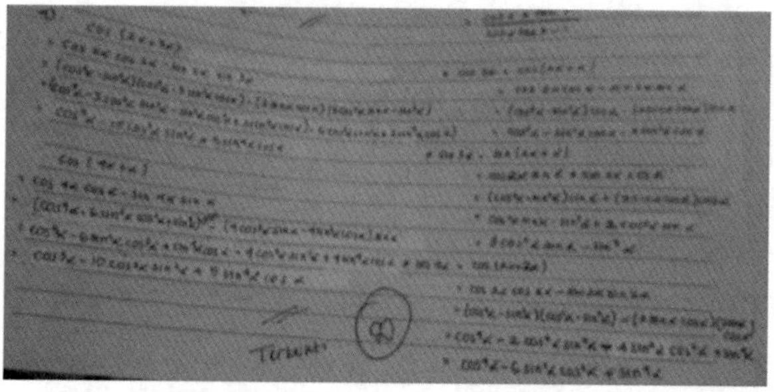

Figure 4. Correct Answer Item No. 3 Experiment Class.

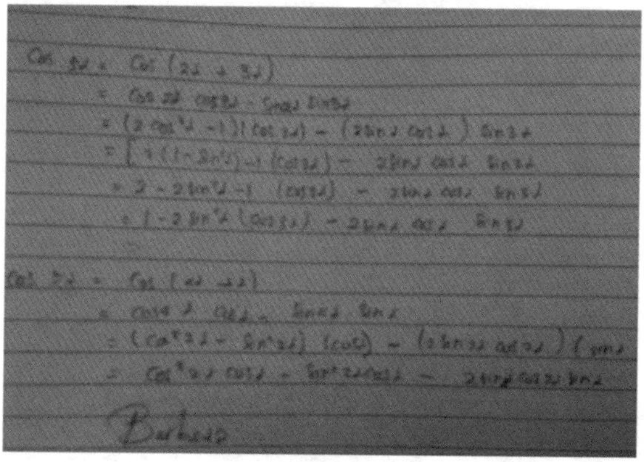

Figure 5. Wrong Answer Item No. 3 of the Control class.

Figure 4 and 5 is an example of the completion of the third indicator, namely direct proof, where students first describe the results of various versions based on trigonometric rules. The results should be the same if spelled out in several versions; the above is a real and false example indirectly proving trigonometric material.

In the experimental class, 11 students were able to prove right away directly without the slightest error, whereas in the control class, only five students were able to answer correctly, with a score of 4. Meanwhile, students who received a score of 3 in the experimental category were five students; for the control class, 15 students get a score of 3. Error in rating 3 is the student can describe the rules correctly, but the student is wrong in answering that is the student proves that the formula is different in each version, so this is what reduces the student's score while score 2 in the experimental class were eight students, while in the control class there were six students.

The student's mistake in proving is that the student is not careful in describing the rules of the formula directly. Four students in the experimental class received a score of 1 and in the control class, there were two students. The innovative and control level contained five students who could not prove the third indicator directly. That is following the findings of (Jingga et al., 2017) that the mistakes made describe each relation of trigonometric comparisons resulting in calculations to be complicated; the cause is the inability of students to determine the relationship between formulas on trigonometric identities.

Based on the analysis of the students' answers to the third indicator, it appears that the experimental class is superior to the control class, where more preparatory courses can complete the third indicator very well even though only 33%. That proves that the discovery method can further improve mathematical proof ability based on the analysis of student answers.

The Fourth Indicator Constitutes Proof

The fourth indicator is composing evidence. Arranging the evidence referred to in the proof ability indicator is the ability of students to find the angle with known trigonometric equations with zero value, with the previous sign the student has been able to prove directly followed by the student being able to compile the evidence from the information that has been found in advance so that the student can collect proof well.

Below is a picture of the correct student answers in the experimental class and the control class.

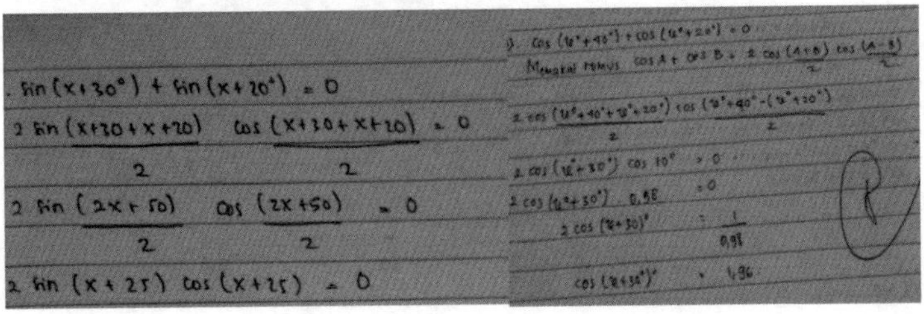

Figure 6. Wrong Answer item No. 4 Control class.

Based on Figure 6, it can how students have not been able to arrange evidence well; students can only derive from the informed problem formula. However, it is still wrong to operate algebra in trigonometric rules. That is by the findings (Hadi & Faradillah, 2019) that in serving algebra students still have difficulty understanding concepts. In the experimental class, ten students were able to arrange the evidence correctly, while in the control class, there were only two students who were able to arrange the evidence correctly and got a score of 4. In the experimental and control level, there was only one student who got a rating of 3. in the trial and control class there were three people and 22 people in sequence who got a score of 2. In the experimental type, 15 students and four students in the control class got a score of 1, while those who got a score of 0 there were four innovative class students and two control class students. It is clear that for indicators compiling proof, students still have difficulty answering this fourth indicator.

Based on the results of the analysis of the answers to the fourth indicator, that the experimental class outperformed the fourth indicator even though there were still many students who were wrong or incorrect in answering the fourth indicator instrument. The innovative level is higher than the control class.

The Fifth Indicator Validates the Form of Proof

The fifth indicator is validating the form of proof; the purpose of validating here is students can verify the structure of trigonometric inequality with the evidence that has known previously. Students can confirm the form of evidence in the way of results that have found; if students can validate the proof well, it means students can have mathematical proof ability well. Below is a picture of the effects of student answers invalidating the form of evidence on the fifth indicator.

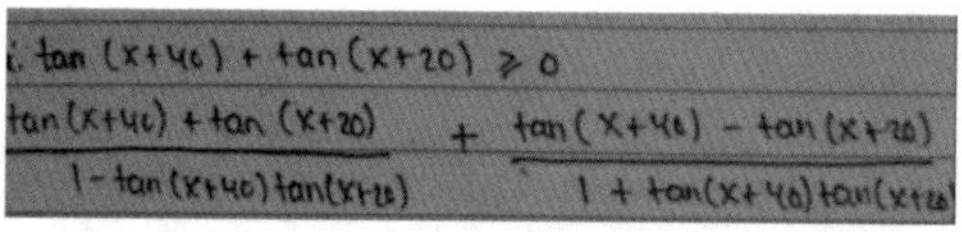

Figure 7. Wrong Answer item No. 5 experimental class.

Based on Figure 7, it found that students are still not able to validate correctly, on the answer sheet the students only describe the trigonometry rules that are known beforehand, if students are not able to prove from the available information means that students cannot validate the form of the problem is true or false, so the indicators The fifth is a combination of the first to fourth indicators and ends by confirming the results that have been obtained by students. That is consistent with the findings (Minggi et al., 2016) states that there are two categories of difficulties for students in proving, namely the lack of understanding of mathematical evidence and the lack of knowledge of concepts and principles in mathematics.

In the experimental class only four students were able to validate the form of proof well, while in the control class, there was one student who could verify the type of research. In the experimental category, there is one student, and two control class students who get a score of 3, students who get a score of 3 are capable students who have been able to prove it's just wrong invalidating the form of proof. In the experimental class of 2 students and 21 control class students who got a score of 2, the student's error in answering the fifth indicator problem was that the student was wrong in proving so that even the results would be validated even wrong. In the experimental class, there were 18 students and two control class students who got a score of 1, and in the innovative type, there were eight students and seven control class students who were unable to answer the fifth indicator.

Based on the analysis of students' answers to the fifth indicator that students still have difficulty validating the form of evidence, it how the lowest score obtained by many experimental and control class students. Based on the analysis, it appears that the control class is superior to the innovative level; it can see how the answer with a score of 3 is more, even though the results are also incorrect and there are still errors.

Based on the first to fifth indicators, it is clear that students still have difficulty in terms of mathematical proof ability so that this causes no effect of discovery methods on mathematical proof ability. Analysis of the answers showed that the experimental class was better than the control class, as seen in how students were able to get a higher score than the control class. Not too prominent differences that also influence the results of statistical analysis in the calculation of hypothesis testing. That also causes the discovery method not to have a significant influence on increasing the mathematical proof ability. Also, seen from the average results of the experimental class and the control class, even the control class is slightly superior to the innovative level, this also causes the discovery method to have no significant effect on mathematical proof ability.

The thing that causes the average experimental class is less than the control class is the mathematical proof ability of preliminary class students who are unable to complete all instruments of mathematical proofing ability; the student does not answer the instrument question with his knowledge, there is no writing described by the student. Students are emptying all the answers of the instruments that have been given by researchers. That causes the average experimental class to be less than the average control class. So the control class is superior to the innovative level based on average results.

CONCLUSION AND SUGGESTION

Based on the results of the study, it can conclude that the average score of students' mathematical proof ability by using discovery learning methods is no better than those not using discovery learning methods.

Based on the results of research using the t-test, it can conclude that there is no significant effect of mathematical proof ability that uses discovery learning methods with those that do not use discovery learning methods.

REFERENCES

1. Dereu, E. (2019). Student ' s Self Discovery of Right Triangle Trigonometry.
2. Fajri, N., Ikhsan, M., & Subianto, M. (2018). Mathematical reasoning abilities of students through a model of discovery learning in senior high school. Proceedings of The 8th Annual International Conference (AIC) on Social Sciences, (2000), 123–132.
3. Hadi, W. (2016). Meningkatkan kemampuan penalaran siswa SMP melalui pembelajaran discovery dengan pendekatan saintifik (Studi kuasi eksperimen di salah satu SMP Jakarta Barat). Kalamatika, I(1), 93–108.
4. Hadi, W., & Faradillah, A. (2019). The Algebraic Thinking Process in Solving Hots Questions Reviewed from Student Achievement Motivation. Al-Jabar: Jurnla Pendidikan Matematika, 10(2), 327–337. https://doi.org/10.1017/CBO9781107415324.004
5. Himmi, N. (2017). Korelasi Self Efficacy Terhadap Kemampuan Penalaran Matematis Mahasiswa Semester Pendek Mata Kuliah Trigonometri Unrika T.a. 2016/2017. PYTHAGORAS: Jurnal Program Studi Pendidikan Matematika, 6(2), 143–150. https://doi.org/10.33373/pythagoras.v6i2.941
6. Ilmi, M. B., & Rosyidi, A. H. (2016). MATHE dunesa. Jurnal Ilmiah Pendidikan Matematika, 1(5), 21–29.
7. Jingga, A. A., Mardiyana, & Setiawan, R. (2017). Analisis Kesalahan Siswa dalam Menyelesaikan Identitas Trigonometri pada Siswa Kelas X Semester 2 SMA Negeri 1 Kartasura Tahun Ajaran 2015/2016. Keywords in Qualitative Methods, 1(5), 48–62. https://doi.org/10.4135/9781849209403.n73
8. Minggi, I., Paduppai, D., & Assagaf, S. F. (2016). Penyebab Kesulitan Mahasiswa dalam Pembuktian Matematika. Jurnal Penelitian Pendidikan INSANI, 19(1), 18–2.
9. Tuna, A., & Kacar, A. (2012). 1 ST CYPRUS INTERNATIONAL CONGRESS OF Full Text Book. Prospective Mathematics Teachers' Misconceptions About Trigonometry, 92–97.
10. Walsh, R., Fitzmaurice, O., & O'Donoghue, J. (2017). What Subject Matter Knowledge do second-level teachers need to know to teach trigonometry? An exploration and case study. Irish Educational Studies, 36(3), 273–306. https://doi.org/10.1080/03323315.2017.1327361

Chapter

12

ORGANIZING THE MATHEMATICAL PROOF PROCESS WITH THE HELP OF BASIC COMPONENTS IN TEACHING PROOF: ABSTRACT ALGEBRA EXAMPLE

Aysun Yeşilyurt Çetin[1] and Ramazan Dikici[2]

[1]Atatürk University, Erzurum, Turkey
[2]Mersin University, Turkey

ABSTRACT

The aim of this study is to identify the basic components of the mathematical proof process in abstract algebra and to organize the proof process into phases with the help of these basic components. A basic component form was prepared by arranging a draft basic component form, which was created as a result of a document analysis in accordance with the opinions of three academicians, who were experts in algebra. The data obtained as a result of both document analysis and expert examination were analyzed by the

Citation: (APA): Çetin, A. Y., & Dikici, R. (2021). Organizing the mathematical proof process with the help of basic components in teaching proof: Abstract algebra example. LUMAT: International Journal on Math, Science and Technology Education, 9(1), 235-25. (21 pages).

Copyright: © Creative Commons Attribution 4.0 International License (http://creativecommons.org/licenses/by/4.0/).

descriptive analysis method and are explained here in a detailed manner. It is believed that this basic component form will facilitate the step-by-step addressing of such a complex process as proofs in a non-compulsory and non-hierarchical order.

Keywords: Teaching abstract algebra, basic components, mathematical proof process, teaching proof

INTRODUCTION

The aims of mathematics are to understand the rules on which numbers, algebra, and geometry are based; to find new and non-routine ways in these systems; and to explain new situations that are encountered. The world of mathematics is a world of ideas, insights, and discoveries, which will not be realized by people who are only interested in how to separate a function, and it is necessary to use mathematical proof and abstraction to enter this world (Goldberg, 2002). A mathematical proof, which is far more than just one or several examples supporting a mathematical statement (Derek, 2011), is a logical explanation of why a mathematical statement is true (Altıparmak & Öziş, 2005). The best proof is the one that helps understand the proved theorem by showing not only that it is true but also why it is true (Hanna, 2000).

The Purpose of the Study

Proofs constitute the basis of mathematics (CadwalladerOlsker, 2011; Sarı, 2011). Therefore, many studies have been conducted on mathematical proofs. However, the mathematical proof process was not presented step by step in any study as in this study.

Abstract Algebra is one of the most important course of mathematics and mathematics education at the university level. According to Agustyaningrum, Husna, Hanggara, Abadi and Mahmudi (2020), abstract algebra is full of definitions and theorems which all require proof. Therefore, the students need to understand every definition and theorem they learn and be able to organize the concepts needed to proving theorems.

For that to be possible, students should have the ability to organize the required information for proving theorems in abstract algebra. This requires them to have an idea about the structure and stages of the proofs.

Due to these reasons, the mathematical proof process in abstract algebra should be examined and presented step by step. The aim of this study is to classify the mathematical proof process in abstract algebra with the help of basic components. Furthermore, it was also aimed to determine the knowledge that would help students facing problems and would also increase their understanding of mathematical proof. The following two research questions are posed for this purpose:

1. What are the difficulties students face during the mathematical proof process?
2. What are the stages of the mathematical proof process in abstract algebra?

The answer to the first research question was sought by reviewing the literature addressing the difficulties experienced in the mathematical proof process and analyzing lecturers and students' notes in order to see how the mathematical proof process in abstract algebra was experienced by students and teacher candidates and what kinds of difficulties they faced in this process. Similarly, the second research question was replied based on literature and the opinions of experts. Thus, the proof process was staged as basic components and then exemplified.

Unlike previous studies (Boero, 1999; Leron, 1983) in which the structure and stages of proofs were revealed, this study is inspired by the studies that deal with the difficulties students experience in the process of proof and the existing literature. In studies investigating the difficulties of proof, the content of the proof is also addressed, albeit not directly, and this gives us an idea of the nature of the proof.

LITERATURE REVIEW

The proof process is a complicated and hierarchical process in which more than one thinking process and stage are included, and mathematical information is used in an intertwined manner. Students come to a dead end when they do not know what to do when constructing proofs (Selden & Selden, 2008; Weber, 2001). In the proof process, students mentally go through steps of identifying the problem, making an assumption, testing the truth of the assumption by outlining the proof, conclusions, re-checking the proof that has been obtained, and making justification (Faizah, Nusantara, Sudirman, & Rahardi, 2020). Researchers state that pre-service mathematics teachers who encounter a mathematical proof for the first time face problems such as not knowing how or where to start, not understanding the logic of the proof, not being able to decide on what kind of method and conceptual information to use, and not being able to conclude the proof (Güler, 2013; Güler & Dikici, 2014; Moore, 1990; Moore, 1994; Polat & Akgün, 2016; Selden, Selden, & Benkhalti, 2018; Weber, 2001; Yeşilyurt Çetin & Dikici, 2020).

To understand the abstract and conceptual structure of mathematics, it is very important for students to understand the concept of proofs, what a proof is, why proofs are constructed, and the proof process (Sarı, 2011). If students do not know how to construct a proof, they try informal approaches, such as using examples or looking at a graph (Raman, 2002). At the undergraduate level, when mathematics students are expected to construct a proof, sufficient time is not allocated to helping students learn how to do so, and therefore the difficulties that students face cause many of them to quit studying mathematics (Selden & Selden, 2008). Pre-service teachers have difficulty in constructing long, acceptance-based proofs that they encounter for the first time (Polat & Akgün, 2016). Furthermore, most pre-service teachers do not know how to start a proof (Güler & Dikici, 2014; Polat & Akgün, 2016).

According to Weber (2001), the primary cause of failure in constructing proofs is the lack of strategic knowledge and it must therefore be ensured that students gain effective strategic knowledge. Pre-service teachers have difficulty in determining a proper proof method and strategy when constructing proofs (Doruk & Kaplan, 2015; Güler, 2013). According to Karakuş and Dikici (2017), students of secondary school mathematics teaching have difficulty in using proof methods effectively, although they think that proof methods play a significant role in the proof process. Demiray (2013) determined that pre-service teachers are highly successful in refutation and proof by contradiction, but they give incorrect answers to contrapositive proof questions since they have difficulty in realizing and understanding the equivalence of contrapositive expressions, and pre-service teachers further fail to see the difference between an empirical argument and a valid proof. According to Ceylan (2012), the use of examples in the proof process by pre-service teachers may mean that they do not have sufficient logical inferences. Furthermore, according to Güler, Özdemir, and Dikici (2012), pre-service teachers failed to understand the relationship between mathematical induction steps completely and perceived this proof method as a procedure to be followed. The results of that study showed that pre-service teachers understand mathematical induction but fail to generate the induction step by using the induction hypothesis. In another study, Jones (2000) pointed out that pre-service mathematics teachers do not have a sufficient level of skills to construct proofs; that they do not have the mathematical knowledge necessary for effective mathematics teaching, including those in advanced grades; and that they graduate this way. It was also shown that pre-service teachers in advanced grades can construct proofs more smoothly in technical terms, but this does not provide them any benefit in terms of their associated mathematical knowledge in conceptual terms.

Karaoğlu (2010) stated that sufficient conceptual knowledge is needed to complete a proof and to understand how to use such knowledge in the proof of a theorem. Preservice teachers have difficulty with proofs that can be done by using definitions even at the most basic level (Şahin, 2016). A pre-service mathematics teacher's experience with proofs, sufficient conceptual understanding, and skill in following different methods are all important for directing towards conceptual images when encountering a mathematical problem and having the skills to begin a proof (Bayazit, 2009). When pre-service teachers do not have conceptual knowledge about the theorem they want to prove, they cannot start to construct the proof, and so they begin with the help of the concepts obtained by investigating all the concepts related to the theorem in their mind and try to find the one that works. However, if their available knowledge is insufficient, they cannot conclude the proof (Karaoğlu, 2010). Another difficulty encountered in the proof process is incomplete or incorrect preliminary knowledge (Polat & Akgün, 2016; Yeşilyurt Çetin, & Dikici, 2020). Even if the preservice teachers know which property and definition to use in the proof, they have difficulty using them while constructing the proof (Güler, 2013; Yeşilyurt Çetin, & Dikici, 2020).

By using the relevant literature, Moore (1990) addressed the difficulties experienced by undergraduate mathematics students in understanding proofs and the construction of proofs within seven categories. Accordingly, these students:

- do not know or cannot express definitions;
- have an inadequate intuitive understanding of the concepts;
- are not competent in proving conceptual images;
- are unsuccessful or unwilling to create and use their own examples;
- do not know how to use definitions to create the whole structure of a proof;

- are unsuccessful in understanding and using mathematical language and notation; and
- do not know how to start the proof process.

According to Moore (1990), students also experience an inability to coordinate and use all the information simultaneously in constructing mathematical proofs. Güler (2014) addressed the difficulties faced by pre-service teachers in the proof process within five categories as follows: determining how to start the proof, using the mathematical language and notation, using definitions, forming the setup of the proof (fully determining the steps to be followed), and selecting the elements from the set.

Polat and Akgün (2016), who observed the proof process among pre-service teachers, stated that pre-service teachers could not decide on which definition to use when constructing proofs, do not make any plans before starting the proof process, and cannot decide on how to begin the proof. They examined the reasons for the difficulties experienced by pre-service teachers in the proof process under two headings, those originating from the individual and those originating from the mathematical subject, and they charted them as follows:

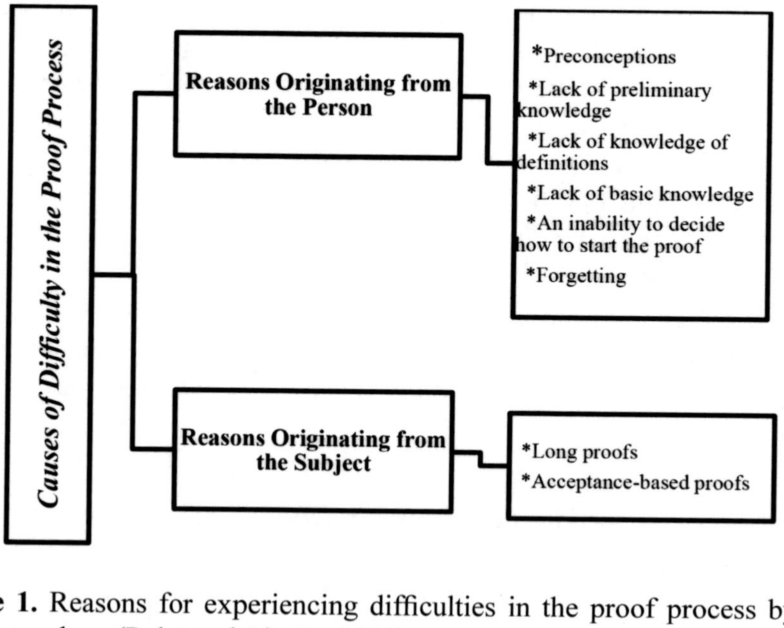

Figure 1. Reasons for experiencing difficulties in the proof process by pre-service teachers (Polat and Akgün, 2016).

According to Stewart and Thomas (2019), the reason why students face so many difficulties during the proof process is that each component in a proof is packed with different conceptual ideas. Therefore, it may be unrealistic to expect students to logically piece together the many definitions, other theorems, and various results that would help in completing a proof.

As can be seen from this literature review, the proof process is a complex process involving many difficulties that must be overcome step by step. In this study, each of these steps that make up a proof is called a basic component, and it is aimed to structure the mathematical proof process in abstract algebra into phases with the help of these basic components. This would make it possible to refine the proof process from relative complexity and address it within the framework of a certain order.

METHOD

A qualitative research method was used in this study, which aimed to phase the mathematical proof process in abstract algebra with the help of basic components by giving explanations about the process in accordance with expert opinions. Qualitative research methods are preferred when there is a theory gap in a subject, or when the existing theory is incapable of explaining the phenomenon. Moreover, the primary tool in data collection and analysis in qualitative research is the researcher himself or herself (Merriam, 2013). In this study, in which the proof process is partitioned into phases, the researchers themselves were the main elements in the collection and analysis of data.

Among qualitative research methods, a case study design was utilized in this study. A case study is a qualitative research method that offers the opportunity to gain indepth knowledge of the meaning of situations and events (Merriam, 1998). In this study, basic components were created in order to understand the proof process in depth, and as these basic components were identified, relevant documents and literature were used together with expert opinions.

Participants

In determining the basic components that make up the mathematical proof process in abstract algebra, three professors who conduct academic studies in the field of algebra were asked to make an examination and provide their opinions. These experts had at least 10 years of experience in teaching abstract algebra, number theory, and linear algebra.

Data Collection

In this study, document analysis was performed, and expert opinions were obtained in order to find the answer to the following questions: "What are the difficulties students face during the mathematical proof process?" and "What are the stages of the mathematical proof process in abstract algebra?". Documents are important sources of information that must be used effectively in qualitative studies. A document analysis involves the examination of written materials about the cases or situations that are being investigated (Yıldırım & Şimşek, 2008).

Mathematical proofs are complex processes with many challenges that consist of several steps. Therefore, the proofs determined by the researchers were divided into steps so that the proof completion process could be addressed step by step with the complexity removed. In addition to literature review, the textbooks (Çallıalp, 2009; Karakaş, 2001; Taşçı, 2010) and course materials for abstract algebra, the notes of the lecturers teaching abstract algebra, the course notes of students were also examined and evaluated to identify the difficulties in the mathematical proof process. Accordingly, a draft basic component form involving those difficulties was constructed. The literature on the difficulties experienced in the mathematical proof process was briefly explained to three experts in algebra together with basic information about the study, and they were asked to share their views on the draft basic component form and evaluate it. And then this draft form was reshaped in the light of their opinions. The final form of the 'basic component form', which was shaped by the document analysis and expert opinions, is presented in detail below in the "Findings" section.

Credibility and Transferability

The credibility and transferability issues for the findings were pursued in the following way: After 'the draft basic component form' was prepared, two mathematicians checked 'the draft basic component form' and after 'the basic component form for proofs' was prepared, five mathematicians checked 'the basic component form for proofs'.

One of the methods used to increase the credibility of a study is to obtain expert opinions. While determining the basic components, the opinions of three experts were obtained, and the process was directed in this direction. In this study, textbooks and course materials for abstract algebra, the notes of the lecturers, the course notes of students, notes taken during the study, and the relevant literature were examined and evaluated in the light of expert opinions. In this way, the credibility of the study was increased through a diversity of methods.

Analysis of the Data

In case studies, the researcher must rely on his or her own instincts and skills (Merriam, 2013). The data for the determination of the basic components that make up the mathematical proof process were analyzed by descriptive analysis in line with the researchers' instincts and skills and are presented with a holistic approach. Purpose in descriptive analysis to present the obtained findings in an organized and interpreted manner (Yıldırım & Şimşek, 2008). In the descriptive analysis of the data, the expert opinions on the draft of the basic component form, which was created as a result of document analysis, were taken as a basis. The opinions of the three expert academicians on the draft of the basic component form are quoted in detail.

The literature addressing the difficulties experienced in the mathematical proof process is explored step by step and analyzed as follows:

Table 1. Difficulties experienced in the mathematical proof process according to literature review

The Challenges Encountered in the Mathematical Proof Process	The Role of the Challenges Encountered in the Mathematical Proof Process in the Draft of the Basic Component Form
Understanding the logic of the proof	Determine the hypothesis and determine the judgement
Knowing how and where to start the proof	Use of hypothesis
• Being able to decide what kind of method and conceptual information to use • Knowing how to use preliminary knowledge, definitions, properties and concepts to create the whole structure of the proof • Understanding and using the mathematical language and notation • Forming the setup of the proof (fully determining the steps to be followed)	Process steps
Being able to conclude the proof	Reaching proof

Figure 2, created based on Table 1, was submitted for expert examination. Figure 3 was created in accordance with the expert opinions.

Ethical Issues

In this study, the purpose and method used are presented to the reader with a detailed explanation. The experts whose opinions were used were informed about the research process and their personal information was kept confidential, but some academic information about the experts was included in order to ensure the credibility and transferability of the study. Furthermore, the researchers in this study did not act biasedly in the process of data analysis and have reported the findings that they obtained without making any changes to them.

FINDINGS

In this section, it is aimed to determine the basic components of the mathematical proof process. To this end, the proof process was schematized as a draft basic component form after examining textbooks about theories and proofs in abstract algebra, the course notes of the lecturers and the course notes of students together with the literature revealing the challenges encountered in the mathematical proof process.

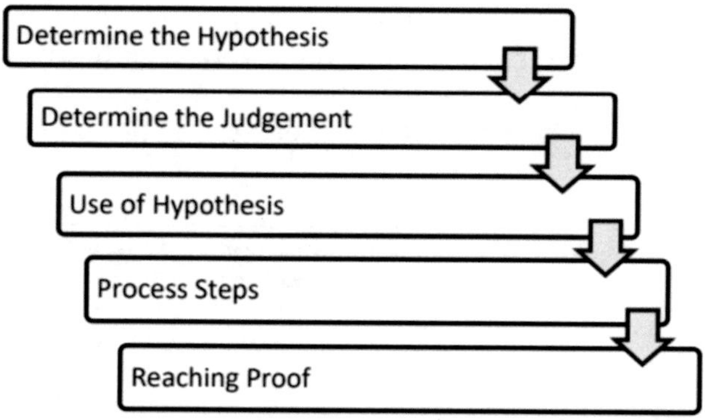

Figure 2. Draft basic component form.

The draft of the basic component form was submitted for the review of the three expert academicians. The literature on the difficulties experienced by students in the mathematical proof process was also provided to the experts, and their opinions on the draft form were obtained. One of the experts (P1) stated that the division of the process steps here should be expanded by being detailed in a non-hierarchical order, such as the use of definitions, use of properties, use of knowledge and theorems, and the performing operations. The second expert (P2) expressed similar opinions and stated that the step of using the hypothesis should be included in this non-hierarchical order. The third expert (P3) stated that the step of determining the method should be added after the step of determining the judgement.

The literature that reveals the difficulties experienced in the mathematical proof process and the expert opinions about the basic component form were analyzed as follows:

Table 2. Developmental process of the basic component form based on literature and expert opinions

The Challenges Encountered in the Mathematical Proof Process	**Expert Opinions**	**The Role of the Challenges Encountered in the Mathematical Proof Process in the Draft of the Basic Component Form**
Understanding the logic of proofs	No opinion	Determine the hypothesis and the judgement
Being able to decide what kind of method and conceptual information to use	*The step of determining the method should be added after the step of determining the judgement (P3),*	Determine the proof method
Knowing how and where to start the proof	*The step of using the hypothesis should be included in a non-hierarchical order (P2)*	Use of hypothesis
Knowing how to use definitions to create the whole structure of the proof	*The step of using definitions should be included in a non-hierarchical order (P1, P2)*	Use of definition
Knowing how to use properties to create the whole structure of the proof	*The step of using properties should be included in a non-hierarchical order (P1, P2)*	Use of property
Being able to decide what kind of conceptual information to use to create the whole structure of the proof	No opinion	Use of conceptual knowledge
Knowing how to use preliminary knowledge and basic information to create the whole structure of the proof	*The step of using knowledge and theorems should be included in a non-hierarchical order (P1, P2)*	Use of knowledge
Forming the setup of the proof (fully determining the steps to be followed)	*The step of performing operations should be included in a non-hierarchical order (P1, P2)*	Perform Operations
Being able to conclude the proof in accordance with mathematical language and notation	No opinion	Complete the proof

Based on the expert opinions and the challenges encountered in the mathematical proof process, the basic component form was reshaped as follows.

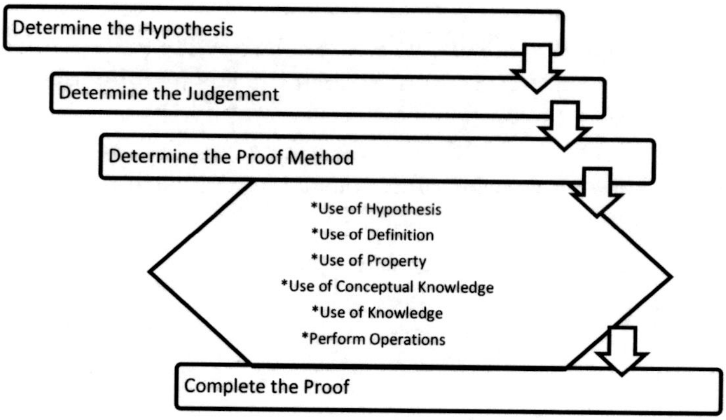

Figure 3. The basic component form for proofs.

These steps are non-hierarchical, and there is no obligation to apply every step in every proof. In other words, it is not necessary to determine the hypothesis in the proof of a theorem that only determines the judgement, such as "Prime numbers are infinite." Similarly, while no definition is used in certain proofs, there is no need to use any conceptual knowledge or properties in some others.

It is expected that an assumption that will provide the basis for a proof, or in other words that will lay the foundation of the proof, will be established in the stage of determining the hypothesis. For example, for the proposition $\Rightarrow$, the expression p is the hypothesis and it is important to be able to determine hypothesis p and write it in mathematical language and notation in order to start proving this proposition.

At the stage of determining the judgement, it is expected that the judgement to be achieved based on the hypothesis will be determined, creating the basis of the proof in this way. For example, for the proposition $p \Rightarrow q$, the expression q is the judgement, and it is important to write judgement q in mathematical language and notation in order to be able to shape the proof of this proposition.

In the basic component form, comprising the process steps of the draft basic component form as finalized by expert opinions, the components are the use of hypothesis, use of definitions, use of properties, use of conceptual knowledge, use of knowledge, and perform operations. These are used according to the content of the proof and in a non-hierarchical order. Therefore, in some proofs, all these basic components are used, while only one or some of them are used in others.

The writing of expressions to help construct a proof based on the hypothesis is expected in the step of using the hypothesis, and it is expected that an auxiliary theorem will be used, which should be known at the time or else should be information gained during the flow of the proof in the step of using knowledge. The use of a property that helps to build the proof is essential in the step of using properties, while the use of an expression that is present in the hypothesis/ judgement or that is mentioned anywhere in the proof and expected to help in the construction of the proof is essential in the step of using definitions. It is expected that the appropriate concepts and the information related to these concepts will be selected and used correctly in the step of using conceptual knowledge. For example, using the definition of subgroups or prime numbers was addressed as the use of definitions, selecting and using the concepts of unit elements and inverse elements in accordance with their properties was addressed as the use of conceptual knowledge, and using group properties when performing an operation was addressed as the use of properties. The component of performing operations aims to reveal the state of taking the proof to a certain level with various algebraic operations. In order to perform an algebraic operation, it may sometimes be necessary to use a definition, sometimes a property, and sometimes conceptual knowledge. At this point, deficiency in one of these areas will make it impossible to successfully complete the proof process.

At the stage of completing the proof, students are expected to complete the proof according to mathematical language and notation by using all the data or information obtained.

It is believed that the basic proof components determined in this form will help students successfully undertake the proof completion process by following a specific order.

Examples of Proofs Divided into the Basic Components

Examples of proofs that are divided into the basic components of the mathematical proof process in abstract algebra are presented below. The proofs below have been taken from the textbooks (Çallıalp, 2009; Karakaş, 2001; Taşçı, 2010) and course materials for abstract algebra, the notes of the lecturers, the course notes of students. These proofs are divided into the basic components by researchers and confirmed by five mathematicians that checked 'the basic component form for proofs'.

An example of a proof in which the components of "use of property", "use of definition", "perform operations", "use of conceptual knowledge", and "complete the proof" are used is given in the following theorem.

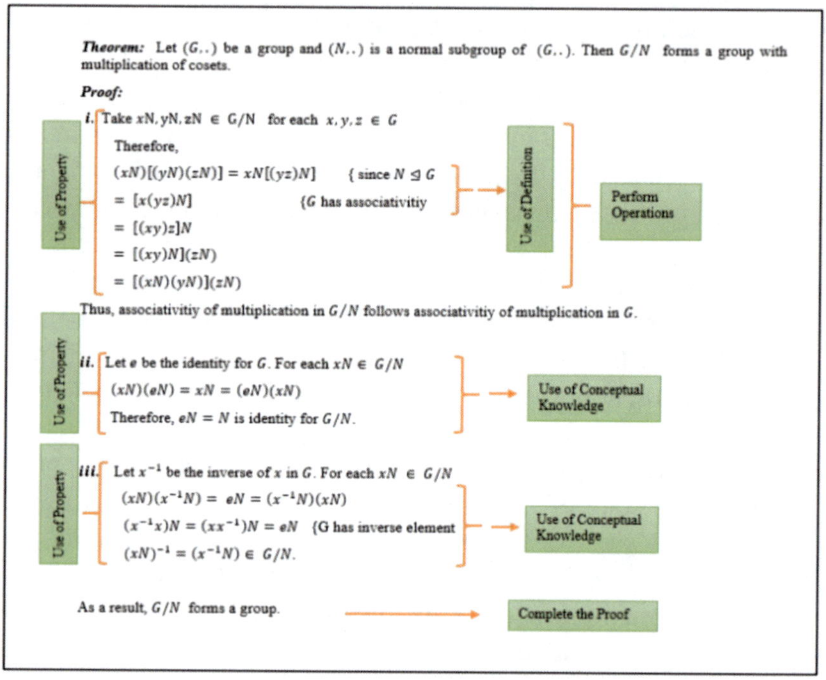

Figure 4. First example for a proof divided into basic components.

In the proof of the above theorem, in cases i, ii, and iii, the subgroup properties are called "use of property". In step i, the process of performing operations is discussed using the definition of the normal subgroup. In step ii, knowledge about the concept of the unit element is used. In step iii, knowledge about the concept of the inverse element is used. Based on all these steps, the completion of the proof is written as the final sentence.

An example of a proof in which the components of "determine the proof method", "determine the hypothesis", "use of hypothesis", "use of definition", and "complete the proof" are used is given in the following theorem.

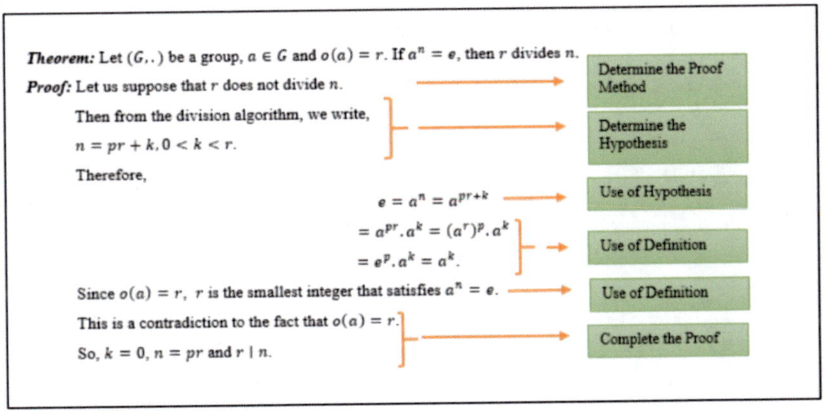

Figure 5. Second example for a proof divided into basic components.

The proof of the theorem presented above was started by determining the method, and then the hypothesis was determined, and an equation was established using this hypothesis. In the established equation, the proof was constructed using the definition of the unit element $(a) = r \Rightarrow a^r = e$ and an inference was made by using the definition of order. All of this knowledge was interpreted together, and the proof was completed.

An example of a proof in which the components of "determine the hypothesis", "determine the judgement", "use of definition", and "complete the proof" are used is given in the following theorem.

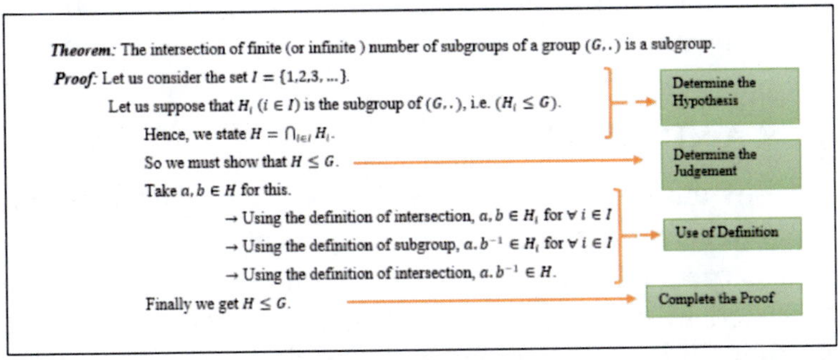

Figure 6. Third example for a proof divided into basic components.

In the proof of the theorem presented above, the hypothesis and judgement were determined based on the statement of the theorem and proof was started in that way. Necessary comments were made using the definitions of intersection and subgroup, and the proof was completed in light of these comments.

The example of a proof in which the components of "determine the hypothesis", "use of hypothesis", "use of definition", "use of conceptual knowledge", "use of knowledge", and "complete the proof" are used is given in the following theorem.

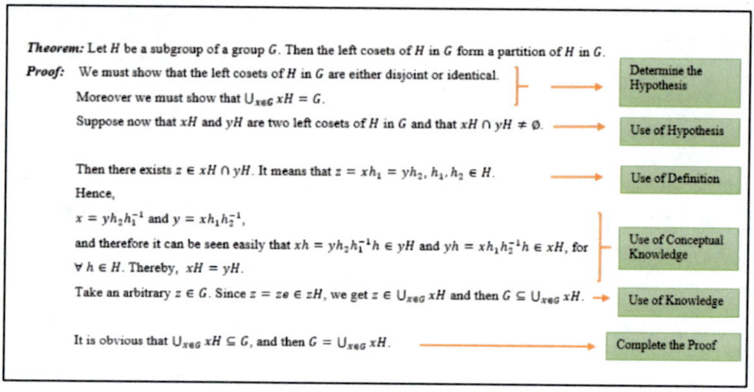

Figure 7. Forth example for a proof divided into basic components.

In the proof of the theorem presented above, the hypothesis was determined based on the statement of the theorem, and by using this hypothesis, left cosets xH and yH, the intersections of which are different from the empty set, were established. Then, using the definition of intersections, the proof was advanced by using conceptual knowledge about the concept of cosets. Finally, it was concluded that $G \subseteq \bigcup_{x\in G} x$H by using the information achieved while determining the hypothesis and the flow of the proof, and the proof was completed using all this information.

RESULTS AND DISCUSSION

The purpose of this study is to stage the mathematical proof process in abstract algebra according to its basic components. Therefore, a literature review was carried out regarding how the mathematical proof process was experienced by students and teacher candidates and what kinds of difficulties they faced in this process. The proof process was then staged as basic components based on the difficulties identified according to the literature and the opinions of experts.

Proofs were addressed step by step with a non-compulsory and non-hierarchical structure with the basic component form (Figure 3) that was created in line with the challenges that students encounter in the mathematical proof process. The final basic component form was established in accordance with the opinions of algebra specialists. It is not necessary for each of these components to be included in a proof. The basic components of a proof may vary according to the statement and the proof structure of the theorems. Similarly to this study, Boero (1999) and Leron (1983) dealt with proof generation with non-linear steps.

One of the difficulties encountered in the mathematical proof process is not knowing how and from where to begin the construction of the proof (Karaoğlu, 2010; Moore, 1990; Moralı, Uğurel, Türnüklü, & Yeşildere, 2006; Polat & Akgün, 2016; Yeşilyurt Çetin & Dikici, 2020). In the basic component form presented in Figure 3, the process of starting the proof is addressed in two sub-steps: determining the hypothesis and determining the judgement. The step of determining the hypothesis includes the establishment of an assumption that constitutes the basis of proof, while the step of determining the judgement includes the determination of the judgement to be achieved based on the hypothesis. Similarly, according to Boero (1999), the first two stages of mathematical proof construction are generating an assumption and formulating the statement according to shared textual conventions.

Another difficulty encountered in the proof process is determining the proof method and strategy (Doruk & Kaplan, 2015; Güler, 2013; Karakuş & Dikici, 2017; Weber, 2001), and it is necessary to check the accuracy of the selected method while examining the proof process. In this study, the step of determining the proof method in the basic component form involves determining the appropriate proof method based on the use of the theorem.

The proof process is found to be influenced by the fact that students have adequate conceptual knowledge (Karaoğlu, 2010; Moore, 1990) and understand mathematical definitions and how to use them (Bayazit, 2009; Moore, 1990; Polat & Akgün, 2016; Şahin, 2016). Students' lack of preliminary basic knowledge also makes this process more difficult (Polat & Akgün, 2016). Therefore, whether students' preliminary knowledge and conceptual knowledge are sufficient and whether they can use their knowledge and the definitions are also factors that shape the mathematical proof process. The component of "use of definition" in the basic component form presented in Figure 3 involves the use of definitions for the purpose of the proof. The component of "use of knowledge" involves the use of preliminary knowledge that must be possessed or the information achieved in the proof process, while the component of "use of conceptual knowledge" involves accurately selecting and using mathematical concepts and information related to these concepts. The component of "use of hypothesis" involves the use of this hypothesis in the flow of the proof. In addition, the component of "use of hypothesis" in this study is similar to the third phase of Boero's (1999) study, "exploration of the content of the conjecture".

It is thought that the components of "use of property" and "perform operations", which were added to the basic component form in accordance with the opinions of three academicians, experts in the field of algebra, shape the proof process. The component of "use of property" involves using a mathematical property that is expected to help construct the proof, while the component of "perform operations" involves carrying the proof to a certain level with various algebraic operations.

The component of "complete the proof" requires that all the information obtained in the proof process is addressed in a certain order, that the necessary inference is made, and that the proof is completed with expressions suitable for mathematical language and notation.

It is thought that structuring the proof of a theorem into phases with the help of basic components and addressing the proof process of students step by step with these components will provide convenience in both the teaching and the investigation of the proof process. It is hoped that the teaching of proofs within a specific non-hierarchical order using the basic component form presented in Figure 3 will facilitate understanding and allow teachers to identify which step in the proof process is difficult for a student. Therefore, it is suggested that including the basic component form and the proofs prepared for this form in textbooks would provide convenience for students in the proof process. In addition, similar studies can be conducted on whether the basic component form revealed in this study can be applied to other mathematics courses other than abstract algebra.

As Selden, Selden, & Benkhalti (2018) suggests, if certain stages of the proof are requested from the students, the success in proving can be increased. Proofs can be staged with the basic components set out in this study. In addition, students may be asked to complete some missing components instead of completing a proper proof. In this case, the teacher can decide about which component will be missing according to the use of this component in the proof. For example, in a proof about homomorphism, the 'use of property' in which the homomorphism property is used is left incomplete and the student can be expected to complete. Or, while teaching such kind of proof, it can be emphasized that the most important thing in making such a proof is the 'use of property'. Thus, students focus on a major part of the proof but not a whole proof, and they are exposed to staged proofs rather than long and intimidating proofs.

ACKNOWLEDGEMENTS

The authors would like to thank the experts who allocated their time to this study, shared their opinions and thoughts with the researchers, and contributed to the final shape of the basic component form.

The authors would like to express their appreciation to the anonymous reviewers and the editor Johannes Pernaa for making useful suggestions regarding the presentation of this paper.

REFERENCES

1. Agustyaningrum, N., Husna, A., Hanggara, Y., Abadi, A. M., & Mahmudi, A. (2020). Analysis of mathematical proof ability in abstract algebra course. Universal Journal of Educational Research, 8(3), 823–834. https://doi.org/10.13189/ujer.2020.080313
2. Altıparmak, K., & Öziş, T. (2005). An investigation upon mathematical proof and development of mathematical reasoning, Ege Journal of Education, 6(1), 25–37. Retrieved from https://dergipark.org.tr/tr/pub/egeefd/issue/4918/67296
3. Bayazit, N. (2009). Prospective mathematics teachers' use of mathematical definitions in doing proof (Doctoral dissertation). Florida State University, Florida.
4. Boero, P. (1999). Argumentation and mathematical proof: a complex, productive, unavoidable relationship in mathematics and mathematics education. International Newsletter on The Teaching and Learning of Mathematical Proof, 7,8. Retrieved from http://www.lettredelapreuve.org/OldPreuve/Newsletter/990708Theme/990708ThemeUK. html
5. CadwalladerOlsker, T. (2011). What do we mean by mathematical proof?. Journal of Humanistic Mathematics, 1(1), 33–60. https://doi.org/10.5642/jhummath.201101.04
6. Ceylan, T. (2012). Investigating preservice elementary mathematics teachers' types of proofs in geogebra environment (Master's thesis). Ankara University, Ankara.
7. Çallıalp, F. (2009). Örneklerle Soyut Cebir [Abstract Algebra with Examples]. İstanbul: Birsen Publisher.
8. Demiray, E. (2013). An investigation of pre-service middle school mathematics teachers' achievement levels in mathematical proof and the reasons of their wrong interpretations (Master's thesis). Middle East Technical University, Ankara.
9. Derek, M. (2011). Teaching and learning of proof in the college curriculum (Master's thesis). San Jose State University, Washington.
10. Doruk, M., & Kaplan, A. (2015). Prospective mathematics teachers' difficulties in doing proofs and causes of their struggle with proofs. Journal of Bayburt Education Faculty, 10 (2), 315– 328. Retrieved from https://dergipark.org.tr/tr/pub/befdergi/issue/17275/180470
11. Faizah, S., Nusantara, T., Sudirman, S., & Rahardi, R. (2020). Exploring students' thinking process in mathematical proof of abstract

algebra based on mason's framework. Journal for the Education of Gifted Young Scientists, 8(2), 871–884. http://dx.doi.org/10.17478/jegys.689809

12. Goldberg, A. (2002). What are mathematical proofs and why are they important?. Retrieved from http://www.math.uconn.edu/~hurley/math315/proofgoldberger.pdf
13. Güler, G. (2013). Investigation of pre-service mathematics teachers' proof processes in the learning domain of algebra (Doctoral dissertation). Ataturk University, Erzurum.
14. Güler, G. (2014). Analysis of the proof processes of pre-service teachers regarding function concept. International Journal of Education and Research, 2(11), 161–176. Retrieved from https://www.ijern.com/journal/2014/November-2014/14.pdf
15. Güler, G., & Dikici, R. (2014). Examining prospective mathematics teachers' proof processes for algebraic concepts, International Journal of Mathematical Education in Science and Technology, 45(4), 475–497. https://doi.org/10.1080/0020739X.2013.837528
16. Güler, G., Özdemir, E., & Dikici, R. (2012). Pre-service teachers' proving skills using mathematical induction and their views on mathematical proving. Kastamonu Education Journal, 20(1), 219–236. Retrieved from https://dergipark.org.tr/en/pub/kefdergi/issue/48696/619520
17. Hanna, G. (2000). Proof, explanation, and exploration: An overview. Educational Studies in Mathematics, 44, 5–23. https://doi.org/10.1023/A:1012737223465
18. Jones, K. (2000). The student experience of mathematical proof at university level, International Journal of Mathematical Education in Science and Technology, 31(1), 53–60. https://doi.org/10.1080/002073900287381
19. Karakaş, H. İ. (2001). Matematiğin Temelleri [Fundamentals of Mathematics]. Ankara: Metu Press.
20. Karakuş, D., & Dikici, R. (2017). The opinions of the students of secondary education mathematics teaching on mathematical proof methods. The Journal of International Education Science, 13, 194–206.
21. Karaoğlu, Ö. (2010). The performance of pre-service mathematics teachers in proving theorems supported by key points and arguments (Master's thesis). Gazi University, Ankara.

22. Leron, U. (1983). Structuring mathematical proofs. The American Mathematical Monthly, 90(3), 174–185. https://doi.org/10.1080/00029890.1983.11971184
23. Merriam, S. B. (1998). Qualitative research and case study applications in education. revised and expanded from: Case study research in education. San Francisco: Jossey-Bass Publishers.
24. Merriam, S. B. (2013). Qualitative Research: A guide to design and implementation (Trans. Ed. S. Turan) Ankara: Nobel Academic Publishing.
25. Moore, R. C. (1990). College students' difficulties in learning to do mathematical proofs (Doctoral dissertation). University of Georgia, Athens.
26. Moore, R. C. (1994). Making the transition to formal proof, Educational Studies in Mathematics, 27, 249–266. https://doi.org/10.1007/BF01273731
27. Moralı, S., Uğurel, I., Türnüklü, E., & Yeşildere S. (2006). The views of the mathematics teachers on proving. Kastamonu Education Journal, 14(1),147–160. Retrieved from https://dergipark.org.tr/en/pub/kefdergi/issue/49106/626665
28. Polat, K., & Akgün, L. (2016). Pre-service mathematics teachers' opinions about proof and difficulties with proving. The Journal of Academic Social Science Studies, 43, 423–438.
29. Raman, M. J. (2002). Proof and justification in collegiate calculus (Doctoral dissertation). University of California, Berkeley.
30. Sarı, M. (2011). Undergraduate students' difficulties with mathematical proof and teaching of proof (Doctoral dissertation). Hacettepe University, Ankara.
31. Selden, A., & Selden, J. (2008) Overcoming students' difficulties in learning to understand and construct proofs. In M. Carlson & C. Rasmussen. (Eds.), Making the connection: Research and teaching in undergraduate mathematics, (pp. 95–110), Mathematical Association of America.
32. Selden, A., Selden, J., & Benkhalti, A. (2018). Proof frameworks: A way to get started. PRIMUS, 28(1), 31–45. https://doi.org/10.1080/10511970.2017.1355858

33. Stewart, S., & Thomas, M. O. (2019). Student perspectives on proof in linear algebra. ZDM, 51(7), 1069–1082. https://doi.org/10.1007/s11858-019-01087-z
34. Şahin, B. (2016). Examination of process of proving on divisibility of mathematics teacher candidates. Journal of Bayburt Education Faculty, 11(2), 365–378. Retrieved from https://dergipark.org.tr/en/pub/befdergi/issue/28762/307847
35. Taşçı, D. (2010). Soyut Cebir [Abstract Algebra]. Ankara: Öziş Typography.
36. Weber, K. (2001). Student difficulty in constructing proofs: The need for strategic knowledge, Educational Studies in Mathematics, 48, 101–119. https://doi.org/10.1023/A:1015535614355
37. Yeşilyurt Çetin, A., & Dikici, R. (2020). Examination of pre-service mathematics teachers' ability to make algebraic proof. Online Journal of Mathematics, Science and Technology Education (OJOMSTE), 1(1), 75–85.
38. Yıldırım, A., & Şimşek, H. (2008). Sosyal bilimlerde nitel araştırma yöntemleri [Qualitative research methods in social sciences]. Ankara: Seçkin Publishing.

Chapter

THE IMPLEMENTATION OF SELF-EXPLANATION STRATEGY TO DEVELOP UNDERSTANDING PROOF IN GEOMETRY

13

Samsul Maarif[(1)], Fitri Alyani[(2)], Trisna Roy Pradipta[(3)]

[(1)] Department of Mathematics Education, Universitas Muhammadiyah Prof. DR. HAMKA

[(2)] Department of Mathematics Education, Universitas Muhammadiyah Prof. DR. HAMKA

[(3)] Department of Mathematics Education, Universitas Muhammadiyah Prof. DR. HAMKA

ABSTRACT

Proof is a key indicator for a student in developing mathematical maturity. However, in the process of learning proof, students have the difficulty of being able to explain the proof that has been compiled using good arguments. So we need a strategy that can put students in the process of clarifying proof better. One strategy that can explore student thought processes in explaining geometric proof is self-explanation strategy. This

Citation: (APA): Maarif, S., Alyani, F., & Pradipta, T. R. (2020). The implementation of self-explanation strategy to develop understanding proof in geometry. JRAMathEdu (Journal of Research and Advances in Mathematics Education), 5(3), 262-275.(14 pages).

Copyright: © Creative Commons Attribution 4.0 International License (http://creativecommons.org/licenses/by/4.0/).

research aimed to analyze the ability to understand the geometric proof of prospective teacher students by implementing a self-explanation strategy in basic geometry classes. This study used a quasi-experimental research type of nonequivalent control group design. The participants of this research were 75 students of mathematics education study programs at one private university in Semarang. This research used four instrument tests of geometric proof. Before being used for research, the instruments were tested for validity and reliability using product-moment and Cronbach's alpha. Data analysis in this study used a two-way ANOVA test. The results showed that: the increased ability to understand the geometric proof of students who used self-explanation strategy was better than those who obtained direct learning; there was a significant difference between the increase of students' mathematical proof ability in a group of students with a high and moderate level of initial mathematical ability; the initial ability (high, medium, low) of mathematics did not directly influence the learning process to improve the ability to understand the geometric proof. Hence, it can be concluded that the self-explanation strategy is effective to be used to improve the understanding of the geometric proof.

Keywords: Self-explanation, proving geometry, understanding the geometric proof

INTRODUCTION

Geometry curriculum in mathematics education contains plane, space, analytical plane and space, and transformation geometry. Specifically, in the study of geometry, higher education curriculum places geometry as studied material up to the axiomatic system leading to formal proof. A formal proof of geometry will be formed if students are able in formulating geometry argumentation works well and write those arguments in a formal proof. These abilities are essential since students who have no ability to abstract geometry concepts and express arguments, and then automatically, they are unable to analyze a proof (Maarif, Wahyudin, Noto, Hidayat, & Mulyono, 2018).

Proof is a key indicator for a student in developing mathematical maturity (Otten, Gilbertson, Males, & Clark, 2014; Maarif, Perbowo, Noto, & Harisman, 2019). In recent years, research has been focusing more on how students can construct proofs compared to students' activities to understand a proof that will be constructed (Samper, Perry, Camargo, Sáenz-Ludlow,

& Molina, 2016; Fiallo & Gutiérrez, 2017; Azrou & Khelladi, 2019). This focus can be a potential cause of students' mistakes in constructing proof because they lack an understanding of the flow of proof according to the axiomatic system.

The proving process will be effective if it is carried out through understanding and explaining activity as a fundamental proving competence (Maarif, 2013; Sommerhoff & Ufer, 2019). Understanding geometric proofs can be done through a process of reading a proof practice. After the proof is read, the student explains it again so that the student can understand the whole idea of the geometric proof. The process of explaining the proof of geometry can get students used to form logical thinking in preparing the proof of geometry ability. Therefore, it is needed as a strategy that can improve students' understanding of geometric proof. In this study, we adopted a pedagogical strategy, a self-explanation that has been proven effective in facilitating students in understanding mathematical proofs (Hodds et al., 2014; Tsitouras, Tsivilis, & Kakali, 2014; Roy & Chi, M.T.H 2005; Chi, De Leeuw, Chiu, & Lavancher, 1994).

A self-explanation strategy is recommended and suggested in learning activities to construct geometry because they can help students develop their ideas based on the knowledge they have (Kumar, 2014; Tsitouras et al., 2014; Conati, 2016; Rittle-Johnson, Loehr, & Durkin, 2017). The study by Rittle-Johnson et. al. concluded that self-explanation strategy is an effective learning technique to support procedural and conceptual transfer in various mathematical topics. Other studies revealed that self-explanation strategy contributes to students' knowledge through automatic feedback in the problem-solving process (McNamara, 2017). Conceptual and procedural knowledge becomes an important part of the student in constructing their ideas with a feedback process to be written in geometric proof.

A self-explanation strategy is a metacognitive strategy to mediate individual thoughts between internal mental models and presenting external information, forming an understanding and revising weakness of the mental model (Ainsworth & Burcham, 2007). Self-explanation strategy in learning is useful for students' understanding of facts, concepts, principles, and procedures to correct misconceptions about the subject matter of mathematics being taught (Tekeng, 2015). Self-explanation aimed to: first, anticipate reasoning to predict the next action; second, a link between goals and actions; third, elaborate problems; fourth understand the coherence between texts, examples, and problems; fifth monitor negative understanding; and sixth, monitor positive understanding (Chi et al., 1994). A Self-explanation strategy has the following steps: comprehension monitoring, paraphrasing, inference bridging, prediction, and elaboration (Kurby et al., 2012).

To proof understanding of geometry, students are determined how far they have their initial mathematical abilities. In the learning process, there will always be differences in the students' high, medium, and low levels due to the ability of students to spread like a normal distribution (Haeruman, Rahayu, & Ambarwati, 2017). One of the best predictors to know the mathematical ability is through previous mathematics learning of students with the process of determining high, medium, and low groups (Vilkomir & O'Donoghue, 2009).

Research of self-explanation strategy has been extensively studied in several previous studies on self-explanation strategy (Wylie, Koedinger, & Teruko, 2004; Al Rumayyan et al., 2018; McNamara, 2017). This research is widely applied in language and outside the fields of mathematics education (Nguyen, Trawiński, & Kosala, 2015; Wylie et al., 2004; Kwon, Kumalasari, & Howland, 2011). Even though several studies have applied self-explanation strategy in the algebra (Kumar, 2014; Tsitouras et al., 2014; Conati, 2016; Rittle-Johnson, Loehr, & Durkin, 2017), however, the strategy has not been widely applied in learning geometry, particularly in understanding geometric proof. In this study, the researchers conducted a self-explanation strategy to develop the ability to understand the geometric proof.

Mathematical proof of students determines how far the initial mathematics ability they have. Yu et al. (2015) revealed from a group of randomly selected students will always students who have high, medium, and low abilities. These levels are due to the ability of students to spread normally. The process of determining high, medium, and low groups involves sorting the scores of previous mathematics learning outcomes (daily tests, midterms, or national examinations. This is in line with the findings of Yu et al. (2015) through their research that one of the best predictors of mathematics ability is the result of previous mathematics learning.

In addition, in the conditions of learning in the classroom, the ability of students is different, so it is necessary to adjust the learning environment. The selection of a suitable learning model is needed to bridge all students' abilities that occur in a lecture. This is expressed by Lestari (2017) that differences in students' ability are not merely innate, but also influenced by the environment. Therefore, we need to examine the interaction between self-explanation strategies and the students' initial abilities.

In learning, students are taught how to interpret a mathematical proof problem so that they can describe mathematical problems and can connect the existing geometrical concepts using theorems and definitions so that they are able to prove a geometry problem in accordance with their initial abilities. The use of reasoning is needed to dctermine and justify the conjecture that has been made based on understanding the geometrical concepts students have. The process of determining the conjecture in the mathematical proof is carried out by exploring the geometrical problems that will be proven, which will then be used as data to be justified in the conclusion in the form of formal evidence. By paying attention to the above, a self-explanation strategy is considered appropriate as an effort to increase the ability to understand the geometric proof. This study was designed to see the ability to understand the geometric proof of prospective teacher-students using selfexplanation strategies.

Therefore, we submitted research questions as follows: first, is the increasing of the geometric proof understanding ability of students who gain self-explanation strategy better than students who gain direct instruction learning?; second, is there a difference in understanding geometric proof in terms of the students' initial mathematical ability?; third, is there an interaction effect between learning factors and the level of initial mathematical ability to improve and achieve proof understanding ability?. This study described the selection of appropriate learning strategies to develop the ability to understand the proof of geometry so that it can help the process of developing the design of learning in lectures on geometry.

RESEARCH METHODS

This study was a quasi-experimental research type of nonequivalent control group design (Sztajn, Wilson, Edgington, Myers, & Dick, 2013). This study used a nonequivalent control group design because it desires to examine how the self-explanation strategy implemented to understand geometric proof. This research was conducted at the Mathematics Education Study Program at one of the private universities in Semarang. The number of participants was 75 students selected using the Cluster Random Sampling technique to determine one experimental class (applying self-explanation strategies) and one control class (direct instruction learning).

Based on the stages of learning using a self-explanation strategy that includes monitoring comprehensive, paraphrasing, bridging inference, predicting, and elaborating, then the steps of learning with a self-explanation strategy are implemented as Table 1.

Table 1. The stages of learning using self-explanation strategy

Steps of Explanation Strategy	Students' Activities
Monitoring comprehensive	Students explore the problems presented in the proof worksheet by collecting data, such as known information. Students try to find out their weaknesses in knowledge by asking questions to themselves: • Am I understand the concept being taught? • Do I not understand the concept being taught? • Am I able to prove theorems being taught? • Am I know theorems being taught? • Do I not know five Euclid axioms?
Paraphrasing	If students find information that is not understood, students rewrite the information in an easily understood language
Bridging inference	Each student makes a conclusion on the results of each thought to be discussed in a group
Predicting	Each student provides a prediction of presented problem solutions
Elaborating	Students discuss the prediction solution of each group member to develop final conclusions from the presented problem solutions.

The instrument used in this study was a test of the ability to understand the geometric proof. There is four ability test of understanding geometric proof. The test of understanding geometric proof uses indicators of choosing one of the two presented theorems used in proving a statement (Arnawa, Sumarno, Kartasasmita, & Baskoro, 2007; Bieda, Ji, Drwencke, & Picard, 2014). The following is one that was used in research by referring to indicators of choosing one of the two presented theorems that to be used in proving a statement.

> Presented the following theorem:
> i. If the P triangle and the Q triangle are congruent, then the ratio of the corresponding sides' lengths is equal.
> ii. If two sides and an angle are enclosed in two congruent triangles, then the triangles are congruent.
>
> Of the two theorems above, which theorem can be used to prove the following statement:
> **If O is the intersection of the three axes of triangle ABC, then the distance of point O to the vertex of the triangle is the same.**
> Write down the formal proof.

To get a good measuring instrument, before it was used in a research class, an instrument test was first conducted to determine the validity and reliability level. The trial was conducted in one class consisting of 35 students. Validity analysis used productmoment with SPSS 21 software application. The validity test showed that the four items of the test items were able to comprehend the geometric proof in a row of 0.611; 0.726; 0.665 and 0.467, all of which are greater than the r product moment value of 0.275. So, it can be concluded that all test items' ability to understand the geometric proof is valid.

Based on the results of the validity test, the next step was to conduct a reliability test. The results of the decisions on each test meet the requirements for reliability testing. The reliability test was carried out using Cronbach's Alpha with the application of SPSS 21 software. The reliability test results of understanding mathematical geometric proof had $r_{count} = 0.425 > 0.275 = r_{critical}$, so it can be concluded that the mathematical proof ability test is reliable.

Data analysis was performed using statistical tests. Quantitative data that has been grouped based on the learning model and the initial students' ability level were testing using parametric statistical analysis requirements as a basis for using the type of hypothesis test. Testing the requirements carried out is a test for normality and homogeneity of data variance. Test for normality and homogeneity of data variance respectively used Shapiro-Wilk test and Levene test. The final part of the quantitative data analysis was directed to determine the effect of mutual interactions between learning models and students' initial abilities. To test the effect of these interactions, researchers planned to use two-way ANOVA test through SPSS 21 software. The reason for using twoway ANOVA test was that the effectiveness of statistical tests with several initial mathematical abilities was tested. In addition, in hypothesis testing, one of the interaction hypotheses was between the learning strategy and the initial mathematical ability factor. So, it will be more effective by using two-way ANOVA test.

Meanwhile, students' initial mathematical abilities were taken from National Examination scores at Senior High School or equivalent level grade. The National exam scores describe students' ability in mathematics because an instrument measures that the government has made nationally through a standardized process. The researcher used data of initial mathematical students' ability in Senior High School or equivalent grade with the reason that for the first time, students took lectures at the university level and describe the mathematical abilities of students in the first semester.

To deepen the meaning of the data, after analyzing statistical data, observation and interviews were with the respondents related to the application of self-explanation strategies. Those respondents represent three levels of students' ability: high (coded R1), moderate (coded R2), and low (coded R3).

RESULTS AND DISCUSSION

The findings of the ability to understand students' geometric proof using selfexplanation strategies are presented in Table 2. It can be seen that the ability to understand students' geometric proof who learned using a self-explanation strategy was better than those who learned using direct instruction for all initial mathematical understanding ability (high, medium, and low). If it was reviewed based on the factors that influence the increase of the understanding ability of geometric proof, it shows that the learning factor and the type of initial mathematical ability affected the development of the ability the understanding of geometric proof.

The data analysis of the ability of geometric proof through a statistical test of average difference was tested to support the description of the understanding ability of geometric proof data that has been explained above. As a prerequisite test, a normality test of the student's mathematical proof ability test was performed in the experimental and control classes. The statistical test used was Shapiro-Wilk test. Based on Shapiro-Wilk test with a significance level of $\alpha = 0.05$, the Sig value of the experimental class was 0.029, while the control value was Sig = 0.007. Since both Sig values (experimental class and control class).

Based on these results, the average difference test above was carried out using Two-Way ANOVA test presented in Table 3.

Table 2. The gain of understanding geometric proof ability based on learning factors and types of Students' initial mathematical knowledge level

Types of Students' initial mathematical knowledge level	Learning	Mean	SD
High	Self-explanation (N=6)	8.50	4.550
	Direct Instruction (N=5)	7.40	2.074
Medium	Self-explanation (N=23)	9.00	3.542
	Direct Instruction (N=23)	3.91	1.929
Low	Self-explanation (N=6)	5.67	2.251
	Direct Instruction (N=7)	3.14	2.410
Total	Self-explanation (N=35)	8.26	3.713
	Direct Instruction (N=35)	4.26	2.393

Table 3. The Summary of Two-Way ANOVA in testing the increase of students' mathematical proof understanding ability based on learning factor and students' initial mathematical knowledge level

Sources	Type III Sum of Squares	df	Mean Square	F	Sig.
Learning strategy (A)	1.056	1	1.056	24.190	0.000
Type of the level of initial mathematical ability (B)	0.329	2	0.164	3.766	0.028
A×B	0.107	2	0.054	1.230	0.299

From the two-way Anova test results in Table 3, it can be seen that the Sig was 0000. Therefore, it can be concluded that there was a significant difference in the increase of geometric proof understanding between students who learned using self-explanation and students who learned using Direct Instruction at a 5% significance level. Based on Table 3, it also obtained the value of sig for the level of initial mathematical ability of 0.028. It can be concluded that there was a difference in increasing mathematical proof ability between students who learned using Self-Explanation strategy and students who learned using Direct Instruction based on the level of students' initial mathematical ability at 5% significance level.

The level of initial understanding ability was tested to identify the difference between groups using Tukey HSD. The results as shown in Table 4. Based on Table 4, it can be concluded that there was a significant difference between the increase of students' mathematical proof ability in a group of students with a high and moderate level of initial mathematical ability at a significant $\alpha = 0.05$. These results indicate that there was a significant difference at the initial mathematical ability level towards increasing students' understanding of geometric proof ability.

In addition, Table 3 also found that there was no significant interaction effect between Self-Explanation strategy and students' initial mathematical ability factors towards the increase of understanding geometric proof ability at a 5% significance level (sig = 0.299). This shows that mathematics's initial ability does not directly influence the learning process to improve the ability to understand the geometric proof. It means that the self-explanation strategy is very effective for all initial mathematical abilities.

Table 4. HSDTukey test results on data of the increase of students' understanding geometric proof ability based on initial mathematical ability level

(I) initial mathematical ability level	(J) initial mathematical ability level	Mean Difference (I-J)	Std. Error	Sig.	95% Confidence Interval	
					Lower Bound	Upper Bound
High	Medium	.08117	.062692	.403	-.06925	.23160
	Low	.17304*	.065750	.028	.01528	.33080
Medium	High	-.08117	.062692	.403	-.23160	.06925
	Low	.09187	.058336	.264	-.04810	.23184
Low	High	-.17304*	.065750	.028	-.33080	-.01528
	Medium	-.09187	.058336	.264	-.23184	.04810

Based on Table 4, it can be concluded that there was a significant difference between increasing mathematical proof ability in the group of students with a high and moderate level of initial mathematical ability at significant $\alpha = 0.05$. These results indicate that there was a significant difference in the level of initial mathematical ability to increase the ability to understand the geometric proof.

The results of the statistical test showed that an increase in the students' geometric proof understanding in the experimental class was better than all students who were in the control class. This result also applies to initial mathematical ability, although only in the high and medium category. These findings are also supported by the results of influence test analysis between learning factors and initial mathematical ability, which indicate that learning factors contribute positively to increased students' mathematical proof ability in the level of initial mathematical ability. This shows that self-explanation strategy has a good impact on improving students' geometric proof understanding ability in a basic geometry course. The study results are in line with the research conducted by Al Rumayyan et al.(2018), which states that students taught using self-explanation strategies can prove that deductive hypotheses from mathematical problems are better than students taught by conventional learning methods.

In the understanding stage of monitoring self-explanation strategy, students are expected to know their weaknesses in initial knowledge. Next, students determine actions to correct these weaknesses. Lecturers provide stimuli to make students aware of their own abilities. The stimulation given by the lecturer is in the form of questions on the students' worksheet. The students' worksheet was designed to meet the principles of selfexplanation. The questions were structured in order to achieve the expected evidence solution. The students feel helped by the questions in the student worksheet, which can be seen from the results of an interview between researchers (R) and one of the students.

R : In worksheets of a student, there are questions that help self-explanation in order to reach conclusions. According to R1 [mentioning the name of respondent], Do those questions help in drawing a conclusion?

R1 : Actually, this really helped me. This is due to the questions presented containing steps to reach a conclusion. So, we will be guided by those questions.

R : In R1 opinion [mentioning the name of respondent], in the geometry learning process, which one is better, using a learning model like this [self-explanation strategy] or conventional learning model?

R1 : It is better than using this [self-explanation strategy] because we can learn by our self and learn to explain. Besides, it can improve the social relationship because we can communicate with friends. There are additional values by using this strategy (selfexplanation strategy].

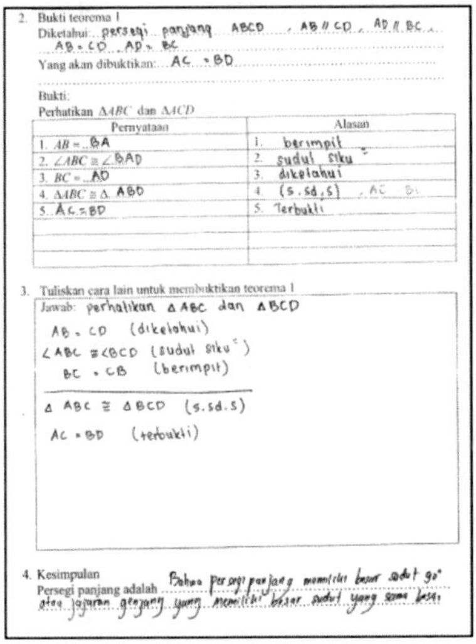
2. Bukti teorema 1
Diketahui: persegi panjang ABCD, AB // CD, AD // BC, AB = CD, AD = BC
Yang akan dibuktikan: AC = BD

Bukti:
Perhatikan ΔABC dan ΔACD

Pernyataan	Alasan
1. AB = BA	1. berimpit
2. ∠ABC ≅ ∠BAD	2. sudut siku"
3. BC = AD	3. diketahui
4. ΔABC ≅ Δ ABD	4. (s. sd. s)
5. AC = BD	5. Terbukti

3. Tuliskan cara lain untuk membuktikan teorema 1
Jawab: perhatikan Δ ABC dan Δ BCD
AB = CD (diketahui)
∠ABC ≅ ∠BCD (sudut siku")
BC = CB (berimpit)
Δ ABC ≅ Δ BCD (s. sd. s)
AC = BD (terbukti)

4. Kesimpulan
Persegi panjang adalah Bahwa persegi panjang memiliki besar sudut 90° atau jajaran genjang yang memiliki besar sudut yang sama besar

Translation of Figure 1:
2. Proof of Theorem 1
Known as a rectangle ABCD, AB//CD, AD//BC, AB=CD, AD=BC
will be proven: AC=BD
Prove:
See ΔABC dan ΔACD

Statement	Reason
1. AB=BA	1. Coincide
2. ∠ABC = ∠BAD	2.Right angle
3. BC=AD	3. Known
4. ΔABC ≅ ΔABD	4. (Side, Angle, Saide)
5. AC=BD	5. Proven

3. Write another way to prove theorem 1
Answer: Known as ΔABC dan ΔBCD
AB = CD (Known)
∠ABC = ∠BCD (Right angle)
BC=CB (Coincide)
ΔABC ≅ ΔABD (Side, Angle, Saide)
AC=BD (Proven)

4. Conclusion
That a rectangle has an angle of 90⁰ or a parallelogram that has two equal angles.

Figure 1. The result of paraphrasing process in student worksheets.

The next step of the self-explanation strategy was paraphrasing, which is the activity of stating and bringing the problem into students' own language to make it easier in deciding the problem solution that will be proven (See Figure 1). In a learning activity, students are faced with problems of proof, which then students discuss to find the solution of proof with guided questions in student worksheets. Then, students were invited to write other solution of proof that has been written by their own language. This refers to the opinion proposed by Rittle-Johnson et. al. who revealed that the repetition of written evidence was very effective in getting used to the process of proof (Rittle-Johnson et al., 2017). By updating the way the student will prove, students will get used to thinking systematically based on logic flow to gain geometric proof that will be achieved, and students will enable them to express their ideas in their own language. It can be seen that learning which used a self-explanation strategy can provide an opportunity for students to express ideas of proof in a group discussion. This refers to the opinion proposed by Boeroe (Reiss & Renkl, 2002) that learning of proof must emphasize students' activeness in expressing ideas in constructing a proof.

One of the mathematical proof abilities is that students evaluate presented proof validity by sorting the steps of proof to obtain valid proof construction. In the selfexplanation strategy, students were assisted by worksheets that were prepared based on the strategy's steps. Students were given proof problems that were complemented by a randomly arranged proof process. Next, students in groups tried to evaluate the presented evidence validity by sorting steps of proof to obtain valid proof construction.

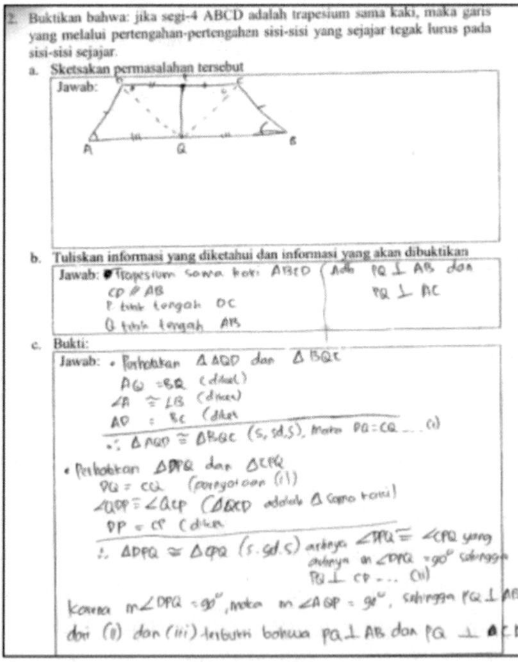

Figure 2. The answer to student's exercise in worksheets of student.

Translation of Figure 2:

2. Prove that: if the rectangle ABCD is an isosceles trapezoid, then the line that passes through the midpoint of the parallel sides is perpendicular to both sides.

a. Sketch out the problem

answer:

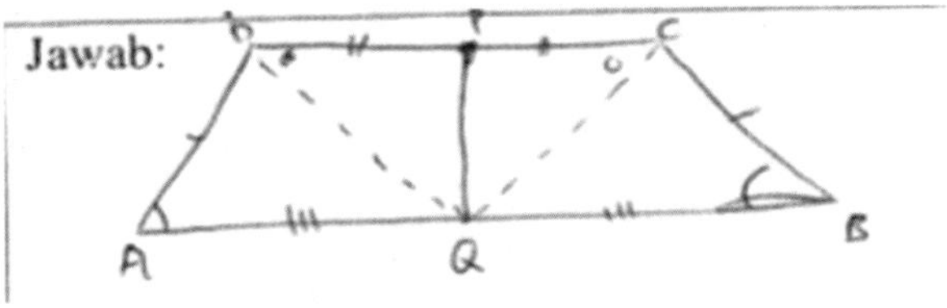

b. Write down the information that is known and information that will be proven

Answer:

Known: isosceles trapezoid ABCD, CD//AB, P is midpoint DC and Q midpoint AB Will be proven that: PQ ⊥ AB and PQ ⊥ AC

c. Proof:

Answer:

- See as ΔABC dan ΔBCD

AQ = BR (Known)

∠A = ∠B (Known)

<u>AD=BC (Known)</u>
ΔABC ≅ ΔABD (Side, Angle, Saide), so DQ=CQ.....(Statment i)

- See as ΔDPQ dan ΔCPQ

DQ = CQ (Statement i)
∠QDP = ∠QCP (ΔCPQ is isosceles triangle)
<u>DP=CP (Known)</u>
ΔDPQ ≅ ΔQCP (Side, Angle, Side), so ∠DPQ = ∠CPQ, that means∠DPQ = 90°, so PQ ⊥ AC.......(Statement ii)
because ∠DPQ = 90° caused ∠AQP = 90°. So, ∠DPQ = 90°.......(statement iii)
for the statement (i), (ii) and (iii) proven that : PQ ⊥ AB and PQ ⊥ AC.

Student worksheets greatly assist students in understanding geometric proof. The practice exercises contained in the worksheet provide experiences for students in practicing geometric proof problems. For example, students are asked to write explanations of the evidentiary strategy by drawing sketches of images, writing information that is known to write the interrelationships of the theorems used, and writing formal proof (See Figure 2).

The self-explanation strategy is a star inference to anticipate reasoning. The anticipation of reasoning can predict the follow-up and coherence of the problem to be proven. In learning of self-explanation strategy, students enable expressing their thoughts, so that strategy ideas in solving proof problems. Based on the interview results, students felt helped by the help questions on the student worksheet. So, through self-explanation strategies, students can develop questions that will create ideas and ideas for evidence. In addition, with self-explanation strategies, students are able to explain to themselves and the group of geometric proof that has been constructed. The results of the interview with students (R1, R2, R3) are the following:

P : Related to self-explanation, in student worksheets, there are questions that help selfexplanation, what do you think about these questions, are they helpful in the learning process?

R1 : Actually those questions are helpful [nod]..because those questions help to obtain a conclusion P : In the self-explanation process, explaining to ourselves, we usually mumble when solving problems [paused for a moment because the conversation was cut off by respondent]

R1 : Yes, I mumbled [cut off the question]

P : Why you did it, R1? [mentioning the name of respondent]

R1 : Firstly, usually, when we say something by words, we will memorize instead of we scribe and mutter in our hearts. Probably it will be remembered faster if we mumble

P : When explaining something in a group, which one is more comfortable, explaining directly or explaining while giving scribbles?

R1 : Yes sir, why if we often explain or are explained, we can absorb knowledge easily from the material?

P : Well, next question, in self-explanation learning, how is the importance of the selfexplanation process? What do you think, R2?

R2 : It is very important because that is not only to add knowledge to ourselves but also to others. So, if we understand the material, we can explain and be more fluent.

P : Related to self-explanation, when we face problems, we will think something, what R3 [mentioning the name of respondent] do when you are facing problems?

R3 : My strategy is to identify the problem carefully, then read the previous material. If there is an obstacle, I will ask my friend.

P : What questions are usually come up in your mind?

R3 : How to draw a sketch P : and then?

R3 : Connecting to suitable theorems

According to the interview results, R1 revealed that self-explanation was very helpful. R1 muttered when solving a proof problem. Muttering activity is meant to explain the answer. According to R1, when expressed an answer by words, it would be better than muttering in the heart. Thus, when solving problems of questions in our minds, it was done by muttering so that the knowledge gained would be easier to remember. Furthermore, R1 also wrote the answers in front of his/her group member while explaining the answers. For R1, this way was easier to remember and understand the discussed material.

Meanwhile, R2 revealed that self-explanation is very important because it is not only adding knowledge to our selves but also to others by explaining. According to R1, when we often explain, we will absorb the explained material easily. In line with R2, R3 also revealed the importance of self-explanation. When facing proof, R3 tried to think of something and ask themselves how to solve it, represent the sketch of the picture, and what suitable theorems are. Thus, a self-explanation process which is helped by guided questions in worksheets of student can help students in solving and understanding geometry proof problems. This is consistent with the results of research conducted by Azrou and Khelladi (2019), which revealed that in the process of proof scaffolding in the form of questions is very helpful for students to express the argument of proof and develop it in the form of statements that point to the proof intended.

The description above shows that the self-explanation strategy is effective in improving the ability to understand the geometric proof. The learning process using selfexplanation strategies emphasizes the active role of students in addressing the problems provided in the sheet. They are presented with guided questions to help students draw conclusions. The results showed that there was no interaction between self-explanation strategy and initial mathematical abilities, which meant that the self-explanation strategy was effective for all levels of students' initial mathematical abilities. This shows that there is an opportunity to apply a self-explanation strategy in developing the ability to understand proof for other material, outside of geometry.

CONCLUSION

The self-explanation strategy is effective learning in supporting the process of understanding geometric proof for students. This research concludes: first, the increased ability to understand the geometric proof of students who used self-explanation strategy was better than those who used direct learning; second, there was a difference in the ability to increase the understanding of geometric proof in terms of the level of initial mathematical ability (high, medium, low); and third, there was no interaction effect of learning factors and initial mathematical ability on increasing the ability to understand the geometric proof. It shows that the initial ability of mathematics does not directly affect the learning process to improve the ability to understand the geometric proof. Hence, the selfexplanation strategy was very effective for all initial mathematical ability. To sum up, the self-explanation strategy is effective for improving the ability to understand the geometric proof, especially for students who have high and medium mathematical initial abilities.

The self-explanation is an alternative learning strategy in geometry, particularly in higher education. Activities in the learning process using self-explanation enable students to complete their provided analysis assignments to develop students' understanding of the geometric proof. Therefore, the researcher recommends further research to focus on developing worksheets based on structured self-explanation questions and the difficulty level of geometry material. It is crucial because it determines the successful implementation of the self-explanation strategy.

ACKNOWLEDGEMENT

Thank you to students of mathematics education department in one of the private universities in Semarang who participated in this research.

BIBLIOGRAPHY

1. Ainsworth, S., & Burcham, S. (2007). The impact of text coherence on learning by self-explanation. Learning and Instruction, 17(3), 286–303. https://doi.org/10.1016/j.learninstruc.2007.02.004
2. Al Rumayyan, A., Ahmed, N., Al Subait, R., Al Ghamdi, G., Mohammed Mahzari, M., Awad Mohamed, T., … Schmidt, H. G. (2018). Teaching clinical reasoning through hypothetico-deduction is (slightly) better than self-explanation in tutorial groups: An experimental study. Perspectives on Medical Education, 7(2), 93–99. https://doi.org/10.1007/s40037-018-0409-x
3. Arnawa, I. M., Sumarno, U., Kartasasmita, B., &Baskoro, E. T. (2007). Applying the Apos Theory To Improve Students Ability To Prove in Elementary Abstract Algebra. Journal of the Indonesian Mathematical Society, 13(1), 133–148. https://doi.org/10.22342/jims.13.1.80.133-148
4. Azrou, N., & Khelladi, A. (2019). Why do students write poor proof texts? A case study on undergraduates' proof writing. Educational Studies in Mathematics, 102(2), 257-274. https://doi.org/10.1007/s10649-019-09911-9
5. Bieda, K. N., Ji, X., Drwencke, J., & Picard, A. (2014). Reasoning-and-proving opportunities in elementary mathematics textbooks. International Journal of Educational Research, 64, 71–80. https://doi.org/10.1016/j.ijer.2013.06.005
6. Chi, M. T. H., De Leeuw, N., Chiu, M. H., &Lavancher, C. (1994). Eliciting self-explanations improves understanding. Cognitive Science, 18(3), 439–477. https://doi.org/10.1016/0364-0213(94)90016-7
7. Conati, C. (2016). Commentary on: "toward Computer-Based Support of MetaCognitive Skills: A Computational Framework to Coach Self Explanation." International Journal of Artificial Intelligence in Education, 26(1), 183–192. https://doi.org/10.1007/s40593-015-0074-8
8. Fiallo, J., & Gutiérrez, A. (2017). Analysis of the cognitive unity or rupture between conjecture and proof when learning to prove on a grade 10 trigonometry course. Educational Studies in Mathematics, 96(2), 145–167. https://doi.org/10.1007/s10649-017-9755-6
9. Haeruman, L. D., Rahayu, W., &Ambarwati, L. (2017). Pengaruh Model Discovery Learning Terhadap Peningkatan Kemampuan Berpikir

Kritis Matematis dan Self-Confidence Ditinjau dari Kemampuan Awal Matematis Siswa SMA di Bogor Timur. Jurnal Penelitian Dan Pembelajaran Matematika, 10(2), 157–168. https://doi.org/10.30870/jppm.v10i2.2040

10. Hodds, M., Alcock, L., & Inglis Loughborough, M. (2014). Self-explanation training improves proof comprehension. Journal for Research in Mathematics Education, 45(1), 62–101. https://doi.org/10.5951/jresematheduc.45.1.0062
11. Kumar, A. N. (2014). An evaluation of self-explanation in a programming tutor. Lecture Notes in Computer Science (Including Subseries Lecture Notes in Artificial Intelligence and Lecture Notes in Bioinformatics), 8474 LNCS, 248–253. https://doi.org/10.1007/978-3-319-07221-0_30
12. Kurby, C., Magliano, J., Dandotkar, S., Woehrle, J., Gilliam, S., & McNamara, D. (2012). Changing how students process and comprehend texts with computer-based self-explanation training. Journal of Educational Computing Research, 47(4), 429–459. https://doi.org/10.2190/EC.47.4.e
13. Kwon, K., Kumalasari, C. D., & Howland, J. L. (2011). Self-explanation prompts on problem-solving performance in an interactive learning environment. Journal of Interactive Online Learning, 10(2), 96–112.
14. Lestari, W. (2017). Pengaruh Kemampuan Awal Matematika dan Motivasi Belajar terhadap Hasil Belajar Matematika. Jurnal Analisa, 3(1), 76. https://doi.org/10.15575/ja.v3i1.1499
15. Maarif, S, Perbowo, K. S., Noto, M. S., & Harisman, Y. (2019). Obstacles in Constructing Geometrical Proofs of Mathematics-Teacher-Students Based on Boero's Proving Model. Journal of Physics: Conference Series, 1315, 012043. https://doi.org/10.1088/1742-6596/1315/1/012043
16. Maarif, S. (2013). Aplikasi Software Cabri Geometri Pada Materi GeometriSebagai Upaya Mengeksplorasi Kemampauan Matematis. In Prosiding Seminar Nasional Matematika Dan Pendidikan Matematika STKIP Siliwangi, Bandung. Vol. 1, pp. 261-271.
17. Maarif, S, Wahyudin, W., Noto, M. S., Hidayat, W., &Mulyono, H. (2018). Geometry Exploration Activities Assisted With Dynamic Geometry Software (Dgs) in a Teacher Education Classroom. Infinity Journal, 7(2), 133. https://doi.org/10.22460/infinity.v7i2.p133-146

18. McNamara, D. S. (2017). Self-Explanation and Reading Strategy Training (SERT) Improves Low-Knowledge Students' Science Course Performance. Discourse Processes, 54(7), 479–492. https://doi.org/10.1080/0163853X.2015.1101328
19. Nguyen, N. T., Trawiński, B., & Kosala, R. (2015). Intelligent information and database systems 7th asian conference, aciids 2015 bali, indonesia, march 23–25, 2015 roceedings, part II. Lecture Notes in Computer Science (Including Subseries Lecture Notes in Artificial Intelligence and Lecture Notes in Bioinformatics), 9012, 1–2. https://doi.org/10.1007/978-3-319-15705-4
20. Otten, S., Gilbertson, N. J., Males, L. M., & Clark, D. L. (2014). The Mathematical Nature of Reasoning-and-Proving Opportunities in Geometry Textbooks. Mathematical Thinking and Learning, 16(1), 51–79. https://doi.org/10.1080/10986065.2014.857802
21. Reiss, K., &Renkl, A. (2002). Learning to prove: The idea of heuristic examples. ZDM - International Journal on Mathematics Education, 34(1), 29–35. https://doi.org/10.1007/BF02655690
22. Rittle-Johnson, B., Loehr, A. M., & Durkin, K. (2017). Promoting self-explanation to improve mathematics learning: A meta-analysis and instructional design principles. ZDM - Mathematics Education, 49(4), 599–611. https://doi.org/10.1007/s11858-017-0834-z
23. Roy, M. & Chi, M. T. . (2005). The Self-explanation Principle. Cambridge Handbook of Multimedia Learning, (0325054), 271–286. Retrieved from http://chilab.asu.edu/papers/MultiM_chapter_final.pdf
24. Samper, C., Perry, P., Camargo, L., Sáenz-Ludlow, A., & Molina, Ó. (2016). A dilemma that underlies an existence proof in geometry. Educational Studies in Mathematics, 93(1), 35–50. https://doi.org/10.1007/s10649-016-9683-x
25. Sommerhoff, D., &Ufer, S. (2019). Acceptance criteria for validating mathematical proofs used by school students, university students, and mathematicians in the context of teaching. ZDM - Mathematics Education, 0(0), 0. https://doi.org/10.1007/s11858-019-01039-7
26. Sztajn, P., Wilson, H., Edgington, C., Myers, M., & Dick, L. (2013). Using Design Experiments to Conduct Research on Mathematics Professional Development. Alexandria: Revista de EducaçãoEmCiência e Tecnologia, 6(1), 9–34. https://doi.org/10.5007/%x

27. Tekeng, N. Y. (2015). Using self-explanation strategy to improve students' understanding of the to be learned material. Auladuna, 2(36), 173–184.
28. Tsitouras, A., Tsivilis, S., &Kakali, G. (2014). A Taguchi Approach for the Synthesis Optimization of Metakaolin Based Geopolymers. In Advances in Science and Technology (Vol. 92). https://doi.org/10.4028/www.scientific.net/ast.92.44
29. Vilkomir, T., &O'Donoghue, J. (2009). Using components of mathematical ability for initial development and identification of mathematically promising students. International Journal of Mathematical Education in Science and Technology, 40(2), 183–199. https://doi.org/10.1080/00207390802276200
30. Wylie, R., Koedinger, K. R., & Teruko, M. (2004). Is Self-Explanation Always Better? The Effects of Adding Self-Explanation Prompts to an English Grammar Tutor Adding Self-Explanation to an Existing Tutor. Human-Computer Interaction, (Figure 1), 1300–1305.
31. Yu, Z., Ding, Y. L., Lu, F., Miao, L. Y., Shen, Z. Y., & Ye, W. X. (2015). Warfarin dosage adjustment strategy in Chinese population. International Journal of Clinical and Experimental Medicine, 8(6), 9904–9910.

Chapter

14

MATHEMATICAL PROOF: THE LEARNING OBSTACLES OF PRE-SERVICE MATHEMATICS TEACHERS ON TRANSFORMATION GEOMETRY

Muchamad Subali Noto[1,2], Nanang Priatna[2] , Jarnawi Afgani Dahlan[2]

[1]Universitas Swadaya Gunung Djati, Jl. Perjuangan No.1 Cirebon, Indonesia

[2]Universitas Pendidikan Indonesia, Jl. Dr. Setiabudi No. 229 Bandung 40154, Indonesia

ABSTRACT

Several studies related to mathematical proof have been done by many researchers on high-level materials, but not yet examined on the material of transformation geometry. The aim of this research is identification learning obstacles pre-service teachers on transformation geometry. This study is qualitative research; data were collected from interview sheets and test. There were four problems given to 9 pre-service mathematics teachers. The results of this research were as follows: learning obstacles related to the difficulty in applying the concept; related to visualize the geometry object; related

Citation: (APA): Maarif, S., Alyani, F., & Pradipta, T. R. (2020). The implementation of self-explanation strategy to develop understanding proof in geometry. JRAMathEdu (Journal of Research and Advances in Mathematics Education), 5(3), 262-275.(14 pages).

Copyright: © Creative Commons Attribution 4.0 International License (http://creativecommons.org/licenses/by/4.0/).

to obstacles in determining principle; related to understanding the problem and related obstacles in mathematical proofs such as not understanding and unable to express a definition, not knowing to use the definition to construct the proof, not understanding the use of language and mathematical notation, not knowing how to start the proof.

Keywords: Mathematical proof; Pre-service mathematics teachers; Geometry

Geometry is an integral part of the learning of mathematics (Fachrudin, Putri, & Darmawijoyo, 2014; Sukirwan, Darhim, Herman, & Prahmana, 2018; Ahamad, Li, Shahrill, & Prahmana, 2018). However, the development of learning geometry at this time is less developed. One reason is the difficulty in forming a real construction student carefully and accurately, the notion that to paint geometry requires precision in the measurement and requires a long time, and not infrequently students experiencing obstacles in the process of evidence (Rizkianto, Zulkardi, & Darmawijaya, 2013; Novita, Prahmana, Fajri, & Putra, 2018). Meanwhile, the painting plays an essential role in teaching geometry at school for painting geometric connection between physical space and theory. If further investigation of the link between the objects with the abstract geometry student obstacles in learning geometry, it will be alleged that in fact there is a problem in teaching geometry at school relates to the formation of abstract concepts. Learn abstract concepts cannot be done only with the transfer of information, but it takes a process of formation of concepts through a series of activities experienced directly by students (Nurhasanah, Kusumah, & Sabandar, 2017). The series of abstract concept formation activity of these are referred to the process of abstraction.

Studying mathematics meant to be also studied branches of mathematics is the science of geometry. Everything in this universe is geometry so that the branches of mathematics through geometry learn about the concepts embodied in the objects that exist in nature through geometric concepts. Thus, assessment of learning geometry must continue to be developed so that each learner can analyze the geometry of objects into a concept of geometry and can construct geometry knowledge with formal proofs. Mata-pereira & Ponte (2017) say that a proof is a connected sequence of assertions that includes a set of accepted statements, forms of reasoning and modes of representing arguments. Stefanowicz & Kyle (2014) say that a proof is a sequence of logical statements, one implying another, which explains why a given statement is true.

However, mathematical proofs in geometry material lately become an obstacle that they seem poorly developed. Difficulties analyzing geometric properties are realized in the form of theorems to create a concept widely experienced by the students. Proof becomes a severe matter in determining the school curriculum in every different country. This is what makes reasoning and proof NCTM enters into one of the standard processes. This means that in every lesson a teacher must enter the elements in each classroom. Maya & Sumarmo (2011) state that possessing mathematical proving ability was certainty ability because it is an essential ability that should be possessed by all students who learn mathematics. Komatsu (2017) state also that proof and proving to play a crucial role in the discipline of mathematics and should be an essential component of mathematical learning.

Several studies have been conducted regarding evidence learning in secondary school (Harel & Sowder, 1998; Mariotti, 2006). Several research methods have also been conducted in proof-learning. Duval's research (1991) has identified the arguments and evidence that explain the difficulties students experience in understanding and making proof. Other researchers such as Balacheff, 1988; Harel & Sowder, 1998; Marrades & Gutiérrez, 2000 focuses more on identifying the types of empirical and deductive evidence generated by students that enable student progress in learning to prove.

Some researchers identify and explain the reasons why students are unwilling or unable to complete deductive evidence from their allegations (Arzarello, Micheletti, Olivero, & Robutti, 1998). Researchers also focus on analyzing learning especially on deductive evidence (Antonini, 2003; Stylianides, Stylianides, & Philippou, 2007; Antonini & Mariotti, 2008). More details can be found in Mariotti's research (2006), Reid and Knipping (2010), and Hanna and De Villiers (2012).

In traditional learning, mathematical proof is only used as a means to eliminate the doubts of students on the concepts taught. However, the evidence is not used as a means of increasing the higher mathematical ability. Just as revealed by Hana (Christou, et al. 2004) that the function of evidence and proof are: verification, explanatory, systematization, invention, communication, construction, exploration, and incorporation. Verification of proving and the proof is regarded as the most fundamental functions in the proof because both are products of the process of the development of mathematical thinking very mature. Verification refers to the truth of the statement while explanations provide insight into why this is true.

As for the role played proof in mathematics, namely: 1) to verify that a statement is true, 2) to explain why a statement can be said to be true, 3) to establish communication mathematics, 4) to find or create new math and 5) To make systematic statement in an axiomatic system (Knuth, 2002). Hanna (Stylianides, Stylianides, and Philippou, 2007) say that there are three main reasons why the ability to prove the need to be improved. First, the proof is crucial to learn to explore mathematics. Second, the ability of students in the proof can improve their math skills more broadly, because the evidence "involved in all situations where the conclusion must be reached in the making of decisions to be made", and the third, the difficulties experienced by high school students and college students will affect their ability to perform mathematical proofs on a broader level again, so it is crucial for students to learn mathematical proof on the level of previous education.

METHOD

This study was conducted to analyze student learning obstacles, especially regarding the difficulty students epistemology regarding materials, both presented in the form of materials or materials in lectures. This research method is descriptive qualitative research that aims to describe obstacles regarding epistemology student learning mathematical proofs related to the subject of transformation geometry. Subjects were nine student teachers Unswagati contracting mathematics courses transformation geometry consisting of 3 students with high prior knowledge mathematically, three students with medium prior knowledge of the mathematically and three students with low prior knowledge of mathematically. The beginning of knowledge is based on the acquisition of student achievement index in the previous semester. For students learning obstacles regarding epistemology student at transformation material using five indicators: concept, visualization, principles, understand the problem, and mathematical proofs.

RESULTS AND DISCUSSION

This research resulted in qualitative data. Learning obstacles students understand the concepts in terms of epistemology transformation geometry in working on the geometry transformation is divided into 5 types, it is in terms of indicators in assessing learning obstacles, namely: a) learning obstacles related to the difficulty in applying the concept; b) learning obstacles related to visualize the geometry object; c) learning obstacles related to obstacles in determining principle; d) learning obstacles related to understanding the problem and e) related obstacles in mathematical proofs.

Related understanding and applying the concept of learning difficulty is the difficulty experienced by students in understanding and applying the concept by the command matter. Examples of these obstacles one student does not understand the concept, students cannot mention the definition of transformation. It happened at the beginning of the mathematical knowledge of students with high, medium or low. Here is one example of the questions and responses of students experiencing barriers to learning. For example, on the following question: write the definition of a transformation in the field of V.

Learning obstacle students with high prior knowledge is described as follows. S1 had trouble with the concept, to define the transformation; these students do not write domain/codomain of a function called transformation. S5 can write correctly and complete the definition of transformation. S8 experienced obstacle of the concept, to define the transformation, these students do not write domain/ codomain of a function called transformation.

Learning obstacle students with prior knowledge is being described as follows. S3 having trouble against the concept, it cannot define the transformation correctly; these students mention that the transformation is a bijective function, but do not write domain /codomain of these functions. S4 in defining transformation by merely mentioning that a transformation is injective functions only. S9 can write the definition of transformation correctly and completely. Obstacles students with low initial knowledge are described as follows. S2 no difficulty is in writing the definition of transformation. S6 and S7 to write the definition of transformation are not complete. Both of these students do not write domain/codomain of a function is the transformation in the field of S6 V. It also found one in writing notation.

Related Learning Obstacles Visualize learning obstacles related visualize the geometry object. The point is that students have obstacles regarding describing the line of the transformation result. Examples of these obstacles include the inability of students in painting properly and appropriately. Here is one example of the questions and responses of students who experience learning obstacles. For example, in Question 2 is as follows: Draw the line $g' = M_h(g)$

if h = {(x,y) |y = x + 1} and $g = \{(x, y) | y = -x\}$.

Learning obstacle students with high prior knowledge is described as follows. S1 is having trouble visualize right lines g and h so that images reflection created false images. S5 can visualize by what is known and questioned, but the images are still made without a ruler. S8 can paint lines mirroring the results appropriately. Obstacles students with prior knowledge are being described as follows.S3 had difficulty in visualizing the line h so that the reflection is illustrated one. S4 in the paint did not use a ruler, but the results are correct reflection depicted. S9 cannot describe all that is known. Obstacles students with low initial knowledge are described as follows. S2 can paint reflection results correctly. S6 and S7 cannot describe all that is known so that the reflection nothing, other than that, the two students were not drawing Cartesian coordinates as a first step in painting a line mirroring results.

Learning obstacles related to the principle of. Learning obstacles is the difficulty experienced by students in terms solves the problem by defining the principles to be used in solving the problem of transformation. Examples of this difficulty are the inability of students in the mentioned properties of isometry, so it cannot be a member of reasons of the questions in the matter. Here is one example of the questions and responses of students who have difficulty learning. For example, the following problem: Given: T and S isometry. Determine the statements below True or False? Give your reason.

a. If g is a line, then g'= (TS)(g) is also a line.

b. If g // h and g'= (TS)(g), h'= (TS)(h) then g'// h'.

c. If S is a reflection of the S is involutory.

Learning obstacle students with high prior knowledge is described as follows. S1 can answer correctly, but the reasons expressed by one. S5 can be answered correctly and the reasons for appropriately. S8 can mention the definition of isometry correctly, but cannot answer questions related to the principle of isometry, so that reason used improperly. Obstacles students with prior knowledge are being described as follows. S3 was having difficulty writing down the definition of isometric and members wrong reasons related to statements given. S4 can write for the right reasons, but they are notational wrong. S9 is important to specify the definition of isometry, giving the wrong reasons. Obstacles students with low initial knowledge are described as follows. S2 is important to specify the definition, and the reason given was also incorrect. S6 and S7 wrong in writing down the definition and does not include the reason (no answer).

Learning obstacles related to understanding the problem. Is the difficulty of learning obstacles experienced by students regarding understanding the problem to solve the problem by using the steps in the completion of the write down what is known and asked about the matter. Examples of this difficulty are the inability of students to solve problems by the steps to completion. Here is one example of the questions and responses of students who have difficulty learning. For example, on the following question: define a line equation $g' = M_h(g)$ if h = {(x,y) |y = x + 1} and $g = \{(x, y) | y = -x\}$.

Learning obstacle students with high prior knowledge is described as follows. S1 was having trouble determining what is known and troubleshooting procedures are still wrong. S5 can understand the problem, find out what is known and asked, can solve the problem by the settlement procedures. S8 can understand the problem and solve it according to the procedure. Obstacles students with prior knowledge are being described as follows. S3 can understand the problem and solve the problem with proper procedures, but there are errors in arithmetic operations. S4 can understand the problem and solve the problem according to the procedure. S9 cannot understand the problem and cannot solve it. Obstacles students with low initial knowledge are described as follows. S2 can understand the problem but cannot finish the correct procedure. S6 and S7 are not able to understand the problems and did not finish.

The difficulty is in proving mathematical. Learning difficulty is the difficulty experienced by students in constructing the proof of the matter. Here is one example of the questions and responses of students who have difficulty learning. For example, the following problem: to prove that the reflection on the line g is an isometry.

Learning obstacle students with high prior knowledge is described as follows. S1 can construct evidence correctly, but there is still incorrect notation. S5 can construct evidence properly. S8 can construct proofs in part, at the end of the part that is wrong in giving reasons. Obstacles students with prior knowledge are being described as follows. S3 and S4 obstacles for constructing proofs can't use the existing definition. S9 trouble is to begin constructing proofs. Obstacles students with low initial knowledge are described as follows. S2 cannot use a definition for constructing proofs. S6 and S7 begin constructing the evidence about be proved, difficulty in starting the construction of the evidence and not be able to use the definition for constructing proofs.

Some mathematical proofs related research shows some of the things that are essential mathematical proofs. Activities considered difficult by students to learn and teachers to teach include justification or proof (Suryadi, 2007). Research studies conducted Dryfus (Jones and Rood, 2001) showed that students always fail to look at the adequacy of the evidence because they are too often asked to prove things that are obvious to them. Students also fail to distinguish between the different forms of mathematical reasoning such as heuristic or argument, explanation or proof. A significant gap in the research literature is still at least main set students "because it looks right" instead of "because he worked on issues" for the argument that believed.

The research result Knuth (2002) showed that teachers recognize the many roles of the play proof in mathematics, in learning the role of evidence should not be abandoned, and the evidence as a tool for learning mathematics. The results also show that many teachers still have limitations in determining the nature of the evidence used in the study of mathematics.

Heinze and Reiss (2003) research results showed that some students found with some of the answers wrong though ideas about the solution are proving correct. This occurs in the empirical argument. Many errors occur experienced by students is related to aspects of the structure of evidence. Most of the students interviewed were mostly already know that the argument does not form empirical evidence.

Study interviews in this study also showed that the three aspects of methodological knowledge of relevant proof when assessing the evidence. It seems that on this aspect conclusion chain is not problematic, because the proof is right mostly depicted as true. However, some cases it was not clear whether the students understand every step of the evidence compiled. The problem with this aspect of the scheme of evidence, in particular, an inductive argument, often trained using inductive argument in elementary school. Students have difficulty bridging the gap between empirical arguments to formal arguments. This is confirmed by a study by Lin (Heinze and Reiss, 2003) showed that Taiwanese students a different problem that is wrong or improper argument transformation in improving formal arguments were observed. Mariotti study (2001) revealed the geometry construction is an essential part of the experience of students who should be organized. Results of the study Mariotti showed that if the geometry is just a pencil and paper geometry theory perspective, it is difficult to understand. When students draw on paper students can only focus on the images being constructed and can't manipulate it.

CONCLUSION

Average-ability students, especially in mathematical constructing proofs of 11.33 (maximum score 28) with details as follows: for students with high mathematical prior knowledge gained an average of 19.67; students with early mathematical knowledge have gained an average of 9.33, and with low prior knowledge mathematical earned an average of 5.00. There arc five kinds of difficulties related to the students regarding epistemology on geometry transformations, namely a) learning difficulties related to the difficulty in applying the concept; b) associated learning difficulties visualizing the geometry object; c) learning difficulties related to difficulties in determining principle; d. Related learning difficulties to understand the problem and e. Related difficulties in mathematical proofs. Specialized in mathematical proofs, students have difficulty, among others: do not know how to start the construction of the evidence, can't use the definition (concept) and the principle already known, and are likely to begin construction of the evidence with what must be proved.

ACKNOWLEDGMENTS

Author would say thank to KEMENRISTEK DIKTI that giving research and doctoral dissertation grant.

REFERENCES

1. Ahamad, S. N. S. H., Li, H. C., Shahrill, M., & Prahmana, R. C. I. (2018). Implementation of problem-based learning in geometry lessons. Journal of Physics: Conference Series, 943(1), 012008.
2. Antonini, S., & Mariotti, M. A. (2008). Indirect proof: What is specific to this way of proving? ZDM International Journal on Mathematics Education, 40(3), 401–412.
3. Arzarello, F., Micheletti, C., Olivero, F., & Robutti, O. (1998). A model for analyzing the transition to formal proofs in geometry. In A. Olivier & K. Newstead (Eds.), Proceedings of the 22nd PME Conference, 24–31. Stellenbosch, Republic of South Africa: PME.
4. Balacheff, N. (1988). Aspects of proof in pupils' practice of school mathematics. In D. Pimm (Ed.), Mathematics, teachers and children, 216–235. London: Hodder & Stoughton.
5. Christou, C., Mousoulides, N., Pittalis, M and Pantazi, D.P. (2004). Proofs Through Exploration In Dynamic Geometry Environments. Proceedings of the 28th Conference of the International Group for the Psychology of Mathematics Education 2004, 215–222.
6. Duval, R. (1991). Structure du raisonnement déductif et apprentissage de la démonstration. Educational Studies in Mathematics, 22(3), 233–261.
7. Fachrudin, A. D., Putri, R. I. I., & Darmawijoyo. (2014). Building Students' Understanding of Quadratic Equation Concept Using Naïve Geometry. Journal on Mathematics Education, 5(2), 192-202.
8. Hanna, G., & de Villiers, M. (Eds.). (2012). Proof and proving in mathematics education. Dordrecht, The Netherlands: Springer.
9. Harel, G., & Sowder, L. (1998). Students' proof schemes: Results from exploratory studies. In A. H. Schoenfeld, J. Kaput, & E. Dubinsky (Eds.), Research in collegiate mathematics education, 234–283. Providence, RI: American Mathematical Society.
10. Heinze, A., & Reiss, K. (2003). Reasoning and proof: Methodological knowledge as a component of proof competence. Proceedings of the third conference of the European Society for Research in mathematics education, 1-10.
11. Jones, K & Rodd, M. (2001). Geometry and Proof. Proceedings of the British Society for Research into Learning Mathematics 21(1), 95-100.

12. Komatsu, K. (2017). Fostering empirical examination after proof construction in secondary school geometry. Educational Studies in Mathematics, 96(2), 129–144.
13. Knuth, E.J. (2002). Teachers' Conception of Proof in the Context of Secondary School Mathematics. Journal of Mathematics Teacher Education 5(1), 61–88.
14. Marrades, R., & Gutiérrez, A. (2000). Proofs produced by secondary school students learning geometry in a dynamic computer environment. Educational Studies in Mathematics, 44(2), 87–125.
15. Mariotti, M. A. (2001). Introduction To Proof: The Mediation Of A Dynamic Software Environment. Educational Studies in Mathematics 44(1), 25–53.
16. Mariotti, M. A. (2006). Proof and proving in mathematics education. In A. Gutiérrez & P. Boero (Eds.), Handbook of research on the psychology of mathematics education. Rotterdam. The Netherlands: Sense.
17. Mata-pereira, J., & Ponte, J. (2017). Enhancing students' mathematical reasoning in the classroom: teacher actions facilitating generalization and justification. Educational Studies in Mathematics, 96(2), 169–186.
18. Maya, R., & Sumarmo, U. (2014). Mathematical Understanding and Proving Abilities: Experiment With Undergraduate Student By Using Modified Moore Learning Approach. Journal on Mathematics Education, 2(2), 231-250
19. Novita, R., Prahmana, R. C. I., Fajri, N., & Putra, M. (2018). Penyebab kesulitan belajar geometri dimensi tiga. Jurnal Riset Pendidikan Matematika, 5(1), 18-29.
20. Nurhasanah, F., Kusumah, Y. S., & Sabandar, J. (2017). Concept of triangle: Examples of mathematical abstraction in two different contexts. International Journal on Emerging Mathematics Education, 1(1), 53-70.
21. Rizkianto, I., Zulkardi, & Darmawijaya. (2013). Constructing Geometric Properties of Rectangle, Square, and Triangle in the Third Grade of Indonesian Primary Schools. Journal on Mathematics Education, 4(2), 160-171.
22. Reid, D. A., & Knipping, C. (2010). Proof in mathematics education. Rotterdam, The Netherlands: Sense.

23. Stefanowicz, A., & Kyle, J. (2014). Proofs and Mathematical Reasoning. University of Birmingham. Retrieved from https://www.birmingham.ac.uk/
24. Stylianides, G. J., Stylianides, A. J., & Philippou, G. N. (2007). Preservice teachers' knowledge of proof by mathematical induction. Journal of Mathematics Teacher Education, 10(3), 145–166.
25. Sukirwan, Darhim, Herman, T., & Prahmana, R. C. I. (2018). The students' mathematical argumentation in geometry. Journal of Physics: Conference Series, 943(1), 012026.
26. Suryadi, D. (2007). Model Bahan Ajar dan Kerangka-Kerja Pedagogis Matematika Untuk Menumbuhkembangkan Kemampuan Berpikir Matematik Tingkat Tinggi [Model of Teaching Materials and Framework of Mathematical Pedagogy for Developing Higher Mathematical Thinking Ability]. Unpublished Research Report. Bandung: SPS UPI.

Chapter

15

STUDENTS' MATHEMATICAL PROBLEM-SOLVING ABILITY BASED ON TEACHING MODELS INTERVENTION AND COGNITIVE STYLE

Aloisius Loka Son [1,2], Darhim[2] , Siti Fatimah [2]

[1]Universitas Timor, Kefamenanu, Indonesia
[2]Universitas Pendidikan Indonesia, Bandung, Indonesia

ABSTRACT

The study aimed to analyze the interaction effect teaching models and cognitive style field dependent (FD)-field independent (FI) to students' mathematical problem-solving ability (MPSA), as well as students' MPSA differences based on teaching models and cognitive styles. Participants in this study were 145 junior high school students, with details of 50 students learning through the Connect, Organize, Reflect, and Extend Realistic Mathematics Education (CORE RME) model, 49 students use the CORE

Citation: (APA): Son, A. L., & Fatimah, S. (2020). Students' Mathematical Problem-Solving Ability Based on Teaching Models Intervention and Cognitive Style. Journal on Mathematics Education, 11(2), 209-222. (10 pages).

Copyright: © Creative Commons Attribution 4.0 International License (http://creativecommons.org/licenses/by/4.0/).

model, and 46 students use the Conventional model. Data collection tools used are the MPSA test, and the group embedded figure test (GEFT). The MPSA test finds out that there are interaction effect teaching models and cognitive styles on students' MPSA, as well as a significant difference in MPSA students who study through the CORE RME model, CORE model, and Conventional model. Based on cognitive style, between students who study through CORE RME model, CORE model, and Conventional model found that there was no significant difference in MPSA between FI students. Furthermore, there were significant differences in MPSA between FD students and also MPSA of FI students better than MPSA FD students. Therefore, teaching models and student cognitive styles are very important to be considered in the learning process, so students are able to solve mathematical problems.

Keywords: Mathematical Problem-Solving Ability; Teaching Models; Field Dependent-Field Independent

Problem-solving is a characteristic of mathematical activity and is a major means of developing mathematical understanding (NCTM, 2000). This statement implies that problem-solving is an integral part of all mathematics learning. Furthermore, students learn to apply their mathematical skills with new ways; they develop a deeper understanding of mathematical ideas and feel the experience of being a mathematician through solving-problems (Badger et al., 2012). Therefore, students can develop new knowledge, solve problems that occur, apply and use various strategies, and also reflect and monitor the problem-solving process.

The problem-solving process requires implementing a certain strategy, which may lead the problem solver to explore multiple ideas by developing and testing hypotheses. Related with the process, NCTM (2000) said that in order to find solutions for any given problem, students should utilize their knowledge, through which they often able to develop a new mathematical understanding. Foshay and Kirkley (2003) said that Bransford's IDEAL model is a common problem-solving model used, consisting of identify the problem, define the problem through thinking about and sorting out relevant information, explore solution through looking at alternatives, brainstorming, and checking out a different point of view, act on the strategies, and look back and evaluate the effects of your activity.

The famous problem-solving steps according to Polya (1957), such as understanding the problem, devising a plan, carrying out the plan, and looking back. Firstly, understanding the problem is the ability to convince yourself that students understand the problem correctly, by describing known and unknown elements, what quantities are known, how they are, whether there are exceptions, and what is asked. Secondly, devising a plan is the ability to find the relationship of information which was given and the unknown that allows students to calculate unknown variables. Thirdly, carrying out the plan is the ability to carry out the plan contained in the second step, by examining each step in the plan and writing it down in detail to ensure that each step is correct. Lastly, looking back is the ability to test the solution that has been obtained by criticizing the results, and giving conclusions correctly.

Although problem-solving is the main goal in learning mathematics, but that goal remains one of the most difficult cognitive abilities for students to understand (Tambychik & Meerah, 2010; Căprioară, 2015). Several evidences show that students still find difficulties in solving mathematical problems as evidenced by a survey by TIMSS and PISA. One of the benchmarks used in the assessment by TIMSS is that students can apply their mathematical knowledge and understanding in solving problems (IEA, 2016), and PISA measures the capacity of students to apply their knowledge and skills in identifying, interpreting, and solving problems in various situations (OECD, 2019). Data from the TIMSS and PISA survey shows that the ability to solve mathematical problems of Indonesian students is still below expectations. The International Association for the Evaluation of Educational Achievement (IEA) reported the result of TIMSS survey in 2015, Indonesia ranked 45th out of 50 participating countries (IEA, 2016). While the result of the PISA study released by the Organization for Economic Cooperation and Development (OECD) shows that in 2018, Indonesia ranked 72 out of 78 participating countries (OECD, 2019).

Difficulties in solving mathematical problems are also experienced by seventh-grade students of Junior High School in North Central Timor Regency, located in the border areas between the Republic of Indonesia and the Democratic Republic of Timor Leste. This is proven through the results of research by Son, Darhim, and Fatimah (2019) about errors made in solving algebraic problems based on Polya's and Newman's theory. The results showed that more than 50% of the participants made errors in solving algebra problems. More students made errors on all indicators, both based on Polya's steps and based on Newman's theory. During interviews with the research participants on the reasons why they made errors in solving the algebra questions given, many students said that these questions were rarely found in the learning process. They are not familiarized with solving math problems. This shows that one of the reasons for the students' inability to solve given problems is that they were not well trained to solve problems during mathematics class.

Therefore, learning mathematics should encourage students to apply mathematics confidently in solving problems. Learning mathematics at school should help students in understanding mathematics, and applying it in solving daily problems both in society and the workplace. The learning program has to enable students to develop new mathematical knowledge through problem-solving, solve mathematics and other problems, implement and adjust various strategies available to solve problems, and monitor and reflect the process of solving mathematical problems (NCTM, 2000).

Students' problem-solving abilities will increase if the teacher uses a student-centered learning model (Wijayanti, Herman, & Usdiyana, 2017). The Connect, Organize, Reflect, and Extend (CORE) model is a student-centered learning model because through CORE students can build their knowledge by connecting and organizing new knowledge and old knowledge, thinking about the concepts being learned and expanding their knowledge during the learning process (Curwen, Miller, Smith, & Calfee, 2010). The CORE model combines four elements: connecting is the stage of linking old information with new information or between concepts, organizing is the stage of organizing the information obtained, reflecting is the stage of rethinking information already obtained, and extending is the stage of expanding knowledge.

Related with making connections between old and new information in mathematics learning, NCTM (2000) asserts that if mathematical ideas are interconnected with real-world phenomena, students will view mathematics as something useful, relevant and integrated and becomes very powerful process in developing students' understanding of mathematics. This NCTM statement illustrates that students' mathematical understanding will be more developed if the learning of mathematics begins by making connections between the subjects studied with the student experience, not only between mathematical concepts but must be connected to real-world phenomena. Mathematics learning that places real context and student experience as a starting point for learning is Realistic Mathematics Education (RME)

(Prahmana, Zulkardi, & Hartono, 2012; Saleh et al., 2018; Apsari et al., 2020).

RME is a learning approach that uses the real-world context as a starting point for learning and views mathematics as a human activity (Freudenthal, 2002; Yilmaz, 2020). Through horizontal and vertical mathematical activities, students are expected to be able to find and construct mathematical concepts (Treffers, 1987). Realistic in this learning can be meaningful: (1) real context that exists in everyday life; (2) formal mathematical contexts in the world of mathematics; or (3) imaginable contexts that do not exist in reality but can be imagined (Heuvel-Panhuizen & Drijvers, 2014).

Many researchers, especially in Indonesia, researched the influence of the CORE model, as well as a realistic mathematical approach to students' mathematical problem-solving abilities. Their results show that there is an increase in students' mathematical problem-solving abilities after learning with the CORE model (Purwati, Rochmad, & Wuryanto, 2018; Wijayanti et al., 2017), and the achievement and improvement of mathematical problem-solving abilities of students who study through RME approach are better than students who learn using conventional approach (Ulandari, Amry, & Saragih, 2019; Huda, Florentinus, & Nugroho, 2020; Chong, Shahrill, & Li, 2019). These previous studies analyzed the effect of the CORE model as well as a realistic mathematical approach to problem-solving abilities, but the implementation was separated. In this study, the CORE teaching model has collaborated with a realistic mathematical approach which is then called the CORE RME teaching model.

CORE RME teaching model is done through the CORE model syntax namely Connect, Organize, Reflect, and Extend. In the Connect stage, given real context problems that have to do with the student experience. Furthermore, at the Organize stage, students are given the opportunity to carry out reinvention and self-developed models of these real problems. Reflect stage is the stage of rethinking and seeing the relationship between the models of which is built by students and the model for the appropriate subject matter. Furthermore, the Extend stage is the stage of expanding knowledge with other real problems. The learning syntax of the CORE RME model can be described in the implementation flowchart as shown in Figure 1.

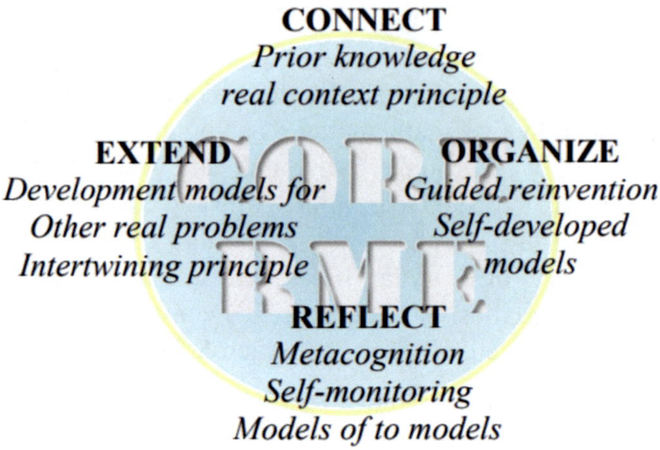

Figure 1. CORE RME models cycle.

Learning through CORE RME syntaxes such as Figure 1 can trigger the development of students' mathematical problem-solving abilities because it is supported by several main principles in RME namely guided reinvention, progressive mathematization, didactical phenomenology, and self-developed models (Gravemeijer, 1994). Mathematical problem-solving ability (MPSA) of students can be seen from several dimensions, one of which is cognitive style. Cognitive style is one of the important variables that can influence student problemsolving (Mefoh, Nwoke, & Chijioke, 2017). Therefore, some researchers throughout the world are very interested in examining the relationship between cognitive style dimensions and mathematical abilities (Chrysostomou, Pantazi, Tsingi, Cleanthous, & Christou, 2012). Cognitive styles are divided into several types, namely fielddependent and field-independent cognitive styles, impulsive and reflective cognitive styles, perceptive and receptive cognitive styles, and intuitive and systematic cognitive styles (Volkova & Rusalov, 2016).

Field-dependent (FD) and field-independent (FI) are the most popular cognitive styles (Mefoh et al., 2017). FI and FD are cognitive styles characteristics that are characterized by general ways of thinking, problem-solving, learning and dealing with others (Abrams & Belgrave, 2013). This definition explicitly illustrates that FI and FD cognitive styles are related to one's problem-solving performance. Pithers (2006) says that there is a strong relationship between FI-FD cognitive style and problem-solving performance, where the solution depends on critical elements utilization in a different context from the original context where it was presented.

FI's cognitive style reflects the students' ability to rely on their knowledge and experience when solving problems, whereas FD's cognitive style describes students' orientation to the outside world when solving problems (Volkova & Rusalov, 2016). This is the difference between FI students and FD students when solving problems, in which FI students tend to be independent and confident, while FD students tend to rely on external influences. Although a lot of researches have been conducted on the FI and FD cognitive styles, there is still less attention given to this type of cognitive style in relation to certain mathematical fields such as problem-solving and mathematical operations (Nicolaou & Xistouri, 2011), so this research was conducted to study MPSA students based on learning model intervention and the FI-FD cognitive styles.

METHOD

The research method used is quantitative research with a quasi-experimental approach because it does not re-group random samples, but uses classes that have been formed by the school that is used as a population. The research design used is the nonequivalent comparison group design which is a better condition for all quasi-experimental research designs. In this research, there are two experimental groups namely a group of students who study through the CORE RME model, and the CORE model, while the control group is a group of students who study through the Conventional model.

Participants in this study were 145 students with details of 50 students who study through the CORE RME model, 49 students who study through the CORE model, and 46 students who study through the Conventional model. These 145 people are Grade VII students in two state junior high schools in Kefamenanu City, Timor-NTT, Academic Year 2018/2019. These two-public junior high schools were selected using a purposive sample of 5 public junior high schools in the city of Kefamenanu, with the reason that the two schools used the 2013 Curriculum for the first time.

The instrument used to obtain data in this study was the Group Embedded Figure Test (GEFT), and a mathematical problem-solving ability test. GEFT is a psychiatric test developed by Witkin (1971) to determine the cognitive style of FI and FD students. The number of GEFT questions is 18 numbers with the assessment criteria is that if the student's final score is in the range of 0-11 then the student has a cognitive style of FD. Whereas, if the final score is in the 12-18 interval, then the student has the FI cognitive style. This GEFT level of reliability has been measured by previous researchers. The value obtained from the Alpha Cronbach reliability of 0.84, meaning that the reliability of GEFT is very high.

MPSA test consists of 4 numbers in the form of a description test, which are arranged through an expert validation process, and then are tested on students to find out the level of validity and reliability. The average validator assessment results are 91.67 which showed that the test questions are in good category and can be used at a later stage. While the results of trials on 19 students obtained Cronbach's alpha value of 0.69 which means the item test was reliable. While the Pearson correlation value of the four questions in a row is 0.73; 0.75; 0, 65; and 0.79, which means all four questions are valid.

Data analysis techniques used were two-way anova statistical analysis, one-way anova, Kruskal Wallis and t-test one-tailed. Two-way anova test was carried out to find out there is an interaction effect between teaching models and cognitive styles on students' mathematical problem-solving abilities, oneway anova test to find out the difference in mathematical problem-solving abilities based on teaching models, Kruskal Wallis test to find out the difference in mathematical problem-solving abilities between FI students and between FD students, and t-test one-tailed to find out the comparison of students' mathematical problem-solving abilities between FI students and FD students. Both the prerequisite test and the hypothesis test in this study were analyzed using IBM SPSS Statistics 22.

RESULT AND DISCUSSION

The Interaction of Teaching Models and Cognitive Styles with Students' Mathematical Problem-solving Abilities

Interaction test between teaching models and cognitive styles on MPSA of students using the twoway anova test, because the significance value of Kolmogorov-Smirnova on standardized residuals is $0.20 > 0.05$ which means the data distribution of interaction between teaching models and cognitive styles on students' MPSA normally distributed. The two-way anova test output is presented in Table 1.

Table 1. Interaction test of teaching models and cognitive styles on MPSA of students

Source	Sum of Squares	df	Mean Square	F	Sig.	Ho
Corrected Model	2075.31[a]	5	415.06	19.94	0.00	
Intercept	60045.00	1	60045.00	2885.12	0.00	
Teaching Models	139.06	2	69.53	3.34	0.04	
Cognitive Style	1564.82	1	1564.82	75.19	0.00	
Teaching Models* Cognitive Style	192.97	2	96.48	4.64	0.01	Reject
Error	2892.861	139	20.81			
Total	66048.00	145				
Corrected Total	4968.17	144				

a. R Squared = 0.418 (Adjusted R Squared = 0.397)

Table 1 shows that Ho is rejected which means there is an interaction between teaching models and cognitive styles on students' mathematical problem-solving abilities. This result is reinforced by the picture that shows the lines that are not parallel but tends to the intersection of lines between the teaching model with the cognitive style of FI and FD shown in Figure 2.

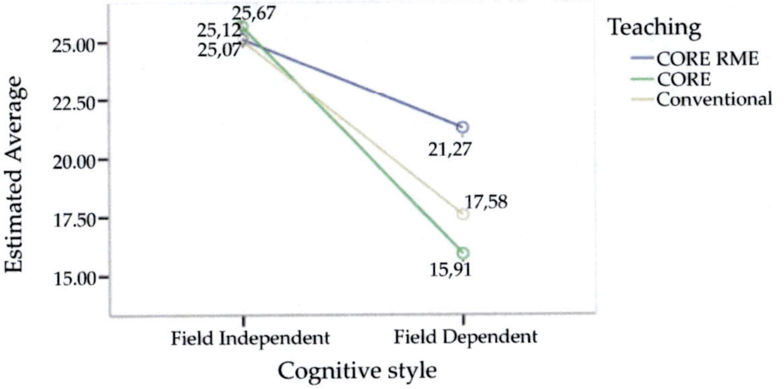

Figure 2. Profile plots teaching models and cognitive style against the MPSA.

Figure 2 shows that there is an interaction effect between teaching models and cognitive styles on student MPSA. This means that teaching models and cognitive styles both influence students' MPSA. MPSA students are not only influenced by the use of teaching models but are also influenced by other factors such as cognitive style. Chinn & Ashcroft (2017) said that if a teacher wants to teach effectively, it should be realized about the different student cognitive style. The realizing of different cognitive styles in teaching can help teachers to percentage the teaching materials effectively.

Cognitive style is very important to be considered to determine the teaching model that is suitable for students to be able to solve mathematical problems (Marwazi, Masrukan, & Putra, 2019). Teaching models are the frame of implementation of a teaching strategy, so this result finding research to implicated for there is an interaction effect between teaching strategy and cognitive style to MPSA students. The statement supported by the research result of Sudarman, Setyosari, Kuswandi, and Dwiyogo (2016) that there are significant interactions between the use of learning strategies and cognitive style on learning outcomes solving mathematical problems.

Significance value at the output of the test of equality of error variances is $0.00 < 0.05$ which means that the data group is not homogeneous, so that differences in students 'mathematical problem-solving abilities both based on learning and students' cognitive style are carried out separately as described below.

The Difference in the Mathematical Problem-Solving Abilities of FI and FD Students

This section analyzes differences in MPSA between FI and FD students who study through the CORE RME model, between FI and FD students who study through the CORE model, and between FI and FD students who study through the Conventional model. Test the difference between FI students and FD students using the t-test one-tailed whose results are presented in the following Table 2. Based on t-test result of Table 2, it could be concluded that MPSA FI students who learn through the CORE RME model, CORE model, or Conventional model better than MPSA FD students.

Table 2. Test difference in MPSA for students FI and FD

Data Source	*df*	t_{count}	$t_{0.05}$	Ho
FI and FD of CORE RME	48	2.33	1.68	Reject
FI and FD of CORE	47	9.64	1.68	Reject
FI and FD of Conventional	44	7.48	1.68	Reject

This is caused by the characteristics of FI students and FD students who tend to be different, namely students with the cognitive style of FD find it difficult to process information, perceptions change easily when information is manipulated in accordance with the context, tend to accept existing structures, due to lack of restructuring. Whereas, FI students who are generally more independent, competitive, and confident (Onwumere & Reid, 2014). The difference in characteristics is what causes the MPSA of FI students to be better than the MPSA of FD students. This is supported by the results of research that says that the problem-solving ability of FI students tends to be better than the problem-solving ability of FD students (Anthycamurty, Mardiyana, & Saputro, 2018; Sudarman et al., 2016).

Differences in Mathematical Problem-Solving Abilities between FI Students

This section analyzes the differences in MPSA between FI students who learn in using the CORE RME model, the CORE model, and the Conventional model. This difference test uses the Kruskal Wallis test because this data group is not homogeneous. The results of the Kruskal Wallis test can be presented in the following Table 3.

Table 3. MPSA difference test among FI students

	MPSA	Ho
Chi-Square	0.82	
df	2	Accept
Asymp. Sig.	0.66	

Table 3 shows that Ho is accepted which means there is no significant difference in the mean rank of MPSA between FI students who study through the CORE RME model, the CORE model, and the Conventional model. The use of these three different teaching models turns out to be found that the FI student MPSA is the same. Whatever the teaching model is used in the teaching and learning process in the classroom does not affect the MPSA of FI students. They have the same tendency in interacting with the environment including in terms of learning so that the use of certain learning models does not interfere with their creativity. FI students have the same characteristics and are general that is more independent, competitive, and confident (Witkin, 1971). Students who have a similar cognitive style will have the same MPSA because they feel more positive and have similar in their learning activities (Carraher, Smith, & De Lisle, 2017).

Differences in Mathematical Problem-Solving Abilities among FD Students

MPSA test differences between FD students who study through CORE RME, CORE models, and Conventional models are done with the Kruskal Wallis test because this data group is not homogeneous. Kruskal Wallis test results can be presented in Table 4.

Table 4. MPSA difference test among FD students

	MPSA	Ho
Chi-Square	14.55	
df	2	Reject
Asymp. Sig.	0.00	

Table 4 shows that there was a significant difference in MPSA between FD students who study through the CORE RME model, the CORE model, and the Conventional model. Because there are significant differences, it is continued with the post-hoc multiple comparisons between treatments. The test results of the multiple comparisons between treatments can be presented in Table 5.

Table 5. Post hoc Test MPSA among FD students

Groups	$\lvert \overline{R_u} - \overline{R_v} \rvert$	Critical value $\lvert \overline{R_u} - \overline{R_v} \rvert$	Ho
FI of CORE RME-FI of CORE	$\lvert \overline{R_1} - \overline{R_2} \rvert = 26.11$	$\lvert \overline{R_1} - \overline{R_2} \rvert = 2.04$	Reject
FI of CORE RME-FI of Conventional	$\lvert \overline{R_1} - \overline{R_3} \rvert = 16.95$	$\lvert \overline{R_1} - \overline{R_3} \rvert = 3.02$	Reject
FI of CORE-FI of Conventional	$\lvert \overline{R_3} - \overline{R_2} \rvert = 9.16$	$\lvert \overline{R_3} - \overline{R_2} \rvert = 3.64$	Reject

Based on the results of the post hoc test in Table 5, it can be concluded that at $\alpha = 5\%$, i.e.:

1. There is a significant difference between MPSA FD students who study through the CORE RME model and FD students who study through the CORE model. Descriptively, the average MPSA of FD students learning through the CORE RME model was 21.27, and the average MPSA of FD students who study through the CORE model was 15.91. Because there are inferential differences, and 21.27 > 15.91 it can be concluded that the MPSA FD students who study through the CORE RME model are better than the MPSA FD students who study through the CORE model.

2. There is a significant difference between MPSA FD students who study through the CORE RME model and MPSA FD students who study through the Conventional model. The average MPSA of FD students who study through the CORE PMR model was 21.27, and the average MPSA of FD students who study through the Conventional model was 17.58. Because inferentially there are significant differences, and 21.27 > 17.58 it can be concluded that the MPSA of FD students who study through the CORE PMR model is better than the MPSA of FI students who study through the Conventional model.
3. There is a significant difference between MPSA FD students who study through the CORE model with the MPSA FD students who study through the Conventional model. Descriptively, the average MPSA of FD students who study through the CORE model was 15.91, and the average MPSA of FD students who study through the Conventional model was 17.58. Because there are inferential differences, and 17.58 > 15.91 it can be concluded that the MPSA FD students who study through the Conventional model are better than the MPSA FD students who study through the CORE model.

This section found that MPSA FD students who study through CORE RME model are better than MPSA FD students who study through CORE model, as well as Conventional models. This result research appears like this because according to the scenario of the teaching CORE RME model, start from the step of connect, organize, reflect, until extend, students sit-down in heterogenic each group, so problem-solving performance FD students improved when the effect of FI students. This situation adjusts with students' FD characteristics more effect by their peer friends. Field dependent students are more likely to desire feedback from their peers in educational settings, which increases their ability to be influenced by their peers (Abrams & Belgrave, 2013). MPSA FD students tend to change if learning in the classroom uses learning models that are appropriate to their characteristics. Although FD students have the same characteristics and tend to find difficulties in processing, their perceptions can change if the information is manipulated according to the context (Witkin, 1971).

Differences in Students' Mathematical Problem-Solving Abilities Based on Teaching Models

The difference in MPSA between students learning through the CORE RME model, the CORE model, and the Conventional model is done using the one-way anova test because it meets the assumption requirements that the MPSA data distribution of students is normally, and the data groups are homogeneous. One-way anova test results can be presented in Table 6.

Table 6. Test the difference of MPSA students based on teaching models

	Sum of Squares	df	Mean Square	F	Sig.	Ho
Between Groups	352.52	2	176.26	5.42	0.01	Reject
Within Groups	4615.65	142	32.51			
Total	4968.17	144				

The One-way anova output in Table 6 shows Ho rejected, which means there is a significant difference in MPSA students who study through the CORE RME model, the CORE model, and the Conventional model. Because there were significant differences in MPSA students, it was continued with the Scheffe post hoc test. It was using the Scheffe post hoc test because the number of participants between classes is different. The results of the Scheffe post hoc test are presented in Table 7.

Based on the post hoc test in Table 7, it can be concluded that at $\alpha = 5\%$, i.e.:

1. There is a significant difference between the MPSA of students who study through the CORE RME model and the CORE model. Descriptively, the average MPSA of students who study through the CORE RME model was 22.58, and the average MPSA of students who study through the CORE model was 18.90. Because inferentially there are significant differences, and $22.58 > 18.90$ it can be concluded that the MPSA of students who study through the CORE RME model is better than the MPSA of students who study through the CORE model.
2. There is no significant difference in MPSA students who study through the CORE RME model and the Conventional model, as well as MPSA students who study through the CORE model and the Conventional model.

Table 7. Post hoc test MPSA students based on teaching models

Teaching Models		Mean Difference	Std.	Sig.	Ho
(I)	(J)	(I-J)	Error		
CORE RME	CORE	3.68*	1.15	0.01	Reject
	Conventional	2.56	1.16	0.09	Accept
CORE	Conventional	-1.12	1.17	0.63	Accept

One of the findings in this section is that MPSA students who study through the CORE RME model are better than MPSA students who study through the CORE model. This happens because in learning the CORE RME model uses the CORE model syntax by applying the principles and characteristics of the RME. By applying the principles and characteristics of RME in CORE, students are given the opportunity to do reinvention, rediscover ideas and mathematical concepts with the guidance of the teacher, experience the same processes themselves when mathematics is discovered, and through guided reinvention students can recognize their experience capacity to think in a way that is depth as a means of solving problems (Abrahamson, Zolkower, & Stone, 2020).

CONCLUSION

Teaching models of CORE RME using the CORE syntax by applying the principles and characteristics of RME. The connecting stage emphasizes the student's prior knowledge and real context principle. In the organizing stage, students interactively conduct reinvention and self-developed models. Stages of reflecting, students do self-monitoring, self-reflect on understanding the relationship the model of with models for, and at the extending stage students develop models for at other real problems. The study found that there are interactions effect between the teaching model and cognitive style on the student MPSA. In terms of the intervention of the teaching models, it was found that there were significant differences in the MPSA of students who study through the CORE RME model, the CORE model, and the Conventional model. This difference is determined by MPSA students who study through the CORE RME model are better than MPSA students who study through the CORE model. Whereas when viewed from the FI's cognitive style, there was no significant difference in MPSA between FI students who study through the CORE RME model, the CORE model, and the Conventional model. Whereas based on the FD's cognitive style, there are significant differences in MPSA between FD students who study through the CORE RME model, CORE model, and Conventional model. This difference is determined by MPSA FD students who study through the CORE RME model better than MPSA FD students who study through the CORE model, as well as the Conventional model. Comparison of MPSA FI students and FD students found that MPSA FI students both who study through the CORE RME model, the CORE model, and the Conventional model were better than the MPSA FD students.

Problem-solving is characteristic of mathematics activity, and mathematics as a human activity. Therefore, the teaching model and student cognitive style are very important to consider in learning so students are able to solve mathematical problems. Through the CORE RME model, students could organize their knowledge through real context, students themselves could be developed mathematical models based on their prior knowledge so could improve the MPSA of students. In addition, mathematics learning systems in school not grouped FI and FD students separately, so it suggested for teachers to use of CORE RME models as one alternative to minimize different of MPSA of them.

ACKNOWLEDGMENTS

The author would like to thank Indonesia Endowment Fund for Education (LPDP) that supported and funded this research. Furthermore, thanks to Dr. Sri Adi Widodo, M.Pd., who has contributed his thoughts, and the management of the Journal on Mathematics Education (JME) who helped publish this article.

REFERENCES

1. Abrahamson, D., Zolkower, B., & Stone, E. (2020). Reinventing realistic mathematics education at berkeley - emergence and development of a course for pre-service teachers. In M. Van Den Heuvel-panhuizen (Ed.), International Reflections on the Netherlands Didactics of Mathematics (pp. 255–277). Nederlands: Springer. https://doi.org/10.1007/978-3-030-20223-1.
2. Abrams, J., & Belgrave, F. Z. (2013). Field dependence. The Encyclopedia of Cross-Cultural Psychology, II(1), 1–3. https://doi.org/10.1002/9781118339893.wbeccp221.
3. Anthycamurty, C. C., Mardiyana, & Saputro, D. R. S. (2018). Analysis of problem solving in terms of cognitive style. Proceeding in The International Conference on Mathematics, Science and Education 2017, pp. 1–5. https://doi.org/10.1088/1742-6596/983/1/012146.
4. Apsari, R. A., Putri, R. I. I., Sariyasa, Abels, M., & Prayitno, S. (2020). Geometry representation to develop algebraic thinking: A recommendation for a pattern investigation in pre-algebra class. Journal on Mathematics Education, 11(1), 45-58. http://doi.org/10.22342/jme.11.1.9535.45-58.
5. Badger, M. S., Sangwin, C. J., Hawkes, T. O., Burn, R. P., Mason, J., & Pope, S. (2012). Teaching Problem-Solving in Undergraduate Mathematics. Coventry, UK: Coventry University https://doi.org/10.1017/CBO9781107415324.004.
6. C?prioar?, D. (2015). Problem solving-purpose and means of learning mathematics in school. Procedia-Social and Behavioral Sciences, 191, 1859–1864. https://doi.org/10.1016/j.sbspro.2015.04.332.
7. Carraher, E., Smith, R. E., & De Lisle, P. (2017). Cognitive styles. In Leading Collaborative Architectural Practice (pp. 179–195). https://doi.org/10.1177/002221947000300101.
8. Chinn, S., & Ashcroft, R. E. (2017). Cognitive (thinking) style in mathematics. In Mathematics for Dyslexics and Dyscalculics (Fourth, pp. 48–61). https://doi.org/10.1002/9781119159995.ch3.
9. Chong, M.S.F., Shahrill, M., & Li, H-C. (2019). The integration of a problem solving framework for Brunei high school mathematics curriculum in increasing student's affective competency. Journal on Mathematics Education, 10(2), 215-228. https://doi.org/10.22342/jme.10.2.7265.215-228.

10. Chrysostomou, M., Pantazi, D. P., Tsingi, C., Cleanthous, E., & Christou, C. (2012). Examining number sense and algebraic reasoning through cognitive styles. Educational Studies in Mathematics, 83(2), 205–223. https://doi.org/10.1007/s10649-012-9448-0.
11. Curwen, M. S., Miller, R. G., Smith, K. A. W., & Calfee, R. C. (2010). Increasing teachers' metacognition develops students' higher learning during content area literacy instruction: Findings from the read-write cycle project. Issues in Teacher Education, 19(2), 127–151. Retrieved from https://eric.ed.gov/?id=EJ902679.
12. Foshay, R., & Kirkley, J. (2003). Principles for teaching problem solving. Plato Learning, 1–16. https://doi.org/10.1.1.117.8503&rep=rep1&type=pdf.
13. Freudenthal, H. (2002). Revisiting Mathematics Education. Dordrecht: Kluwer Publisher. https://doi.org/10.1007/0-306-47202-3.
14. Gravemeijer, K. G. (1994). Educational development and developmental research in mathematics education. Journal for Research in Mathematics Education, 25(5), 443–471. https://doi.org/10.2307/749485.
15. Heuvel-panhuizen, M. V. D., & Drijvers, P. (2014). Realistic Mathematics Education. Encyclopedia of Mathematics Education, 521–534. https://doi.org/10.1007/978-94-007-4978-8.
16. Huda, M. J., Florentinus, T. S., & Nugroho, S. E. (2020). Students' mathematical problem-solving ability at Realistic Mathematics Education (RME). Journal of Primary Education, 9(2), 228–235. https://doi.org/10.15294 /jpe.v9i2.32688.
17. IEA. (2016). The TIMSS 2015 International Results in Mathematics. In TIMSS & PIRLS International Study Center. Retrieved from http://timss2015.org/.
18. Marwazi, M., Masrukan, & Putra, N. M. D. (2019). Analysis of problem solving ability based on field dependent cognitive style in discovery learning models. Journal of Primary Education, 8(2), 127–134. https://doi.org/10.15294/jpe.v8i2.25451.
19. Mefoh, P. C., Nwoke, M. B., & Chijioke, J. B. C. C. A. O. (2017). Effect of cognitive style and gender on adolescents' problem solving ability. Thinking Skills and Creativity, 25, 47–52. https://doi.org/10.1016/j.tsc.2017.03.002.
20. NCTM. (2000). Principles and Standards for School Mathematics. United States of America: NCTM.

21. Nicolaou, A. A., & Xistouri, X. (2011). Field dependence/independence cognitive style and problem posing: an investigation with sixth grade students. Educational Psychology, 31(5), 611–627. https://doi.org/10.1080/01443410.2011.586126.
22. OECD. (2019). PISA 2018 Results: What Student Know and Can Do. https://doi.org/10.1787/5f07c754-en.
23. Onwumere, O., & Reid, N. (2014). Field dependency and performance in mathematics. European Journal of Educational Research, 3(1), 43–57. https://doi.org/10.12973/eu-jer.3.1.43.
24. Pithers, R. T. (2006). Cognitive learning style: A review of the field dependent-field independent approach. Journal of Vocational Education and Training, 54(1), 117–132. https://doi.org/10.1080/13636820200200191.
25. Polya, G. (1957). How To Solve It: A New Aspect of Mathematical Method (Second). https://doi.org/10.2307/j.ctvc773pk.
26. Prahmana, R. C. I., Zulkardi, & Hartono, Y. (2012). Learning multiplication using Indonesian traditional game in third grade. Journal on Mathematics Education, 3(2), 115-132. https://doi.org/10.22342/jme.3.2.1931.115-132.
27. Purwati, L., Rochmad, & Wuryanto. (2018). An analysis of mathematical problem solving ability based on hard work character in mathematics learning using connecting organizing reflecting extending model. Unnes Journal of Mathematics Education, 7(3), 195–202. https://doi.org/10.15294/ujme.v7i1.28977.
28. Saleh, M., Prahmana, R.C.I., Isa, M., & Murni. (2018). Improving the reasoning ability of elementary school student through the indonesian realistic mathematics education. Journal on Mathematics Education, 9(1), 41-54. http://dx.doi.org/10.22342/jme.9.1.5049.41-54.
29. Son, A. L., Darhim, & Fatimah, S. (2019). An analysis to student errors of algebraic problem solving based on Polya and Newman theory. International Seminar on Applied Mathematics and Mathematics Education, 1315(1), 12069. https://doi.org/10.1088/1742-6596/1315/1/012069.
30. Sudarman, Setyosari, P., Kuswandi, D., & Dwiyogo, W. D. (2016). The effect of learning strategy and cognitive style toward mathematical problem solving learning outcomes. IOSR Journal of Research & Method in Education (IOSR-JRME), 6(3), 137–143. https://doi.org/10.9790/7388-060304137143.

31. Tambychik, T., & Meerah, T. S. M. (2010). Students' difficulties in mathematics problem-solving: What do they say? Procedia-Social and Behavioral Sciences, 8, 142–151. https://doi.org/10.1016/j.sbspro.2010.12.020.
32. Treffers, A. (1987). Three dimensions: A model of goal and theory description in mathematics education. In A. J. Bishop (Ed.), Springer Briefs in Applied Sciences and Technology (First). https://doi.org/10.1007/978-94-009-3707-9.
33. Ulandari, L., Amry, Z., & Saragih, S. (2019). Development of learning materials based on realistic mathematics education approach to improve students ' mathematical problem solving ability and self-efficacy. International Electronic Journal of Mathematics Education, 14(2), 375–383. https://doi.org/10.29333/iejme/5721.
34. Volkova, E. V., & Rusalov, V. M. (2016). Cognitive styles and personality. Personality and Individual Differences, 99, 266–271. https://doi.org/10.1016/j.paid.2016.04.097.
35. Wijayanti, A., Herman, T., & Usdiyana, D. (2017). The implementation of CORE model to improve students' mathematical problem solving ability in secondary school. Advances in Social Science, Education and Humanities Research, 57, 89–93. https://doi.org/10.2991/icmsed-16.2017.20.
36. Witkin, H. A. (1971). The role of cognitive style in academic performance and in teacher-student relations. In ETS Research Bulletin Series. https://doi.org/10.1002/j.2333-8504.1973.tb00450.x.
37. Yilmaz, R. (2020). Prospective mathematics teachers' cognitive competencies on realistic mathematics education. Journal on Mathematics Education, 11(1), 17-44. http://doi.org/10.22342/jme.11.1.8690.17-44.

Chapter

GROUNDED AND EMBODIED MATHEMATICAL COGNITION: PROMOTING MATHEMATICAL INSIGHT AND PROOF USING ACTION AND LANGUAGE

16

Mitchell J. Nathan[1] and Candace Walkington[2]

[1] University of Wisconsin-Madison, Educational Sciences Building, 1025 West Johnson Street, Madison, WI 53705, USA

[2] Southern Methodist University, Dallas, TX, USA

ABSTRACT

We develop a theory of grounded and embodied mathematical cognition (GEMC) that draws on action-cognition transduction for advancing understanding of how the body can support mathematical reasoning. GEMC proposes that participants' actions serve as inputs capable of driving the cognition-action system toward associated cognitive states. This occurs through a process of transduction that promotes valuable mathematical

Citation: (APA): Nathan, M. J., & Walkington, C. (2017). Grounded and embodied mathematical cognition: Promoting mathematical insight and proof using action and language. Cognitive research: principles and implications, 2(1), 1-20. (20 pages).

Copyright: © Creative Commons Attribution 4.0 International License (http://creativecommons.org/licenses/by/4.0/).

insights by eliciting dynamic depictive gestures that enact spatio-temporal properties of mathematical entities. Our focus here is on pre-college geometry proof production. GEMC suggests that action alone can foster insight but is insufficient for valid proof production if action is not coordinated with language systems for propositionalizing general properties of objects and space. GEMC guides the design of a video game-based learning environment intended to promote students' mathematical insights and informal proofs by eliciting dynamic gestures through in-game directed actions.

GEMC generates several hypotheses that contribute to theories of embodied cognition and to the design of science, technology, engineering, and mathematics (STEM) education interventions. Pilot study results with a prototype video game tentatively support theory-based predictions regarding the role of dynamic gestures for fostering insight and proof-with-insight, and for the role of action coupled with language to promote proof-with-insight. But the pilot yields mixed results for deriving in-game interventions intended to elicit dynamic gesture production. Although our central purpose is an explication of GEMC theory and the role of action-cognition transduction, the theory-based video game design reveals the potential of GEMC to improve STEM education, and highlights the complex challenges of connecting embodiment research to education practices and learning environment design.

SIGNIFICANCE

Commercially available motion-based programs for improving mathematical reasoning through action (e.g., MATHS DANCE) have long attracted the attention of educators, but with little basis in the psychological theory of how actions can reliably influence cognition. Evaluation of these programs seldom accounts for variations in learning and performance due to variation in influences such as pedagogical support (e.g., explicit prompting). Early work exploring grounded and embodied mathematical cognition (GEMC) through video game performance showed some areas of promise. As predicted, dynamic gestures reliably predicted mathematical insight, even when players were not consciously aware of the mathematical relevance of their in-game actions. GEMC also correctly predicted that proof validity would improve when players' directed actions elicited during game play were explicitly connected to mathematical conjectures through pedagogical language. However, some findings that were at odds with our initial hypotheses reveal the gap between embodied learning theory and using theory to design effective learning environments aimed at promoting science, technology, engineering, and mathematics learning. Generally, learning theories under-constrain designs of effective instruction and learning environments, which fundamentally constricts the impact of research on practice. Engineering-based approaches of iterative design generation and refinement are needed to translate research to practice.

BACKGROUND

Principles of grounded and embodied cognition address the role of the body and body-based resources to shape cognition. Our objective is to present a theory of grounded and embodied mathematical cognition (GEMC) consistent with Stokes' (1997) 'use-inspired research,' by contributing generalizable models of mathematical thinking and learning that support the application of theory to the design of effective, scalable learning experiences in science, technology, engineering, and mathematics (STEM) education. In addition to presenting the theory, we discuss how we used our theory of GEMC to guide the design of a video game that engages players' action systems in order to promote mathematical reasoning. Our specific focus is on understanding and improving mathematical proof skills for geometry. As an example of the application of the GEMC theory, we describe early findings from a small pilot study of middle- and high-school students to understand the influences of action-based interventions on their mathematical insights and proofs. We use this occasion to discuss the inherent challenges of designing effective STEM learning environments derived from cognitive theory.

Geometric proof is a valuable content area for making strides for theories of GEMC. First, geometry is seen as the study of the properties of space and shape, and therefore should be suitable to a GEMC perspective. Second, it is an area of advanced mathematics, typically studied by students planning to attend college and other post-secondary educational programs (Pelavin & Kane, 1990). Third, geometric proof is primarily concerned with universal statements about space and objects, and therefore addresses an important area of abstract thought. Proof is an especially intriguing area of study because of the central role of conceptual understanding rather than using only 'canned' procedures or mathematical algorithms that might enable people to generate a valid answer with little understanding of the mathematics involved (e.g., long division). Finally, geometry plays a profound role in all of the STEM disciplines. Because of these deep connections, there is the potential for advancements in this program of research to impact mathematical reasoning and STEM education more broadly.

Research to Practice for STEM: The Need to Improve Proof Education

Justification and proof are central activities in mathematics education (National Council of Teachers of Mathematics, 2000; Yackel & Hanna, 2003). In fact, 'proof and proving are fundamental to doing and knowing mathematics; they are the basis of mathematical understanding and essential in developing, establishing, and communicating mathematical knowledge' (Stylianides, 2007, p. 289).

Research has long revealed that students struggle to construct viable and convincing mathematical arguments and provide valid generalizations of mathematical ideas (Dreyfus, 1999; Healy & Hoyles, 2000; Martin, McCrone, Bower, & Dindyal, 2005). Students tend to be overly reliant on examples when exploring mathematical conjectures and often conclude that a universal statement is true on the basis of only checking examples that satisfy the statement (e.g., Healy & Hoyles, 2000; Knuth, Choppin, & Bieda, 2009; Porteous, 1990). When presented with deductive proofs, students frequently find them unconvincing (Chazan, 1993), and fail to appreciate the utility of deductive reasoning for communicating generalized arguments based on logical inference (Harel & Sowder, 1998). Interviews by Coe and Ruthven (1994) showed that even advanced college mathematics students held restricted manners and attitudes toward proof. These students typically looked for standardized routines to guide their investigations, rather than seeking out methods suited to the specific conjectures at hand. Furthermore, these advanced mathematics students seldom sought out explanations that would illuminate or give them insights into the general rules and patterns, and rarely attempted to connect these patterns to broader mathematical ideas or frameworks.

In reaction, some mathematics education scholars call for more innovative approaches to proof instruction that focus on the construction and negotiation of mathematical meaning (Stylianides, 2007). Harel and Sowder (1998) define proving as 'the process employed by an individual to remove or create doubts about the truth of an observation' (p. 241). Thus, the process of proving encompasses a wide range of activities where students reason critically about mathematical ideas rather than focus only on an abstract, concise end product disconnected from situated reasoning.

Grounded and Embodied Mathematical Cognition

We view mathematical communication as a multimodal discourse practice (e.g., Edwards, 2009; Hall, Ma, & Nemirovsky, 2015; Radford, Edwards, & Arzarello, 2009; Roth, 1994; Stevens, 2012), rather than a formal, written, propositional form. When constructing valid proofs, individuals often communicate a logical and persuasive chain of reasoning using descriptive language, verbal inference, and gestures. Research on mathematicians' proving practices has suggested that proof 'is a richly embodied practice that involves inscribing and manipulating notations, interacting with those notations through speech and gesture, and using the body to enact the meanings of mathematical ideas' (Marghetis, Edwards, & Núñez, 2014, p. 243). Observations show that both teachers and students use multimodal forms of talk using speech-accompanied gestures as a way to track the development of key ideas when exploring mathematical conjectures (Nathan, Walkington, Srisurichan, & Alibali, 2011). We refer to these kinds of arguments and communications as 'informal proofs.' While they relay the key ideas and transformations needed to explain why properties do or do not hold, they are not always organized in the propositional, deductive, and meticulous manner of formal proofs.

Grounded and embodied cognition

Mathematical thinking and communication, like other forms of cognitive behaviors, are of interest to the growing research program on grounded and embodied cognition (Shapiro, 2014). Grounded cognition (Barsalou, 2008, p. 619) is a broad framework that posits that intellectual behavior 'is typically grounded in multiple ways, including simulations, situated action, and, on occasion, bodily states.' When the focus is on the grounding role of the body, scholars typically use the more restricted term, 'embodied cognition'. Grounded cognition is contrasted with models of cognition based on 'AAA symbol systems' that are abstract, amodal, and arbitrarily mapped to the concepts to which they refer (Glenberg, Gutierrez, Levin, Japuntich, & Kaschak, 2004). Yet working with symbolic notational systems and making general claims about idealized entities (such as perfect circles) through logical deduction is at the heart of mathematical proof construction. Several scholars have provided accounts of thinking about abstract entities and relationships that we never actually see or touch based on principles of grounded and embodied cognition (Casasanto & Boroditsky, 2008; Lakoff & Nunez, 2000). Thus, a central goal in this work is to explicate how a GEMC perspective accounts for seemingly abstract forms of reasoning.

Directed actions, gestures, and learning

One form of GEMC intervention explores the effects of directed actions on reasoning. Here we define 'directed actions' as physical movements that learners are instructed to formulate by some kind of pedagogical agent (Thomas & Lleras, 2009). 'Gestures' can be distinguished from directed actions in that they are spontaneously generated movements, often of the hand, that accompany speech and thought (Chu & Kita, 2011; Goldin-Meadow, 2005; Nathan, 2014). Our review of the literature on directed actions, gestures, and learning reveals four empirically based findings of note: (1) gesture production predicts learning and performance; (2) directed actions can influence mathematical cognition; (3) directed actions from earlier training opportunities leave a historical trace, or legacy, expressed through gestures during later performance and explanation; and (4) mathematical reasoning and learning are further enhanced when actions are coupled with task-relevant speech, leading to coordinated action-speech events that are the hallmark of contemporary gesture research. Taken together, these findings support the assertion that actions serve a valuable role in addition to language in both fostering and conveying mathematical ideas.

The first finding - that gesture production predicts learning and performance - includes content areas such as mathematics (Cook, Mitchell, & Goldin-Meadow, 2008; Valenzeno, Alibali, & Klatzky, 2003) and language (Glenberg et al., 2004; McCafferty & Stam, 2009), as well as broad influences such as general problem solving (Alibali, Spencer, Knox, & Kita, 2011; Beilock & Goldin-Meadow, 2010), inference-making (Nathan & Martinez, 2015), and cognitive development (Church & Goldin-Meadow, 1986). Conversely, when gesture production is controlled, gesture inhibition often disrupts performance and learning (Hostetter, Alibali, & Kita, 2007; Nathan & Martinez, 2015).

For example, the likelihood that students produced valid proofs for mathematical conjectures was positively associated with the presence of 'dynamic depictive gestures' (Donovan et al., 2014). Depictive gestures are gestures through which speakers directly represent objects or ideas with their bodies (e.g., forming an angle with their two hands; McNeill, 1992).

Dynamic depictive gestures (which we will often refer to as 'dynamic gestures') are defined as those that show a motion-based transformation of a mathematical object through multiple states (Walkington et al., 2014). The odds of generating valid proofs were 4.14-times greater (95% confidence interval 1.57–10.92) for participants who produced dynamic gestures than those who did not (Donovan et al., 2014). Figure 1 shows a student making a dynamic depictive gesture while proving the statement that 'The sum of the lengths of any two sides of a triangle must be greater than the length of the remaining side.' The gesture uses motion-based transformations to show two sides of the triangle being unable to meet. Pier et al. (2014) demonstrated that the benefits of dynamic gestures for predicting performance on verbal mathematical proofs are over and above the effects of variations in speech.

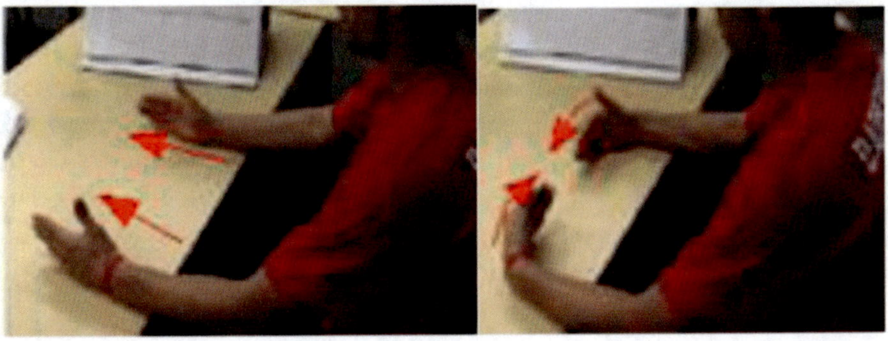

Figure 1. A student's dynamic depictive gestures for the Triangle conjecture used to prove the statement that 'The sum of the lengths of any two sides of a triangle must be greater than the length of the remaining side'.

The second finding comes from a growing body of empirical literature indicating that directed actions can influence learning. Superior problem-solving performance has been demonstrated when students follow directions to perform specific actions hypothesized to foster effective problem-solving strategies (Goldin-Meadow, Cook, & Mitchell, 2009). Thomas and Lleras (2007) showed that manipulating eye-gaze patterns can, unbeknownst to participants, affect the success of solving Dunker's classic Radiation Problem. In mathematics, Abrahamson and Trninic (2015; Abrahamson, 2015) increased primary grade children's awareness of mathematical proportions by engaging their hand and arm motions in order to achieve a particular goal state of the system (a green illuminated screen rather than a red one) once they enacted the appropriate (but tacit) proportions with their relative rates of dual hand movements. It may be said that their hand movements constituted a form of problem solving or epistemic action (Kirsh & Maglio, 1994). According to the authors (Abrahamson & Trninic, 2015), participants' movements did not elicit proportional reasoning, tacitly or otherwise, because, in their view, such forms did not exist yet for these young children, who were engaged in an activity that could later give rise to the concept of proportion. Rather, students were engaged in manipulating objects in the spatial-dynamical problem space. Students could later reflect on their own emergent manipulation strategies, discern motion patterns, and then model these patterns mathematically. Thus, these physical experiences helped children's subsequent performance symbolizing otherwise elusive multiplicative relationships. Fischer, Link, Cress, Nuerk, and Moeller (2015) used digital dance mats, Kinect sensors, and interactive whiteboards to promote physical experiences that children could use to understand the mental number line through embodied training. A mixed-reality environment developed by Lindgren (2015) fostered body 'cueing' that led to higher achievement and more positive attitudes toward learning by providing grounding for students' understanding of physics principles. Enyedy and Danish (2015) also used a mixed-reality environment to support students' understanding of Newtonian force using motion-tracking technology. They found that verbal and physical reflection on embodied activity and first-person embodied play allowed students to engage deeply with challenging concepts.

In the domain of geometry, Petrick and Martin (2012) describe an intervention where high-school students physically enacted (versus observed) dynamic geometric relations, and found that enactment improved learning gains. Shoval (2011) describes an intervention in which students made 'mindful movements' to kinesthetically model angles with their bodies, and demonstrated improved understanding of angles at post-test compared to a control group receiving traditional instruction. Smith, King, and Hoyte (2014) used the Kinect platform to engage students in making different types of angles with their bodies. They found that making conceptual connections between physical arm movements and the grounding metaphor 'angles as space between sides' allowed students to demonstrate greater understanding of estimating and drawing angles. However, they found that in order to benefit from the intervention, it was critical for students to connect their physical actions to the canonical geometric representation and to engage in dynamic movements in which they tested different hypotheses.

In our prior work, Nathan et al. (2014) showed that directing participants' ($N = 120$) body actions affected the generation of appropriate mathematical intuition, insight, and informal proof for two different tasks. They looked at intuition and informal proof for a conjecture on properties of the lengths of sides of all triangles. They also looked at insight and informal proof for a task involving parity for a train of gears. Mathematical insights are defined as partial understandings of the key ideas underlying a mathematical system. Insight is related to but distinct from intuition (e.g., Zander, Öllinger, & Volz, 2016; Zhang, Lei, & Li, 2016): intuition draws on unconscious information to make a judgment (often Yes/No), without leaving a reportable trace of the decision-making process, whereas insights use conscious retrieval processes applied to both unconscious and conscious knowledge to report on one's thoughts about a solution or to provide a partial solution. One of the challenges of insight processes is overcoming unhelpful associations (e.g., when conjectures about triangles inappropriately activate Pythagoras' theorem).

Nathan et al. (2014) found that trials in which participants performed the directed actions were associated with significantly more accurate intuitive judgments on the Triangle conjecture, and more accurate insights on the Gear conjecture, than the trials that used control actions of comparable complexity that were less relevant to the mathematics. Participants who performed relevant directed actions were also significantly more likely to generate an accurate intuition on a transfer task for geometry (i.e., as extended to other polygons) than participants who performed irrelevant actions. However, participants were not more likely to have the key insight for a transfer task involving numerical parity of gears. Thus, directed actions can facilitate mathematical intuition and insight, though transfer appears to be highly task dependent.

While performing mathematically relevant directed actions facilitated key mathematical insights and intuitions for two tasks (Triangle and Gear), directed actions on their own did not lead to superior informal proofs compared to irrelevant actions. Rather, adding pedagogical language in the form of prompts (prospective statements) and hints (retrospective statements) that explicitly connected the directed actions to the tasks significantly enhanced proof performance on the Triangle task. The authors interpret the findings as raising questions about the reciprocal relations between action and cognition: actions on their own facilitate insight, while actions coupled with appropriate pedagogical language explicitly connecting the actions to the mathematical ideas foster informal proofs.

Our third finding is that actions from earlier training opportunities leave a historical trace that is evident when people later solve problems in new, related contexts. Body-based training on the Tower of Hanoi task led participants to integrate their motor experiences into their mental encoding of the objects and their subsequent solution processes (Beilock & Goldin-Meadow, 2010). Donovan et al. (2014) showed that directed actions influence later performance, and found that trained actions can leave an observable 'legacy' in learners' subsequent gestures during proof production. Third and fourth graders learned to solve equivalent equations in one of two ways that involved different manual actions for the assigned condition. Participants using the two-handed 'bucket' strategy were significantly more likely to use both hands when solving post-test and transfer problems, and more likely to exhibit a relational understanding of the equal sign than control group participants who used no manipulatives.

In the fourth finding, we note that reasoning and learning with directed actions appears to be enhanced when the solver's actions are coupled with task-relevant speech, such that action and language become coordinated. As noted in the proof research reviewed earlier, gestures and speech each make independent and significant contributions to predicting performance (Pier et al., 2014). Goldin-Meadow et al. (2009) performed a mediational analysis of students' speech in their study of how directed actions influence performance on equivalent equations. Their results show that students come to apprehend the grouping strategy for solving the equivalence equations even though the strategy was depicted only through directed actions and never explicitly vocalized or gestured to the participants. Students who added grouping in their speech along with the directed actions had increased post-test performance. Their analysis showed that the action condition by itself predicted whether participants added the verbal grouping strategy to their repertoire, while verbalizing the grouping strategy more strongly predicted post-test performance. The authors propose that student-generated speech mediates learning from actions. Carrying out specific actions directs learner attention to solution-relevant features of the task, which helps students confer meaning to the actions.

Several summary points emerge from this literature: gesture production predicts learning and performance; in reciprocal fashion, directed actions are a malleable factor that may influence cognition; actions on their own tend to promote insight and intuition that is not well articulated verbally; reasoning and learning with actions is further enhanced when actions are coupled with task-relevant speech; and actions from earlier training opportunities leave a legacy trace shown in future problem solving. Together, these empirical findings form the basis for a set of hypotheses for ways to promote reasoning in STEM by exploring the mutual influences between action and cognition, as moderated by speech.

Action-cognition transduction

Evidence is mounting that sensorimotor activity can activate neural systems, which can in turn alter and induce cognitive states (Thomas, 2013). Recent work has also identified two critical modes of thinking: System 1 processes that are automatic, effortless, nonverbal, and largely unconscious (e.g., orienting to a sudden sound); and System 2 processes that are effortful mental activities involving agency, choice, and concentration (e.g., checking the validity of an argument; Kahneman, 2011). Whereas the influence of directed actions on cognition may largely be on automatic and unconscious System 1 processes, gestures, which are more intimately bound to language, may influence the verbal, deliberative processes of System 2 (Nathan, in press).

As reviewed above, an emerging literature on cognition and education shows that concepts can be learned through motoric (System 1) interventions. Specifically, 'action-cognition transduction' (ACT; Nathan, in press) explores the bidirectional relationship between cognition and action. ACT theory draws inspiration from reciprocal properties of electromechanical and biological systems relating input-output behavior. Physical devices, such as motors, acoustic speakers, light-emitting diodes (LEDs), and so forth, can run both 'forward' and 'backward'; so input energy, often in the form of electric current, can be transduced when forcibly cranking the rotor of a motor (we call the 'reverse' motor a generator), shining a light on an LED (making a photoreceptor), or singing into a speaker (making a microphone).

Transduction behavior is evident in biological systems as well, with reported influences on cognitive processes. Niedenthal (2007) illustrates how affective state is surreptitiously induced through manipulations in the facial muscles to form specific facial expressions, which in turn influence the cognitive processing of emotion information when presented in writing, speaking, and images. Havas, Glenberg, Gutowski, Lucarelli, and Davidson (2010), in a similar vein, showed that Botox injections affect cognitive processing of emotion-laden sentences through paralysis of facial muscles. Niedenthal (2007) references the 'reciprocal relationship between the bodily expression of emotion and the way in which emotional information is attended to and interpreted' (p. 1002). As noted, interventions directing arm movements (e.g., Nathan et al., 2014; Novack, Congdon, Hemani-Lopez, & Goldin-Meadow, 2014) and eye gaze (Thomas & Lleras, 2007) have led to superior performance in mathematics and general problem solving.

ACT theory offers several testable hypotheses about thinking and learning that have implications for STEM education. One hypothesis states that directed actions, body movements that learners are instructed to formulate, can induce cognitive states that activate relevant knowledge. A second hypothesis is that action-based interventions by themselves are expected to induce cognitive states around ideas that are not propositionally encoded. In this way, actions can foster insights that may be nonverbal, and therefore unavailable for immediate verbalization. In this manner, action-transduced knowledge may operate outside of the awareness of the learner.

Consistent with these two hypotheses, Nathan et al. (2014) found that experimental participants who performed directed actions that were selected for their relevancy to mathematical tasks showed improved intuition and insight, even though participants were largely unaware of their mathematical relevance or influence. This work raises three important question about the epistemological basis for claims about body-based interventions influencing cognition. The first question is how actions absent any action-based goals can conjure something meaningful. The second question is whether there is evidence that thoughts induced by actions can contain entirely new ideas, or if the conjured ideas are simply due to priming effects of pre-existing knowledge. The third question is how movements performed in response to directions but without inherent meaning can contribute to specific meaning-making.

Several empirical studies speak to this first question, and show that presenting stimuli can activate motor systems even in the absence of any motoric goals to act. Skilled *kanji* writers, for example, demonstrate motor-system activation in areas associated with writing these characters, even without any intention to write (Kato et al., 1999). Isolated word presentation of action words (e.g., pick, lick, kick) can induce motor responses in the associated muscle systems (fingers, tongue, legs; Pulvermüller, 2005). Beilock and Holt (2007) showed that people reported preferences for letter dyads (FJ over FV) without cuing any action-based goals because these were less demanding to type, but this held only for skilled typists. These studies illustrate ways that people invoke action-based meaning for presented stimuli even when action-oriented goals are not explicitly cued.

On the second question, Leung et al. (2012) provide evidence across several experiments that embodied interventions can increase the generation of entirely new ideas, rather than only priming prior knowledge. Here, interventions that embodied creative and alternative viewpoints (changing hands, being outside of a box, freely wandering) led to more creative responses on a number of convergent and divergent thinking tasks.

In addressing the third issue, we note that actions performed in response to directions can generate a specific meaning by evoking many multiple meanings that undergo real-time selection. One way that actions may generate new ideas is through mental simulation (e.g., Barsalou, 2008). Mental simulation processes literally 'run' or 're-run' multimodal enactments of external sensory and motoric signals along with internally generated introspective events. This offers one account for why we perceive similarities between enacting, observing, and recalling specific behaviors; and understand the minds and behaviors of others (Decety & Grèzes, 2006). The GAME framework proposed by Nathan and Martinez (2015) provides a computational account of how actions that are initiated without specific meaning contribute to specific meaning-making through mental simulation. The GAME framework builds off the MOSAIC architecture, which provides an account of movement regulation in uncertain environments (Haruno, Wolpert, & Kawato, 2001; Wolpert & Kawato, 1998). In this model, as a movement commences, it simultaneously initiates the parallel production of multiple, paired predictor-controller modules. Each module is intended to anticipate one of the myriad of plausible next states of the motor system. Each predictor-controller pair receives feed-forward signals of expected movement and feedback signals of the actual movement, which provides rapid access to the difference between the projected state of the motor system in the simulated mental model and its actual state as movement occurs. This coupling between actual and simulated motor activity establishes a pathway for transductive influences of actions on the cognitive state of the agent that may start out as nonspecific, and ultimately induce specific, contextually relevant cognitive states. As the movement progresses, there is continuous competition among these predictor-controller modules, each serving as a potential future state of the mental simulation. The system favors those modules that most closely track the external influences from the environment and the internal influences from the current cognitive states. Selection of the most helpful predictor-controller modules is used to update the reader's current mental simulation, enabling idea generation, while the current action potentially alters current cognitive processes. Those specific predictor-controller pair modules that are found to most accurately predict both the state of the world and the state of mind receive greater activation for the future, thus improving body response and action-cognition alignment. Nathan and Martinez (2015) provide evidence in support of the prediction that the execution of motor control programs during movements such as gesture production can influence simulated mental model construction processes, and enable the generation of novel inferences. In this way, even nonspecific movements can induce specific mental states through ACT that can lead to novel cognitive processing, and support the generation of insights through nonverbal means.

A GEMC THEORY OF PROOF-WITH-INSIGHT

Nathan et al. (2014) found that the effect of directed action on producing an informal proof was significantly enhanced when pedagogical hints were introduced to engage participants' language systems. Without language activation, solvers may experience their insights through nonverbal means, but still be unable to verbally articulate a proof. Consequently, we hypothesize the need for co-activated language and motor systems during task performance for achieving valid proofs-with-insight. It follows from our theory that processes coordinating language and motor systems will, in turn, produce a legacy of semantically rich co-speech gestures, which reveal students' abstract and generalizable mathematical thinking (Hostetter & Alibali, 2008). GEMC theory posits that dynamic gestures mediate the generation of correct mathematical insight during proof production. Simulated actions that drive performance are specifically evident in students' generation of dynamic gestures and transformational speech (Pier et al., 2014). Dynamic gestures, to review, are those that manually depict and transform an object. Researchers have identified the importance of dynamic gestures during mental rotation tasks (Göksun, Goldin-Meadow, Newcombe,

& Shipley, 2013; Newcombe & Shipley, 2012; Uttal et al., 2012). Our usage of the term aligns best with Garcia and Infante's (2012) characterization of gestures produced when solving calculus problems, as 'moving the hands to describe the action that occurs in the problem or movements made to represent mathematical concepts' (p. 290). Following Harel and Sowder's (2005) framework, 'transformational speech' describes those utterances that indicate (1) logical inferences, where conclusions are drawn from premises; (2) generalization of relevant mathematical relationships between mathematical entities; and (3) operational thought, such that a cohesive argument progresses through a systematic chain of goals. Pier et al. (2014) identified transformational speech patterns as particularly important to valid proofs. Transformational speech was defined as goal-directed manipulations of mathematical objects through conditional statements ('if ... then ...') and language that repeated key mathematical terms as inference was performed. They found that transformational speech and dynamic gesture independently accounted for unique variance in their model predicting proof validity.

In our theory (Fig. 2), we show, on the far right, that both dynamic gestures and transformational speech must be coordinated to generate proof-with-insight; that is, to articulate a proof that is mathematically valid, intuitively satisfying, consciously understood by the learner, and available for explanation and reflection (Systems 1 and 2). The theory hypothesizes that activating learners' action systems without engaging language systems to instill propositional meaning to these actions can lead to the generation of intuitions (System 1) and insights about the conjectures (top pathway of Fig. 2), but that this by itself will not yield a valid proof that presents the chain of analytical reasoning for why the insight holds. Achieving state of insight, while meaningful to the solver, is also not likely to be persuasive to others without an accompanying chain of justification (Harel & Sowder, 2005) This provides an account for why action-based interventions can yield proficiency in automated procedures and perceptual recognition (i.e., engagement of System 1) but fail to support the abstract understanding and reflection that enables verbal explanations, generalization, and far transfer (i.e., engagement of System 2; Evans, 2003).

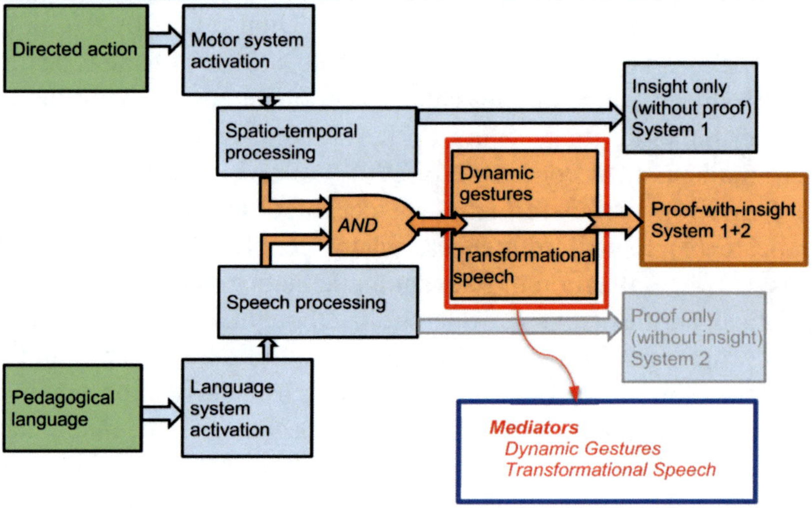

Figure 2. Grounded and embodied mathematical cognition (GEMC) theory. Actions (top pathway) combine with language (bottom pathway) to generate a proof with intuition, which is hypothesized to be mediated by simulated action, as exhibited by the speaker's dynamic gestures and transformational speech.

Prompted speech (e.g., following an authoritative script) can yield the recitation of valid proofs (lower pathway of Fig. 2), but we hypothesize that it does so without generating perceptuo-motor forms of knowing that are characteristic of intuitive understanding (e.g., Kellman & Massey, 2013; Kellman, Massey, & Son, 2010; Koedinger, Corbett, & Perfetti, 2012). Actions and speech production can be independently manipulated to contribute to proof performance, and each may make contributions to students' mathematics skill and knowledge.

Our central claim is that the coordination of action and language is necessary for students to perform simulated actions of the appropriate mathematical entities. With coordination, students produce dynamic gestures along with concurrent transformational speech that serve as mediators of their mathematical thinking. This enables students to formulate an insightful and explicable chain of reasoning that constitutes a mathematical proof that is both externally valid and internally meaningful.

RESEARCH TO PRACTICE VIA LEARNING ENVIRONMENT DESIGN

One of the great challenges for developing theoretically driven interventions is that learning theories markedly under-constrain the design of instruction and learning environments. In terms of implementation, there are myriad design decisions that will have consequences for learning and engagement that must practically be settled, yet that fall outside of the prescribed learning theory. It is in this sense that some Learning Sciences scholars have argued that learning environment development is more closely aligned with engineering than science (Nathan & Sawyer, 2014) and depends on iterative cycles of learning environment design that inform the specific design decisions as well as the overarching theory.

Based on our theory and the findings reviewed above, we offer novel predictions that guide the design of GEMC-inspired learning environments: directed actions and language prompts are hypothesized to each act as viable, independent, malleable factors for enhancing proof performance by fostering mathematical insight and verbalizable proof production. GEMC theory, along with findings from Nathan et al. (2014), served as the basis for the design of a scalable video game environment that would guide directed actions in service of players' proofs and justifications for mathematical conjectures. In accordance with GEMC, game play elicited directed actions thought to (unconsciously) foster dynamic gestures that would facilitate insight, while verbal supports were used to provide pedagogical hints that activate language systems and encourage the integration of actions and language. We narrowed our domain of inquiry to focus on high-school planar geometry and expanded the task set within that domain to include multiple geometry conjectures. We recruited age-appropriate participants to investigate how interactive game play that elicited actions and coordinated language could be used to further develop the emerging theory for promoting proof-with-insight and serve as a prototype for a scalable, embodied game for promoting pre-college mathematical reasoning.

Theoretically Motivated Hypotheses

We offer testable hypotheses about the nature of GEMC. As illustrated in Fig. 3, H1 offers a novel theoretical prediction of a positive association between dynamic depictive gesture production and valid mathematical insights relating to problem tasks. H2 and H3 offer predictions about the influence of directed actions as interventions designed to directly or indirectly influence insight performance. As such, H2 and H3 are a step further from the process account central to GEMC, because implementation choices for when and how to elicit directed actions within the intervention may bear on their influence on outcome measures. H2 predicts that inducing directed actions leads to the production of dynamic gestures. H3 predicts that inducing directed actions improves the generation of mathematical insights. H3 establishes the overall effect that directed actions are a malleable factor for enhancing students' mathematical insight. Establishing the meditational role of dynamic gestures to produce insights (the H2-H1 path) paves the way for a class of embodied interventions within GEMC to promote STEM reasoning.

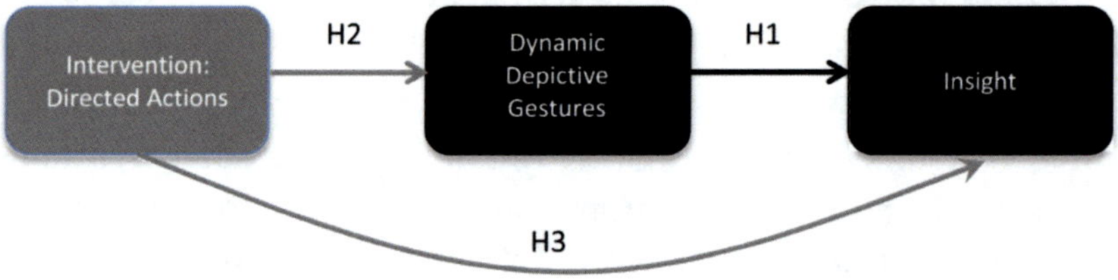

Figure 3. Illustration of hypotheses H1, H2, and H3 for the pilot study. *Black shapes and links* denote predictions about the process account in accord with grounded and embodied mathematical cognition theory. *Gray shapes and links* denote predictions about the intervention designed to improve performance.

- H1. Production of dynamic gestures during proof and justification will predict student performance on mathematical insights.
- H2. Performing mathematically relevant directed actions (without pedagogical language) will predict the production of dynamic gestures.
- H3. Performing mathematically relevant directed actions (without pedagogical language) will predict student performance on mathematical insights.

As illustrated in Fig. 4, H4 and H5 examine the influence of interventions that coordinate language with action systems to enhance transformational proof production. H4 predicts that combining pedagogical language prompts with the directed actions (such as those identified in H2) will elicit more dynamic gestures. H5 predicts that pedagogical language prompts combined with directed actions will lead to improved proof performance. H5 establishes the overall effect that mathematically relevant directed actions coupled with language promotes transformational proofs-with-insight, while H4 considers the meditational role of dynamic gestures on proof performance. H6 (as with H1, above) examines the theoretical process account that proof-with-insight performance is associated with dynamic gesture production.

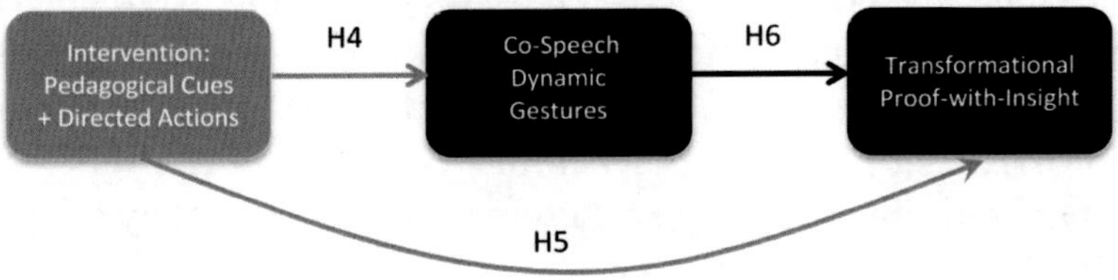

Figure 4. Illustration of hypotheses H4, H5, and H6. *Black shapes and links* denote predictions about the process account in accord with grounded and embodied mathematical cognition theory. *Gray shapes and links* denote predictions about the intervention designed to improve performance.

- H4. A language intervention, in the form of pedagogical cues about the mathematical relevance of game-based directed actions, predicts the production of dynamic gestures during proof production that co-occur during transformational speech production.
- H5. An intervention consisting of language cues coordinated with mathematically relevant directed action predicts the performance on production of valid transformational proof-with-insight.
- H6. Co-speech dynamic gestures predict the production of transformational proof-with-insight.

Pilot Study

A pilot study was conducted using a prototype version of a new video game, *The Hidden Village*, designed as a platform to investigate these hypotheses. The pilot study is provided as an initial, illustrative example of how the GEMC framework can be empirically tested and iterated upon. Given the small sample size and early operationalization of the relevant outcome measures, we make only tentative conclusions from these findings, and offer them as a means to refine our ongoing empirical investigation.

Methods

The Hidden Village utilizes a standard built-in laptop camera and specialized image processing software developed by Extreme Reality, Ltd. (xtr3d.com) to determine in real time whether the player successfully executes each action sequence. The software generates a wireframe skeleton (invisible to players) to track the coordinates of the user's joints (Fig. 5) to determine if player actions match the elicited directed actions of in-game avatars. The game playing procedure (depicted in Fig. 6) begins with setup/calibration instructions. The storyline starts when players, lost in the Hidden Village, encounter an imposing tribe whose culture they must accept by copying arm movements at the welcoming ceremony. Movements are designed to either be task-relevant, capturing some key relation of the subsequent geometry conjecture (Table 1), or task-irrelevant. Action relevance is manipulated randomly across conjectures within player. Sequences of movements must be successfully repeated five times for a player to progress through each location in the village. The player is then presented with a geometric conjecture as a challenge from the tribe and instructed to speak aloud with a proof as to why the conjecture is true or false. All speech and actions were recorded during game play.

Figure 5. Wireframe skeleton that shows how the software in *The Hidden Village* tracks players' body movements using images from the laptop camera.

Figure 6. The procedure for playing *The Hidden Village*. The *top left* image shows the flow of the game through five stages, with the circular cycle repeated for each conjecture. The remaining images show screenshots from each of the stages of the game: tutorial (*top right*), storyline (*middle left*), directed action sequence (*middle right*), player's free response to a geometric conjecture (*bottom left*), player's multiple choice response to a geometric conjecture (*bottom right*). Multiple choice responses are not considered in the present paper for space.

Table 1. Relevant directed actions and associated conjectures

Conjecture	Relevant directed action sequences	intended relevance of directed actions
1. Given that you know the measure of all three angles of a triangle, there is only one unique triangle that can be formed with these three angle measurements. **[FALSE]**		Triangle growing outward, forming similar triangles that had the same angles, but longer side lengths.
2. The area of a parallelogram is the same as the area of a rectangle with the same length and height. **[TRUE]**	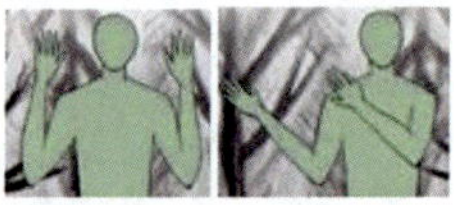	Area of a parallelogram can be formulated from the area of a rectangle when the area is rearranged (tilted).
3. If one angle of a triangle is larger than a second angle, then the side opposite the first angle is longer than the side opposite the second angle. **[TRUE]**		Experience that as an angle opens up, the opposite side necessarily becomes longer.
4. If you double the length and width of a rectangle, then the area is exactly doubled. **[FALSE]**	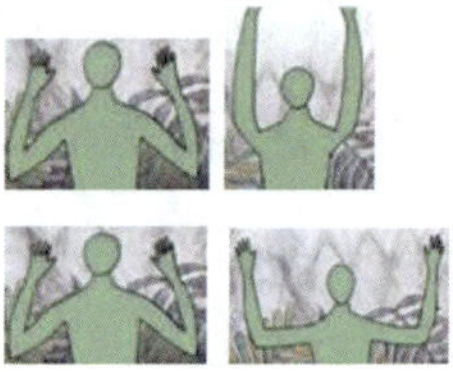	Convey that when the width doubles, the rectangle becomes twice as wide (top) and that when the length doubles, the rectangle becomes twice as long (bottom), resulting in an area that is four times greater.
5. For a given perimeter, a square always has a larger area than a triangle.**[TRUE]**	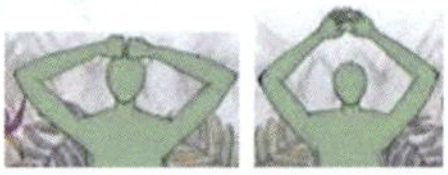	Relevant motions intended to convey that a triangle (left) makes inefficient use of its perimeter, while a square (right) allows more space to fit inside.
6. The measure of a central angle of a circle is twice the measure of any inscribed angle intersecting the same two endpoints on the circumference. **[TRUE]**	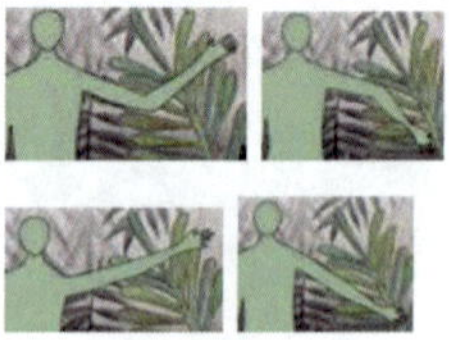	Experience the 2:1 relationship between the larger central angle (top, bent arm) and the smaller inscribed angle (bottom, straight arm).

In the first round of data collection, 18 middle- and high-school students (grades 6 to 11; 16 male, 2 female) attending a video game design summer camp on a university campus individually tested the game; three had previously taken a geometry course. In the second round of data collection, 17 high-school students from the same city (7 male, 10 female) in 9th or 10th grade enrolled in a geometry class at a private school participated in the study. The population at the high school has a mean college admissions American College Testing score of 26, matching the national average.

Students were given a pre-assessment measuring their knowledge of geometric properties of circles, triangles, and quadrilaterals. Students then played the game, experiencing six conjectures in random order. Upon finishing, the interviewer revisited two to four conjectures (depending on time remaining) that the student had received relevant motions for, and revealed that the motions had been relevant by displaying an image that showed the game motions and the conjecture. Participants were asked to think about how their motions might have been connected to the conjecture, and given an additional opportunity to provide a justification following this pedagogical hint.

For scoring student responses, we adopted Harel and Sowder's (2005) notion of transformational proofs, which are part of a broader category of deductive proof schemes. Harel and Sowder argue that deductive proof schemes constitute 'the essence of the proving process in mathematics' (2005, p. 23) and involve operating upon mathematical objects, observing the result, and building upon the proof. As such, transformational proofs have three defining characteristics. First, they are , that is, they show the argument is true for a class of mathematical objects. Second, they involve operational thought, so that the prover progresses through a goal structure, anticipating the results of transformations. Finally, they involve logical inference, such that conclusions are drawn from valid premises. However, as described earlier, we consider such proofs to be 'informal' when they are presented verbally to make conceptual sense to a listener, rather than as a mathematically exhaustive logio-deductive written argument.

Videos of students playing the game were divided into segments of each student proving each conjecture, and were transcribed. Students' responses to each of the six conjectures were scored either 0 or 1 along four dimensions: (1) whether they made any spontaneous depictive gestures while attempting their proof, (2) whether they made any spontaneous dynamic depictive gestures while attempting their proof, (3) whether they recognized the key mathematical insight behind the proof, and (4) whether they formulated a valid informal transformational proof. As noted above, mathematical insights involve an intuitive yet not fully elaborated understanding of the key ideas behind a conjecture. By coding for insight in addition to informal proof, we sought to capture instances where students seemed to have some limited understanding of how the geometric system worked, but were unable to fully formulate and articulate their thinking.

Each of these 0/1 codes was used as a dependent measure in a mixed-effects logistic regression model. This analysis technique was chosen to allow for repeated measures (random effects) to be included for both participant and conjecture. Predictors included gender, pre-test score, highest mathematics course (pre-algebra or lower, algebra, geometry or higher), how many conjectures the participant had proved previously as an ordered factor variable (to account for tiring), and sample (video game camp or private high school). *D*-type effect sizes were calculated using the method in Chinn (2000). In Cohen (1988), effect sizes of 0.2, 0.5, and 0.8 are considered small, medium, and large, respectively.

Results

Results of the regression analyses support the trends that are visually apparent from the descriptive statistics in Table 2. When considering H1 (the effect of gesture on insight and proof), producing any depictive gestures (not necessarily dynamic) predicted that participants would formulate the mathematical insight (odds ratio=3.0, $d=0.6$, $p=0.007$), but not formulate an informal proof ($p=0.11$). However, making dynamic depictive gestures predicted both insight (odds ratio=8.1, $d=1.2$, $p<0.001$) and proof (odds ratio=11.5, $d=1.3$, $p<0.001$). This suggests that producing any depictive gesture may help students glean some key ideas behind the conjecture; however, dynamic gestures may be associated with reasoning about relationships between geometric objects. This supports H1, a novel process claim, which stated that dynamic gestures would predict mathematical insight.

Table 2. Descriptive statistics showing average incidence of gesture, insight, and proof for different experimental conditions

Condition	Depictive gestures	Dynamic depictive gestures	Mathematical insight	Transformational proof
Irrelevant actions (N=81)	16.0%	9.9%	28.4%	14.8%
Relevant actions (N=128)	28.1%	10.0%	25.8%	14.1%
Relevant actions+pedagogical language (N=100)	58.0%	29.3%	43.4%	27.0%

Note: Standard deviations are not provided because when the outcome is binary (0/1), the sample size and mean give all necessary information

The relevance of the directed actions cued by the video game was manipulated within subjects. Participants were more likely to make depictive gestures on trials for which they had performed relevant (versus irrelevant) directed actions (odds ratio=4.2, d=0.8, p=0.008). Counter to H2 and H3, in trials where participants performed relevant directed actions they were not more likely to make dynamic gestures (H2), or demonstrate mathematical insight (H3), than in those trials in which they performed irrelevant directed actions (p-values>0.1). Relevant directed actions seemed to activate participants' motor systems and leave a trace in the form of depictive gestures during proof, but these gestures were often not coded as dynamic gestures. As described above, depictive gestures only predicted insight (not proof), whereas dynamic depictive gestures predicted both insight and informal proof. Note that for these comparisons, most participants (n=27) reported not being consciously aware of the relationship between the directed actions they performed and the conjectures. When participants who did report some conscious awareness were omitted from the analysis (n=8), the pattern of results was the same. In sum, we see the predicted association between dynamic gestures and insight (H1). However, while video game trials cuing mathematically relevant actions were more likely to elicit overall gesture production among players, cuing relevant actions as an in-game intervention did not reliably increase players' production of the all-important dynamic gestures (H2) or lead to a significant increase in mathematical insights (H3).

As noted, players often had no awareness that the actions they were asked to perform related to the mathematical conjectures. When comparing participant performance before and after the intervention of a pedagogical hint explicitly linking actions to the conjectures (H4 and H5), findings emerged that support GEMC. Following the pedagogical hint, participants were more likely to make depictive gestures (odds ratio=5.4, d=0.9, p<0.001) and more likely to make dynamic depictive gestures (odds ratio=4.0, d=0.8, p=0.001), supporting H4 which stated that directed actions combined with pedagogical language would promote dynamic gesture production. In support of H5, when coupling directed actions to pedagogical language, participants were more likely to produce favorable mathematics outcome measures, including expressing the insight (odds ratio=3.1, d=0.6, p<0.001) and formulating a valid informal proof (odds ratio=4.7, d=0.9, p<0.001). In addition, we saw support for our theoretical claim that dynamic gesture production reliably predicts the performance of formulating a valid proof (odds ratio=5.52, d=0.94, p<0.001), supporting H6. Overall, receiving pedagogical language connecting the directed actions to the conjecture was highly beneficial for players' mathematics performance, both for insight and for proof production.

Our results suggest that performing relevant directed actions during game play fosters depictive gestures (versus irrelevant actions; d=0.8), and that making depictive gestures predicts mathematical insight (d=0.6). However, surprisingly, performing relevant actions (versus irrelevant actions) did not predict insight, as would be expected. Supplementary analyses showed that the 'extra' depictive gestures induced by relevant directed actions tended to be static, non-moving gestures that did not display transformational reasoning. Although trials where participants performed irrelevant actions were less likely to have gestures, responses from irrelevant action trials showed a greater proportion of the all-important dynamic gestures (62.0% of gestures from irrelevant trials versus 36.1% of relevant trials).

Qualitative analysis of contrasting cases

To better illustrate students' reasoning in the context of mathematical proof and directed action, we show two transcripts of how directed actions with pedagogical language influenced participants' reasoning. These transcripts were selected to illustrate contrasting cases where pedagogical language was effective versus ineffective at allowing students to link their directed motions to a conjecture and to transform the directed motions into their own, personal co-speech dynamic gestures. The transcripts also illustrate how spontaneous gestures, verbal reasoning processes, pedagogical language, and the directed motions all come together to allow for successful or unsuccessful proof attempts.

Figure 7 shows a photo transcript of a student formulating a new proof for the conjecture that only one unique triangle can be formed from three angle measurements (Conjecture 1 in Table 1) after receiving a hint that his directed actions had been relevant. Initially, the student had incorrectly said the conjecture was true because angle measurements are unique to a triangle. After the hint, his verbal proof showed a growing triangle as a means of disproving the conjecture. To make his argument, he utilized co-speech spontaneous dynamic gestures of a triangle growing outwards, indicating the directed motions may have left a legacy (Donovan et al., 2014).

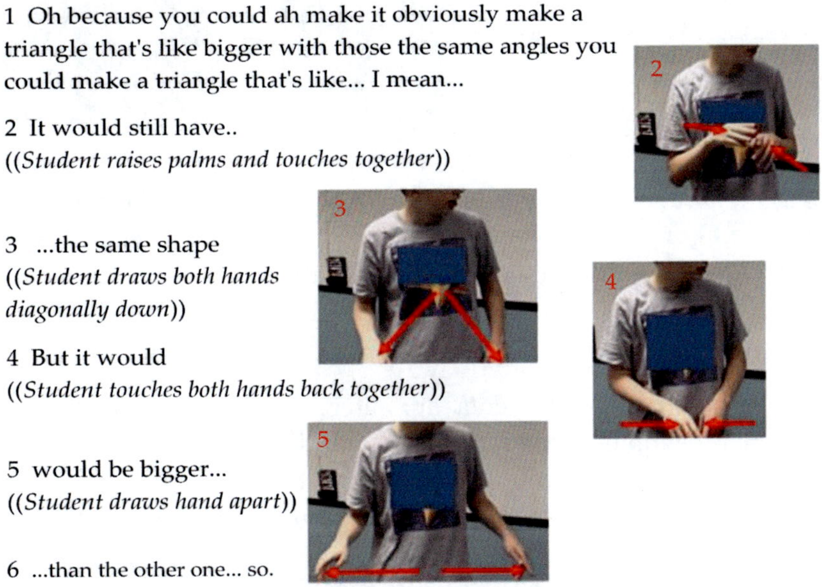

Figure 7. Photo transcript of student proving Conjecture 1 in Table 1 after receiving a hint that his directed actions are relevant to the conjecture. *Red lines* show motion and inferred shapes.

However, relevant directed actions with pedagogical language were not always effective for students. Figure 8 shows a contrasting case where a student's reflection upon the directed motions was unsuccessful for Conjecture 3 in Table 1 (side opposite largest angle is largest). He seems to catch on to the intended insight that the angle in the triangle was growing, and imitated the dynamic arm movement he had been asked to perform (Line 1). Later, when he formulated his proof, he did not use the intended relation, however. He maintained that the conjecture was false because angles needed to add up to 180° (Line 2) and made a non-dynamic depictive gesture that inaccurately indicated a large angle corresponding to a smaller side length (Line 3). This may be an example of an unhelpful insight garnered by a strong association between triangles and angles adding up to 180°.

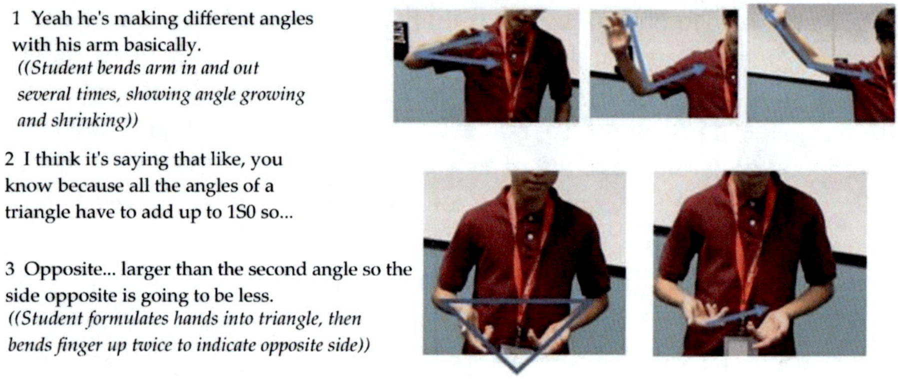

Figure 8. Photo transcript of student providing incorrect proof to Conjecture 3 in Table 1 after receiving a hint that his directed motions are relevant to the conjecture. *Blue lines* show motion and inferred shapes.

Discussion

Our findings from the pilot data are interesting, to say the least. And while our pilot sample sizes are modest, and there are reasons to question whether the two student samples can be combined, the combined results provide valuable early information that can be used to inform our prototype intervention and the theory.

The model of insight performance shown in Fig. 3 explores important theoretical and practical claims for advancing GEMC in an area of advanced mathematics. In support of H1, dynamic gestures led to improvements in insight, which is a novel contribution to theories of embodied cognition for promoting STEM. However, findings contrary to H2 show that is an understanding of how to design the interventions that induce those dynamic gestures that is lacking. Although the directed action cues did significantly influence students' gesture production, most of the gestures were not dynamic and did little to support students' transformational proof production. The increased production of gestures overall following game play trials with relevant directed actions is itself a curiosity given that, as we noted, most players had no idea these directed actions were related to the conjectures. This suggests that the intervention is influencing unconscious reasoning associated with System 1 processes.

H3 incorrectly predicted that relevant directed actions elicited without pedagogical hints that signaled their mathematics relevance would promote insight. H3 makes a claim that bypasses the hypothesized mediational role of dynamic gestures in shaping mathematical reasoning. Thus, at this preliminary stage, we were not able to identify a simple way to improve insight solely by manipulating the relevance of the directed actions within an intervention. A *post hoc* analysis did show, to our surprise, that irrelevant action cues were actually more efficient at eliciting dynamic gestures than were task-relevant cues, in that a higher proportion of the total gestures that were produced were dynamic. In essence, students who received relevant directed actions were gesturing more, but this improvement in gesture rate was not impacting the production of the more influential dynamic gestures. Indeed, for both the relevant and irrelevant conditions, the overall tendency to produce dynamic gestures on the first trial was exactly the same - around 10% of trials showed dynamic gestures (see Table 2). These findings bring to light the chasm between learning theories and their application by highlighting ways that design decisions for the intervention are under-constrained: we see academic benefits when students produce dynamic gestures, but we do not yet know how best to elicit them.

The findings concerning the effects of explicit language connecting the actions to the conjectures also provide important information for the emerging theory and game design. Consistent with the predictions made by H4 and H5, players were significantly more likely to make dynamic gestures (H4) and more likely to generate correct insights and valid informal proofs (H5) after receiving verbal hints. This provides tentative support for the GEMC theory and its call for coordinating language and action systems in service of mathematical reasoning. It raises a question of why the pedagogical hints improved insights, which we take as nonverbal. One indicator is that pedagogical hints were found to influence insight performance directly, and to increase the production of dynamic gestures, which, as hypothesized earlier (H1), promotes mathematical insight and the formulation of valid informal proofs (H6) through a mechanism such as transduction. As predicted, we also found support for H5, the applied path that bypasses the meditational influence of dynamic gestures and predicted a direct relationship between pedagogical language coordinated with directed action and our primary outcome measure of interest, proof-with-insight. Although these results need to be replicated and generalized across a broader set of mathematical conjectures and student populations, these initial findings are encouraging for future game designs that instantiate this direct pathway.

A qualitative analysis of contrasting cases highlighted the significant challenges involved in transforming a directed motion into a meaningful dynamic gesture that makes sense to the learner in the context of the task. This transformation has the potential to support students' proving activities and give them powerful new body-based resources with which to confront the task, as shown in the first transcript. However, making the mapping between directed actions and the sequence of key mathematical ideas needed to prove a conjecture is fraught with difficulty, as shown in the second transcript. This reveals a central challenge in the use of directed actions as an embodied intervention for learning: how to select and design the actions such that the mapping is as sensible and accessible to the learner as possible.

Summary of Findings

The current study provides empirical support for an emerging theory of cognition-action transduction for embodied mathematical cognition. The production of dynamic depictive gestures is strongly predictive of mathematical insight (H1) and informal mathematical proofs (H6), providing support for the GEMC process account. While game-initiated directed actions increased depictive gesture production, this did not translate to improved mathematical reasoning because these depictive gestures were often not dynamic (H2), suggesting that the directed motions utilized by the intervention may need further development.

This highlights the challenges of designing theory-based learning technologies, and reinforces the importance of iterative design approaches for effective educational games.

Our findings also support the hypothesis that pedagogical language directing students to the relevance of the game-directed actions improves both dynamic gesture production (H4) and proof performance (H5), which suggests that the language aspects of the intervention were effective. This supports the view that integrating language systems with nonverbal motor-based forms of knowledge can enhance analytic forms of mathematical reasoning.

CONCLUSIONS

This work offers several contributions, both theoretical and practical, for promoting mathematics education. By focusing on geometry and proof, we contribute to an area of mathematics that makes substantial interconnections across the STEM fields, including classic fields such as physics, chemistry, and biology, as well as more recent advancements, such as nanotechnology and fractals.

From a theoretical perspective we have begun to articulate and investigate a theory of embodied cognition. Our theory makes an overt distinction between insight, or System 1 thinking, as a nonverbal, unconscious form of knowing of mathematical relationships, and conscious, analytic formulation of propositional knowledge, or System 2 thinking, that supports the production of logically valid transformational proofs. While action systems and language systems each contribute to both insight and proof, proof-with-insight, it is hypothesized, depends on a coordination of these two systems. On these matters, we showed partial success. However, the shortcomings offer up valuable direction for future models and research.

On a practical level, our work contributes to issues concerning the design of theory-motivated interventions for improving STEM education, and the proliferation of commercial body-based programs in STEM and language education. One important take-away is that GEMC is not about thoughtless movement. Rather GEMC considers the integration of nonverbal and verbal forms of (mathematical) thinking.

Deriving Design Principles from Theory and Research

Lee (2015) observes that as new theoretical perspectives, like embodied cognition, emerge in education, so do new ways of using technology to support the difficult tasks of teaching and learning. Technology can support the construction of mathematical meanings (National Council of Teachers of Mathematics, 2000), allowing students to 'play with' mathematics (Fey, 1989) and explore justification and proof in a visual, interactive environment (Hanna, 2000). Technology also offers novel opportunities for the embodiment of mathematical ideas (Lee, 2015), allowing students to enact mathematical activities related to visualization, symbolization, intuition, and reasoning. We are witnessing a new genre of educational technologies and interventions for promoting STEM, routed in theories of embodied cognition.

GEMC offers theoretical guidance for the design of effective learning environments. Working from this paradigm, Lindgren and Johnson-Glenberg (2013) offer six 'precepts' for the design of embodied and mixed-reality learning environments. Their recommendations emphasize 'gestural congruency,' where learners' actions are structurally or analogically related to the symbols and their meaning" (p. 446). From this general recommendation, we draw out two design principles for promoting mathematical proof: (1) eliciting dynamic depictive gestures through relevant directed actions that are congruent with the geometric relations that are present in the conjecture tasks; and (2) activating language systems that are explicitly coordinated with the relevant action systems, through pedagogical cues. *The Hidden Village* served as an intervention medium for these design principles, and attained partial success on the impact of language cuing. We were also able to add further empirical support for the theoretically inferred role of dynamic gestures on insight. However, we were not successful at effectively eliciting dynamic gestures on demand through elicitation of relevant directed actions by copying the motions of in-game characters.

One of the key design challenges when creating an environment that directs motions for a sophisticated conceptual area like geometric proof is determining which actions to direct to best facilitate students' reasoning. One idea is to examine the gestures, particularly the dynamic gestures, that competent problem-solvers tend to make, and then turn these gestures into directed actions given by the game for particular conjectures. However, this is complicated by limitations in current technology for accurately detecting certain types of motions. While subtle hand gestures are more difficult for our prototype video game to identify, large arm movements are less problematic. Thus our general approach is to examine the gestures that competent problem-solvers make when they prove our geometric conjectures, and then re-imagine these gestures as large arm movements that the technology can reliably capture. However, some of our motions designed in this way certainly worked better than others.

One finding in the intervention used in Nathan et al. (2014) that has particular relevance was that students who performed directed actions for the Gear task were more likely to have the mathematical insight, but when explicitly prompted to connect their directed actions to the conjecture through pedagogical language, their reasoning and proof performance actually declined. This is because when subject to explicit inspection, the directed actions did not make sense to participants in the manner in which they were intended. Participants would try to imagine how the directed actions were intended to embody the problem solution, and would in some cases guess incorrectly and be led down the wrong path as a result. The same thing occurred occasionally in the pilot study of the video game. For example, some participants who made the two sets of doubling motions for the false conjecture about how doubling the length and width of a rectangle doubles the area actually considered the sets of motions separately, and gave an incorrect proof relating to doubling rather than quadrupling. This is an interesting design challenge - the very motions that promote insight may be problematic when subjected to explicit reflection in order to promote proof. Perhaps prompting players to think about certain unconscious actions and perceptions instills a 'verbal overshadowing effect' (Schooler, Ohlsson, & Brooks, 1993), where explicit attention interrupts critical, nonverbal processes involved in insight formation by attempting to propositionalize and verbalize them. Directed motions and pedagogical prompts need to be carefully selected such that they promote both insight and proof, without making learners overly attentive to inappropriate aspects of the task domains.

Pedagogical language connecting relevant motions to the mathematical task at hand are clearly critical to the success of such an embodied learning environment to promote valid reasoning, but little is known about how the pedagogical prompts should be delivered. Nathan et al. (2014) explored the timing of verbal cues and found that prompts delivered before students had the opportunity to prove the conjecture were far less effective than hints delivered after one attempt at proof had already been made. The timing of the pedagogical cues appears to be an important consideration.

The challenges of deriving effective designs to promote learning is, in part, a limitation of the scientific method as a means to translate theory to practice. An analogy to the physical world makes this plain. Newton's second law ($F=ma$) can help analyze why a bridge stands or falls, but does little to inform bridge design. That is the purview of engineering; for it is within the carefully monitored process of the iterative design cycle that one can rapidly explore, test, and converge on successful designs. Research methods such as design-based research (e.g., Barab & Squire, 2004) offer methodological guidance for the process of applying evidence-based research to the design of learning environments. It is within this space we intend to continue to explore how best to elicit dynamic gestures, and harness their cognitive potential for mathematical insight.

Our future research is informed by the findings and limitations of the current study. One immediate action item is to replicate our findings with a larger, more uniform sample of participants from an age/grade-appropriate context. To this end, we will conduct classroom studies with $N=150$ high-school geometry students across two sites to investigate how directed action and pedagogical language influence insight and proof validity. The findings regarding the lack of an effect of task-relevant directed actions on dynamic gesture production raise questions of the selection of directed actions for instilling dynamic gestures. Two competing hypotheses arise here. One is that our understanding of the principle of gestural congruency, that is, structurally matching learners' directed actions to geometric concepts, is inadequate in this domain, and we need to develop a more valid analytic process to identify the underlying mathematical ideas for each conjecture and translate them to 'congruent' directed actions. Another hypothesis is that we have chosen the appropriate task-relevant directed actions, but have overworked the motor system for these specific actions (by making players match the actions five times before progressing) and thereby inhibited the associated conceptual and spatial relations. Irrelevant actions would, by design, be unrelated to the geometric relations, and therefore may be easier to ignore when learners are confronted with a statement to prove using their already-taxed motor system. Our remedy here is to vary the number of times players have to match the directed actions and see how this impacts insight performance.

Our preliminary results suggest that directed actions in combination with pedagogical language promoted valid informal proofs, and that dynamic gestures played a key, mediating role. While pedagogical prompting yielded some of the strongest findings in support of GEMC, we still understand very little of how these cues foster valid proof production. In future studies we will vary the kinds of pedagogical cues, self-cueing, and the timing of cues in order to understand the robustness of this for promoting mathematical reasoning. We also plan to look at the impact of providing language input alone, without directed actions, on proof performance. Finally, our near-term interest is to introduce this video game to high-school geometry classrooms to support collaborative proof production and collaborative authoring of new mathematical conjectures to provide a generative learning environment for exploring mathematical reasoning about objects and space.

Commercial Programs for Embodied Learning

One final point addresses the proliferation of motion- and body-based interventions for promoting mathematics learning. There is currently no definitive compendium of these programs, or a systematic inventory of their claims. But programs such as The Action Based Learning™ Lab, MATHS DANCE, and Math in Your Feet generally share intertwined goals of incorporating movement into mathematics instruction for intellectual, physical education and health, and interest purposes. Most of these programs lack the research reporting that would support the claims made about advancing academic goals. However, they are tapping into a growing awareness of the role the body plays in learning and engagement. As our current work matures, we hope to provide sound, empirically tested, and theoretically motivated design guidelines for programs that use embodiment to advance STEM learning. With such a vast design space before us, scholars working in embodied cognition for education will find value and inspiration in programs developed by mathematics educators, dancers, biomedical engineers, and mathematicians, as well as those designed by psychologists and educational researchers. Theoretical advancements in GEMC have the potential to provide a common framework for future efforts to design embodied learning experiences to enhance STEM education.

REFERENCES

1. Abrahamson, D. (2015). The monster in the machine, or why educational technology needs embodied design. In V. R. Lee (Ed.), *Learning technologies and the body: integration and implementation in formal and informal learning environments* (pp. 21–38). New York: Routledge.
2. Abrahamson, D., & Trninic, D. (2015). Bringing forth mathematical concepts: signifying sensorimotor enactment in fields of promoted action. *ZDM Mathematics Education, 47*(2), 1–12.
3. Alibali, M. W., Spencer, R. C., Knox, L., & Kita, S. (2011). Spontaneous gestures influence strategy choices in problem solving. *Psychological Science, 22*(9), 1138–1144.
4. Barab, S., & Squire, K. (2004). Design-based research: Putting a stake in the ground. *The journal of the learning sciences, 13*(1), 1–14.
5. Barsalou, L. W. (2008). Grounded cognition. *Annual Review of Psychology, 59*, 617–645.
6. Beilock, S. L., & Goldin-Meadow, S. (2010). Gesture changes thought by grounding it in action. *Psychological Science, 21*(11), 1605–1610.
7. Beilock, S. L., & Holt, L. E. (2007). Embodied preference judgments: can likeability be driven by the motor system? *Psychological Science, 18*(1), 51–57.
8. Casasanto, D., & Boroditsky, L. (2008). Time in the mind: using space to think about time. *Cognition, 106*(2), 579–593.
9. Chazan, D. (1993). High school geometry students' justification for their views of empirical evidence and mathematical proof. *Educational Studies in Mathematics, 24*(4), 359–387.
10. Chinn, S. (2000). A simple method for converting an odds ratio to effect size for use in meta-analysis. *Statistics in Medicine, 19*(22), 3127–3131.
11. Chu, M., & Kita, S. (2011). The nature of gestures' beneficial role in spatial problem solving. *Journal of Experimental Psychology: General, 140*(1), 102.
12. Church, R. B., & Goldin-Meadow, S. (1986). The mismatch between gesture and speech as an index of transitional knowledge. *Cognition, 23*(1), 43–71.

13. Coe, R., & Ruthven, K. (1994). Proof practices and constructs of advanced mathematics students. *British Educational Research Journal, 20*(1), 41–53.

14. Cohen, J. (1988). *Statistical power analysis for the behavioral sciences* (2nd ed.). Hillsdale: Lawrence Erlbaum.

15. Cook, S. W., Mitchell, Z., & Goldin-Meadow, S. (2008). Gesturing makes learning last. *Cognition, 106*(2), 1047–1058.

16. Decety, J., & Grèzes, J. (2006). The power of simulation: imagining one's own and other's behavior. *Brain research, 1079*(1), 4–14.

17. Donovan, A., Boncoddo, R., Williams, C. C., Walkington, C., Pier, E. L., Waala, J., et al. (2014). *Action, gesture and abstraction in mathematical learning*. San Diego, CA: Thematic Panel presented at the Sixth Conference of the International Society for Gesture Studies.

18. Dreyfus, T. (1999). Why Johnny can't prove. *Educational Studies in Mathematics, 38*(1–3), 85–109.

19. Edwards, L. D. (2009). Gestures and conceptual integration in mathematical talk. *Educational Studies in Mathematics, 70*(2), 127–141.

20. Enyedy, N., & Danish, P. (2015). Learning physics through play and embodied reflection in a mixed reality learning environment. In V. R. Lee (Ed.), *Learning technologies and the body: integration and implementation in formal and informal learning environments* (pp. 97–111). New York: Routledge.

21. Evans, J. S. B. (2003). In two minds: dual-process accounts of reasoning. *Trends in Cognitive Sciences, 7*(10), 454–459.

22. Fey, J. T. (1989). Technology and mathematics education: a survey of recent developments and important problems. *Educational Studies in Mathematics, 20*(3), 237–272.

23. Fischer, U., Link, T., Cress, U., Nuerk, H.-C., & Moeller, K. (2015). Math with the dance mat: on the benefits of numerical training approaches. In V. R. Lee (Ed.), *Learning technologies and the body: Integration and implementation in formal and informal learning environments* (pp. 149–166). New York: Routledge.

24. Garcia, N., & Infante, N. E. (2012). Gestures as facilitators to proficient mental modelers. In L. R. Van Zoest, J.-J. Lo, & J. L. Kratky (Eds.), *Proceedings of the 34th Annual Meeting of the North American Chapter of the International Group for the Psychology of Mathematics*

Education (pp. 289–295). Kalamazoo: Western Michigan University.

25. Glenberg, A. M., Gutierrez, T., Levin, J. R., Japuntich, S., & Kaschak, M. P. (2004). Activity and imagined activity can enhance young children's reading comprehension. *Journal of Educational Psychology, 96*(3), 424–436.
26. Göksun, T., Goldin-Meadow, S., Newcombe, N., & Shipley, T. (2013). Individual differences in mental rotation: what does gesture tell us? *Cognitive Processing, 14*(2), 153–162.
27. Goldin-Meadow, S. (2005). *Hearing gesture: how our hands help us think*. Cambridge, MA: Harvard University Press.
28. Goldin-Meadow, S., Cook, S. W., & Mitchell, Z. A. (2009). Gesturing gives children new ideas about math. *Psychological Science, 20*(3), 267–272.
29. Hall, R., Ma, J. Y., & Nemirovsky, R. (2015). Rescaling bodies is/as representational instruments in GPS drawings. In V. R. Lee (Ed.), *Learning technologies and the body: integration and implementation in formal and informal learning environments* (pp. 112–131). New York: Routledge.
30. Hanna, G. (2000). Proof, explanation and exploration: an overview. *Educational Studies in Mathematics, 44*, 5–23.
31. Harel, G., & Sowder, L. (1998). Students' proof schemes. In E. Dubinsky, A. Schoenfeld, & J. Kaput (Eds.), *Research on collegiate mathematics education* (Vol. III, pp. 234–283). Providence: American Mathematical Society.
32. Harel, G., & Sowder, L. (2005). Toward comprehensive perspectives on the learning and teaching of proof. In F. Lester (Ed.), *Second handbook of research on mathematics teaching and learning*. Reston, VA: National Council of Teachers of Mathematics.
33. Haruno, M., Wolpert, D. M., & Kawato, M. (2001). Mosaic model for sensorimotor learning and control. *Neural Computation, 13*(10), 2201–2220.
34. Havas, D. A., Glenberg, A. M., Gutowski, K. A., Lucarelli, M. J., & Davidson, R. J. (2010). Cosmetic use of botulinum toxin-a affects processing of emotional language. *Psychological Science, 21*(7), 895–900.
35. Healy, L., & Hoyles, C. (2000). A study of proof conceptions in algebra. *Journal for Research in Mathematics Education, 31*(4), 396–428.

36. Hostetter, A. B., & Alibali, M. W. (2008). Visible embodiment: gestures as simulated action. *Psychonomic Bulletin & Review, 15*(3), 495–514.
37. Hostetter, A. B., Alibali, M. W., & Kita, S. (2007). Does sitting on your hands make you bite your tongue? The effects of gesture prohibition on speech during motor descriptions. In *Proceedings of the 29th Annual Meeting of the Cognitive Science Society* (pp. 1097–1102). Mawhah, NJ: Erlbaum.
38. Kahneman, D. (2011). *Thinking, fast and slow*. New York, NY: Macmillan.
39. Kato, C., Isoda, H., Takehara, Y., Matsuo, K., Moriya, T., & Nakai, T. (1999). Involvement of motor cortices in retrieval of kanji studied by functional MRI. *NeuroReport, 10*, 1335–1339.
40. Kellman, P. J., & Massey, C. M. (2013). Perceptual learning, cognition, and expertise. *The Psychology of Learning and Motivation, 58*, 117–165.
41. Kellman, P. J., Massey, C. M., & Son, J. Y. (2010). Perceptual learning modules in mathematics: enhancing students' pattern recognition, structure extraction, and fluency. *Topics in Cognitive Science, 2*(2), 285–305.
42. Kirsh, D., & Maglio, P. (1994). On distinguishing epistemic from pragmatic action. *Cognitive Science, 18*(4), 513–549.
43. Knuth, E., Choppin, J., & Bieda, K. (2009). Middle school students' production of mathematical justifications. In D. Stylianou, M. Blanton, & E. Knuth (Eds.), *Teaching and learning proof across the grades: a K–16 perspective* (pp. 153–170). New York: Routledge.
44. Koedinger, K. R., Corbett, A. T., & Perfetti, C. (2012). The Knowledge-Learning-Instruction framework: bridging the science-practice chasm to enhance robust student learning. *Cognitive Science, 36*(5), 757–798.
45. Lakoff, G., & Núñez, R. E. (2000). *Where mathematics comes from: How the embodied mind brings mathematics into being.* New York, NY: Basic Books.
46. Lee, V. R. (2015). *Learning technologies and the body: integration and implementation in formal and informal learning environments*. New York: Routledge.
47. Leung, A. K. Y., Kim, S., Polman, E., Ong, L. S., Qiu, L., Goncalo, J. A., et al. (2012). Embodied metaphors and creative "acts". *Psychological Science, 23*(5), 502–509.

48. Lindgren, R. (2015). Getting into the cue: embracing technology-facilitated body movements as a starting point for learning. In V. R. Lee (Ed.), *Learning technologies and the body: integration and implementation in formal and informal learning environments* (pp. 39–54). New York: Routledge.
49. Lindgren, R., & Johnson-Glenberg, M. (2013). Emboldened by embodiment six precepts for research on embodied learning and mixed reality. *Educational Researcher, 42*(8), 445–452.
50. Marghetis, T., Edwards, L. D., & Núñez, R. (2014). More than mere handwaving: gesture and embodiment in expert mathematical proof. In L. D. Edwards, F. Ferrara, & D. Moore-Russo (Eds.), *Emerging perspectives on gesture and embodiment in mathematics* (pp. 227–246). Charlotte, NC: Information Age Publishing.
51. Martin, T. S., McCrone, S. M. S., Bower, M. L. W., & Dindyal, J. (2005). The interplay of teacher and student actions in the teaching and learning of geometric proof. *Educational Studies in Mathematics, 60*(1), 95–124.
52. McCafferty, S. G., & Stam, G. (Eds.). (2009). *Gesture: second language acquisition and classroom research*. New York: Routledge.
53. McNeill, D. (1992). *Hand and mind: what gestures reveal about thought*. Chicago, IL: Chicago University Press.
54. Nathan, M. J. (in press). One function of gesture is to make new ideas: Evidence for reciprocity between action and cognition. In R. B. Church, M. W. Alibali, & S. D. Kelly, (Eds.), *Why gesture? How the hands function in speaking, thinking and communicating*.
55. Nathan, M. J. (2014). Grounded mathematical reasoning. In L. Shapiro (Ed.), *The Routledge handbook of embodied cognition* (pp. 171–183). New York: Routledge.
56. Nathan, M. J., & Martinez, C. V. (2015). Gesture as model enactment: the role of gesture in mental model construction and inference making when learning from text. *Learning: Research and Practice, 1*(1), 4–37.
57. Nathan, M. J., & Sawyer, K. (2014). Foundations of Learning Sciences. In K. Sawyer (Ed.), *The Cambridge handbook of the learning sciences* (2nd ed., pp. 21–43). Cambridge, England: Cambridge University Press.
58. Nathan, M. J., Walkington, C., Boncoddo, R., Pier, E. L., Williams, C. C., & Alibali, M. W. (2014). Actions speak louder with words: the

roles of action and pedagogical language for grounding mathematical proof. *Learning and Instruction, 33*, 182–193. doi:10.1016/j.learninstruc.2014.07.001

59. Nathan, M., Walkington, C., Srisurichan, R., & Alibali, M. (2011). Modal engagements in pre-college engineering: tracking math and science concepts across symbols, sketches, software, silicon, and wood. In *Proceedings of the 118th American Society of Engineering Education Annual Conference and Exposition.* Vancouver, CA: American Society for Engineering Education.
60. National Council of Teachers of Mathematics. (2000). *Principles and standards for school mathematics*. Reston, VA: National Council of Teachers of Mathematics.
61. Newcombe, N. S., & Shipley, T. F. (2012). Thinking about spatial thinking: new typology, new assessments. In J. S. Gero (Ed.), *Studying visual and spatial reasoning for design creativity*. New York: Springer.
62. Niedenthal, P. M. (2007). Embodying emotion. *Science, 316*(5827), 1002–1005.
63. Novack, M. A., Congdon, E. L., Hemani-Lopez, N., & Goldin-Meadow, S. (2014). From action to abstraction using the hands to learn math. *Psychological Science, 25*(4), 903–910.
64. Pelavin, S. H., & Kane, M. (1990). *Changing the odds ratio: factors increasing access to college*. New York: College Board Publications.
65. Petrick, C., & Martin, T. (2012). *Mind your body: learning mathematics through physical action.* Vancouver, Canada: Paper presented at the Annual meeting of the American Educational Research Association.
66. Pier, E. L., Walkington, C., Williams, C., Boncoddo, R., Waala, J., Alibali, M. W., et al. (2014). Hear what they say and watch what they do: predicting valid mathematical proofs using speech and gesture. In W. Penuel, S. A. Jurow, & K. O'Connor (Eds.), *Learning and becoming in practice: Proceedings of the Eleventh International Conference of the Learning Sciences* (pp. 649–656). Boulder, CO: University of Colorado.
67. Porteous, K. (1990). What do children really believe? *Educational Studies in Mathematics, 21*, 589–598.
68. Pulvermüller, F. (2005). Brain mechanisms linking language and action. *Nature Reviews Neuroscience, 6*(7), 576–582.

69. Radford, L., Edwards, L., & Arzarello, F. (2009). Introduction: beyond words. *Educational Studies in Mathematics, 70*(2), 91–95.

70. Roth, W. M. (1994). Thinking with hands, eyes, and signs: multimodal science talk in a grade 6/7 unit on simple machines. *Interactive Learning Environments, 4*(2), 170–187.

71. Schooler, J. W., Ohlsson, S., & Brooks, K. (1993). Thoughts beyond words: when language overshadows insight. *Journal of Experimental Psychology: General, 122*(2), 166–183.

72. Shapiro, L. (2014). *The Routledge handbook of embodied cognition.* New York: Routledge.

73. Shoval, E. (2011). Using mindful movement in cooperative learning while learning about angles. *Instructional Science, 39*(4), 453–466.

74. Smith, C. P., King, B., & Hoyte, J. (2014). Learning angles through movement: critical actions for developing understanding in an embodied activity. *The Journal of Mathematical Behavior, 36*, 95–108.

75. Stevens, R. (2012). The missing bodies of mathematical thinking and learning have been found. *Journal of the Learning Sciences, 21*(2), 337–346.

76. Stokes, D. E. (1997). *Pasteur's quadrant*. Washington, DC: Brookings Institution Press.

77. Stylianides, A. J. (2007). Proof and proving in school mathematics. *Journal for Research in Mathematics Education, 38*(3), 289–321.

78. Thomas, L. E. (2013). Spatial working memory is necessary for actions to guide thought. *Journal of Experimental Psychology: Learning, Memory, and Cognition, 39*(6), 1974–1981.

79. Thomas, L. E., & Lleras, A. (2007). Moving eyes and moving thought: on the spatial compatibility between eye movements and cognition. *Psychonomic Bulletin & Review, 14*(4), 663–668.

80. Thomas, L. E., & Lleras, A. (2009). Swinging into thought: Directed movement guides insight in problem solving. *Psychonomic Bulletin & Review, 16*(4), 719–723.

81. Uttal, D. H., Meadow, N. G., Tipton, E., Hand, L. L., Alden, A. R., Warren, C., et al. (2012). The malleability of spatial skills: a meta-analysis of training studies. *Psychological Bulletin, 13*(2), 352–402. doi:10.1037/a0028446

82. Valenzeno, L., Alibali, M. W., & Klatzky, R. (2003). Teachers' gestures facilitate students' learning: a lesson in symmetry. *Contemporary Educational Psychology, 28*(2), 187–204.

83. Walkington, C., Boncoddo, R., Williams, C., Nathan, M., Alibali, M., Simon, E., et al. (2014). Being mathematical relations: dynamic gestures support mathematical reasoning. In W. Penuel, S. A. Jurow, & K. O'Connor (Eds.), *Learning and becoming in practice: Proceedings of the Eleventh International Conference of the Learning Sciences* (pp. 479–486). Boulder, CO: University of Colorado.

84. Wolpert, D. M., & Kawato, M. (1998). Multiple paired forward and inverse models for motor control. *Neural Networks, 11*, 1317–1329.

85. Yackel, E., & Hanna, G. (2003). Reasoning and proof. In J. Kilpatrick, W. G. Martin, & D. Schifter (Eds.), *A research companion to principles and standards for school mathematics* (pp. 227–236). Reston, VA: National Council of Teachers of Mathematics.

86. Zander, T., Öllinger, M., & Volz, K. G. (2016). Intuition and insight: two processes that build on each other or fundamentally differ? *Frontiers in Psychology, 7*, 1395.

87. Zhang, Z., Lei, Y., & Li, H. (2016). Approaching the distinction between intuition and insight. *Frontiers in Psychology, 7*, 1195.

INDEX

G

H

I

L

M

Printed in the United States
by Baker & Taylor Publisher Services